FOR LOVE OF INSECTS

For Love of Insects

THOMAS EISNER

The Belknap Press of Harvard University Press

Cambridge, Massachusetts, and London, England ■ *2003*

Designed by Annamarie Why

Library of Congress Cataloging-in-Publication Data
Eisner, Thomas, 1929–
   For love of insects / Thomas Eisner.
     p. cm.
   Includes bibliographical references (p.).
   ISBN 0-674-01181-3 (alk. paper)
   1. Insects.   2. Entomology—Case studies.   I. Title.
QL463.E38 2003
595.7—dc21
2003044399

To Maria

Contents

Foreword

EDWARD O. WILSON

For Love of Insects is an account of a nascent field of biology. It is also a memoir about some of the remarkable discoveries in that field made over the past half century by a world-class biologist who is also a naturalist of such exceptional talent, intensity, and breadth as to deserve the title "the modern Fabre," after France's great pioneer observer of insect life.

I have been privileged throughout that time to count Thomas Eisner among my closest friends. As fellow Harvard graduate students in 1952, we shared the formative experience of an epic field trip, circling through the lower 48 states in an ancient automobile that may have set a national record in breakdowns per week. I have since followed his unbroken and prodigious flow of original discoveries, in conversations and in his hundreds of articles in research journals.

The key to Eisner's success, in addition to his dedication and a marathon capacity for hard work, is his excellence as both a field biologist and a laboratory experimentalist. He also had the wisdom from the start to apply his golden touch to a huge diversity of insects and other invertebrates that had never before been studied in any serious manner. To date, about a million species of insects, spiders, millipedes, and other arthropods have been discovered and given a scientific name. In addition, systematists estimate, millions more species of arthropods, perhaps as high as 10 million, remain undiscovered and nameless. Of those known, fewer than a thousand have been studied in depth. Into this little-known world Eisner entered with state-of-the-art microscopes, cameras, micromanipulators, and chemical analytic equipment.

Another reason for Eisner's success has been his consistent choice of first-rate collaborators, both students and peers. Most notable has been his career-long collaboration in many of the studies with the distinguished natural products chemist Jerrold Meinwald. Both Cornell professors were fortunate in beginning their work about the time that potent new methods of chemical microanalysis were perfected, including especially coupled gas chromatography and mass spectrometry. They were thus able to identify organic chemicals in microgram quantities—the amounts carried in the bodies of individual insects and other small invertebrates.

Thomas Eisner approaches natural environments to conduct research as a benevolent juggernaut. I remember vividly the day in the spring of 1968 when he visited me and my wife, Renee, in Marathon, in the Florida Keys, during one of my sabbatical semesters. A sizable van pulled into the driveway of a house we had rented, and out tumbled Eisner and two assistants, from amidst an assortment of field and laboratory equipment. The next day we explored nearby Lignumvitae Key, site of the least disturbed tropical hardwood forest in the coterminous United States. Eisner soon homed in on brightly colored tree snails that dotted a stand of holywood lignum vitae trees. Peering closely, he lightly tapped the shell of one of the mollusks, which obediently oozed out a droplet of viscous liquid. He mused: Was this secretion a defense against snail-eating predators? If so, against which predators might a droplet at the base of the shell be effective? Maybe not birds or mammals, animals big enough simply to pluck the victim off its perch, crunch its shell, and swallow its soft parts whole. What about ants? They are the ubiquitous all-purpose predators of invertebrates. It happened that on the same trees were foraging workers of the large, red-and-black carpenter ant *Camponotus floridanus,* a very common and aggressive native species. On the spot Eisner devised an experiment that brought workers into close contact with the snails. Out came the mucus droplets, and away ran the ants. It was the first such test of the role of such secretions in these arboreal mollusks. Samples of the mucus were collected by the assistants for later analysis.

That night Eisner composed the draft of a short report for a technical journal. Because I had identified the ants, and just happened to be along, Tom insisted I be a coauthor. Thus was born the scientific study of tree-snail defensive secretions.

Soon afterward we collaborated as part of the successful effort to have

Lignumvitae Key, with its snails, ants, and tropical forest, set aside as a state reserve. I like to imagine future Eisners coming ashore there forevermore.

Although the incident is a nice example of Eisnerian élan, it would be decidedly misleading to let my anecdote suggest that Tom's experiments were generally of a quick and dirty nature. On the contrary, he used such preliminary experiments in a myriad ways, depending on the natural history of particular species and the opportunities it offered in field and laboratory, to discover phenomena previously unknown and unsuspected. If the results were promising, he followed up with detailed field and laboratory research. As a rule he and his collaborators also identified the key substances employed by the species.

It would be equally misleading to conclude that the countless contributions of Thomas Eisner to the biology of so many species were somehow haphazard. On the contrary, he has painted nature as a pointillist, sector by sector of arthropod biology; and from the many local depictions, from the spray of nature's colored dots as it were, he has delineated patterns of evolutionary adaptation, molecular evolution, behavior, and life cycles that likely would not have been revealed by other means. Biology is a science with a high level of particularity. Many of its phenomena make no sense unless profusely illustrated by induction. There are other natural sciences that grow and mature in such a manner. How, for example, could astrophysics exist without a map of the stars?

The discipline that came into existence as a result of the pointillist studies by Eisner, Jerrold Meinwald, and a few other pioneers was chemical ecology. Its importance arises from the fact that the vast majority of organisms—surely more than 99 percent of all species when plants, invertebrates, and microorganisms are thrown in—orient, communicate, subdue prey, and defend themselves by chemical means. Eisner's research has added significantly to our understanding of chemical communication, especially the use of pheromones in courtship, but his most numerous and important contributions have been in the defense mechanisms used by insects and other arthropods. The many behaviors that he has discovered and explained, and their implementation by life around us, amazing in variety and precision, are the worthy focus of this book.

What makes things baffling is their degree
of complexity, not their sheer size . . .
a star is simpler than an insect.

Martin Rees, "Exploring Our Universe and Others,"
Scientific American, December 1999

Prologue

This book is about the thrill of discovery. It is a narrative about insects, a retrospective of a lifetime of exploration of a group of animals that truly can be said to have conquered the planet. Humans, it is claimed, are fundamentally biophilic. In me, love of nature is expressed as an affection for insects. I am an incorrigible entomophile. I view insects with curiosity and without resentment of their success. Quite to the contrary, figuring out exactly what secrets they might have hidden up their chitinous sleeves that enabled them to achieve evolutionary supremacy has been the driving force of my professional life. Insects survive because they have special strategies for doing so. How it is that one goes about deciphering these strategies, and how serendipity, group effort, and sheer good luck combine to provide momentum in such research, are what I have attempted to convey in the pages that follow.

Taken as a group, and viewed on a grand scale, insects have achieved a great deal. They metamorphose—or at least most of them do—growing up as larvae and taking to the wing as adults. They have opted for direct sperm transmittal through insemination, thereby relinquishing forever dependence on water for spawning. And as a consequence of acquisition of an exoskeleton—an external skeleton consisting of a hard cuticle—they have attained quickness of motion, resistance to desiccation, and the capacity to achieve dominance on land. They have succeeded in one major respect where humans have failed. They are practitioners of sustainable development. Although they are the primary consumers of plants, they do not merely exploit plants. They also pollinate them, thereby providing a secure future, both for themselves and for their plant partners.

But it is at the level of the individual species that the beauty of insect adaptation really manifests itself. Each of the 900,000 insects that have been described, not to mention the countless others that remain to be discovered, is a storybook in itself. Insects are the most versatile of evolutionary innovators. Pick an insect at random, and chances are there is something about the way it feeds, or defends itself, or reproduces, that is unique. This book focuses on the process of discovery by which feats of insectan achievement are recognized. It is a personal account, dealing for the most part with findings made by me or my associates in the course of our research. It presents a series of case studies, documented in such fashion as to reveal the process that leads from initial discovery to eventual revelation and experimental proof. The purpose is to involve the reader in the process of inquiry, in hopes not only of strengthening the conviction of the entomophile, but of changing the attitude of the entomophobe.

Included among the examples are some in which the protagonists are not insects, but spiders or other arachnids. There too I have hopes that the narrative will help reverse an affliction. Arachnophobia, if anything, appears to be on the rise.

Photography has been one of my passions over the years, and I have made a special effort to provide photographic documentation of each of the case studies presented. Among the examples that were most thrilling to investigate are:

Beetles that have a protective spray that they eject explosively at 100°C.

Green lacewing larvae that feed on wooly aphids and masquerade as aphids by donning the aphid's wool, thereby causing the ants that guard the aphids to mistake the lacewings for aphids.

Sawfly larvae that feed on pine needles and defend themselves by spitting up pine resin when attacked.

Caterpillars that feed on carnivorous sundew plants by imbibing the glue droplets ordinarily used by the plants to trap insects.

Caterpillars that feed on flowers while disguising themselves as flowers by fastening pieces of petals onto their backs.

A male moth that bestows a plant poison upon a female for eventual defensive incorporation into the eggs, and that announces to the female during courtship, by means of a chemical signal, how much of the poison he holds in store for her, a signal by which the female then determines whether or not to mate with him.

A beetle that clings with its feet when attacked, using 60,000 oil-tipped bristles, which enables the beetle to withstand pulling forces of up to 200 times its body weight.

Termite soldiers that use a spray in defense, a viscous glue that also serves as a social rallying call, to attract further soldiers to the site of trouble.

An aquatic beetle that defends itself with a secretion that it doles out slowly while in the mouth of a fish, thereby assuring rejection by the fish despite the fish's initial effort to rinse the beetle clean by flushing water in and out of the oral cavity.

And there is more.

I am primarily a field biologist and it is in nature that I find the leads to my research. I became interested in insects from the moment I could walk—or so my parents told me—and I never lost the habit of being on the lookout for insects at all times. When walking I constantly scout the ground ahead of me, with the result that in urban settings, on sidewalks, I find a lot of coins. As luck would have it, I have never been without opportunity to commune with nature. I have done fieldwork in the most diverse settings, in Uruguay, Australia, Panama, Europe, Florida, Arizona, not to mention my own patch of woods in Ithaca, New York. The following chronology provides some idea of when I was where, and why.

I was born in Berlin in 1929. When Hitler rose to power in 1933 my family left Germany for Spain, where I befriended caterpillars and learned that they can be kept in boxes, provided one supplied them with leaves of their food plant. I also learned that some of them made munching sounds in the night.

The outbreak of the Spanish Civil War in 1936 forced us to flee to France, where we briefly established residence in Paris. I learned French but hardly anything new about insects. But we did take a trip to Holland to visit my uncle Curt Eisner, who showed me his large collection of *Parnassius* butterflies. Utterly fascinated, I determined that I too would some day have an insect collection.

In 1937 we emigrated to South America, first to Argentina and then to Uruguay, where we lived until 1947. Uruguay was an entomological paradise. I spent summers collecting butterflies. I learned that some insects have powerful stinks and wondered whether these kept them from getting eaten. Fascinated also by the odors of vegetation, I began toying with the notion that plants might be chemically protected as well.

Within a year after I graduated from high school, we moved to the United States. My butterfly collection did not survive the rough handling to which our baggage was subjected on arrival in Boston harbor.

Following rejection by a series of colleges, including Cornell (my letter of rejection now hangs framed in my Cornell office), I went to secretarial school to learn shorthand and typing, and then for 2 years to Champlain College in Plattsburg, New York, from where I transferred to Harvard. I flirted briefly with chemistry at Harvard, but ultimately settled on biology, after taking an entomology course, beautifully taught by Frank Carpenter and his graduate student assistant, Kenneth Christiansen. It finally dawned on me that studying insects was serious business. Earlier, in the summer of 1948, while working as a volunteer in the American Museum of Natural History, I had already been urged to give entomology serious thought. My supervisor, the noted entomologist Charles Michener, had recommended four books—Julian Huxley's *New Systematics,* V. B. Wigglesworth's *Insect Physiology,* R. E. Snodgrass's *Insect Morphology,* and E. B. Ford's *Butterflies*—that made an enormous impression on me. But it was Carpenter's course, and particularly Christiansen's compelling enthusiasm, that persuaded me once and for all to study insects. I graduated from Harvard in 1951, and remained for graduate studies under Frank Carpenter, with whom I obtained my Ph.D. in 1955. My thesis was on a tiny organ in the gut of ants, and on how this organ, called the proventriculus, evolved to enable individual ants to store large amounts of food in the crop for distribution to the remainder of the colony.

In 1951 I met Maria Loebell, the love of my life and best of friends, at the time a student social worker in South Boston. We met in December and were married by June. To have asked for the privilege of sharing her life was the wisest decision I ever made. The three daughters we raised together, Yvonne, Vivian, and Christina, all studied at Cornell and are now professionally active with children of their own. With more time on her hands, Maria is now my constant research partner. She has come to share my love for insects, and in addition to her exquisite mastery of scanning electron microscopy, has made herself indispensable in all aspects of experimental design and data acquisition. And she is the best of scouts in the field. There is little that we miss when we roam the outdoors together on the lookout for wonders. She contributed in countless ways to the production of this book.

In 1951 another graduate student joined Carpenter's group, a multi-talented expert on ants from Alabama of immense personal appeal. Edward O. Wilson was a visionary evolutionist and all-around genius, whose eventual ascendancy to the ranks of the truly great in science was then already foreseeable. We shared interests and working habits and were quick to become friends. Convinced that it would be to our great advantage, and emboldened by receipt of a $200 grant from the Society of the Sigma Xi, we decided in the summer of 1952 to undertake a major exploratory venture. Our intent was to get acquainted with the American landscape and its insects, and we succeeded. We went north from Boston to Ontario, from there across the Great Plains to Montana and Idaho, then to California, Nevada, Arizona, and New Mexico, and finally back east through the Gulf states and home. We crisscrossed every conceivable habitat and by the time of our return 2 months later had added some 12,000 miles to the 100,000 or so already on the odometer of my geriatric Chevrolet. I had named the car Charrúa II, after the fierce Amerindian tribe, but should instead have named it Beagle II, after the ship Darwin sailed on, for it was on that trip that I gained my first genuine appreciation for the vastness of the biological spectacle and the interdependence of life. We stopped at countless locations and were then usually silent as we wandered off in different directions to do our respective explorations, but we found ample time to converse during the endless drives, and it was then that I found myself drawn into discussions of the unknown, fueled by Wilson's formidable intellect. I never regretted that Charrúa II could not safely be driven above 40 miles per hour. It was the intellectual miles covered that mattered. To this day, whenever I wish to check on the validity of an idea, I put the notion to the test in a telephone call to Ed.

Our trip also awakened the conservationist in both of us. Witnessing at first hand, at so many locations nationwide, the destructive consequences of human expansionism convinced us that we would need to assume active roles in the conservation movement. Over the years Ed and I have collaborated on a number of conservation projects, including one that led to the preservation of Lignumvitae Island in the Florida Keys. He himself, through his exquisite writings, has become the chief spokesman for biodiversity preservation worldwide.

I remained as a postdoctoral fellow at Harvard for 2 years, during which I became increasingly interested in insect chemistry and chemical communication. I came face to face with my first bombardier beetle, and concluded that there was no limit to the chemical ingenuity of insects. When in the summer of 1957 I began looking into the availability of academic posts, I advertised myself as a student of insect chemistry. I should have called myself a chemical ecologist, but the term had not yet been coined.

In the late summer of 1957 I moved my young family to Ithaca, New York, home of Cornell University, where I had been hired as an assistant professor in the Entomology Department, to teach the introductory biology course. Today, 45 years later, I am still at Cornell, although in a different department, teaching different courses, and holding a second appointment as director of CIRCE, the Cornell Institute for Research in Chemical Ecology. I am still working on insects, supported since 1959 uninterruptedly by the same grant from the National Institutes of Health.

The years at Cornell were happy ones from the start. In 1958 I met Jerry Meinwald, a Cornell colleague and expert chemist, who was to become my close friend and chief research partner.

Exploration meant travel, and I was soon to discover two study sites upon which I became dependent. One, the Archbold Biological Station, near Lake Placid, Florida, has become a home away from home, which I have been visiting almost yearly since I discovered it in 1958. The other, which I first encountered in 1959, is the Chiricahua region near Portal, Arizona, an area now well known to naturalists, which I have visited on eight occasions. I have also spent time at the Huyck Preserve near Albany, New York, and at the Mountain Lake Biological Station near Blacksburg, Virginia, as well as in the Florida Keys, the Big Thicket of Texas, and, briefly, the Rocky Mountain Biological Station near Crested Butte, Colorado. Foreign ventures include a 1968 visit to the Smithsonian Tropical Research Station on Barro Colorado Island, in Panama, and a year's sabbatical in Australia (1972–73) with the Commonwealth Scientific and Industrial Research Organization in Canberra. I found each site to be rich in opportunity. The thought that fieldwork might be on the decline and that biological field stations might suffer as a result is utterly disconcerting. As a university-based naturalist explorer, I derive my inspiration not from the library, or the classroom, or from interaction with colleagues, but from the times spent observing events outdoors. If this book contributes in any way toward bolstering the preservationist spirit, as I hope it might, it will have fulfilled its purpose.

1
Bombardier

Most naturalists keep good diaries. I don't. Therefore I will never be able to pinpoint the day I came upon my first bombardier beetle (*Brachinus* species; see facing page). It was in Lexington, Massachusetts, and I remember the meadow well, except that it probably doesn't exist as a meadow any more. But that is another story.

It must have been in the summer of 1955, the summer I was writing my doctoral thesis, and probably in early June. I was on my knees, uncovering rocks, and ready for any find, particularly if it involved an unfamiliar insect with unusual chemical talents. My thesis had dealt with the anatomy of ants, but I was on the lookout for something new. All my life I had been passionately interested in insects so there was no question that I'd stick it out with bugs. But I also had a very genuine interest in chemistry, and somewhere in the back of my mind was the thought that these two interests could be combined. I didn't realize it at once, but in stumbling upon those little beetles I had struck gold. Bombardier beetles were precisely the sort of champion chemists I was looking for. My having encountered them when I did was the luckiest of breaks.

At that time, while I was a graduate student, there had been some fascinating developments in the interface of entomology and chemistry. Although the term pheromone to designate a chemical signal had not yet been coined, it was becoming increasingly clear that insects flirt by means of chemicals. Female moths, when ripe and ready, announce that fact by emitting a volatile secretion that attracts the male. Word had it that German chemists were hot on the trail of one such attractant, with the intent of identifying the molecule or molecules involved. And in addition there was the exciting research on insect hormones, those remarkable internal chemical messengers that, operating at infinitesimal concentrations, control growth and the transformations in body form known as metamorphosis. An insect hormone, ecdysone, had just been isolated in pure form, and I remember being tremendously taken by a seminar given by Peter Karlson, the German investigator who had been responsible for the isolation. It was now within the realm of possibility, I thought, to decipher the chemical language of insects. Secretly, I wanted to become one of the cryptographers.

Being at the Biological Laboratories at Harvard at that time was inspiring. Two flights down from my room was the laboratory of Carrol M. Williams, one of the great pioneers of insect endocrinology. Carrol, who was to become a close friend later in life, had been my undergraduate adviser at Harvard and teacher in comparative physiology. I learned about bioassays in his course, about ways for testing quantitatively for the biological activity of a chemical substance. And then there was Ed Wilson, fellow graduate student and most inspiring of friends, with whom I had published my first technical paper and shared countless interests.

Ed was himself becoming interested in chemical communication, and he had begun to explore the role of pheromones in ants. He had discovered the gland responsible for producing the trail substance of certain ants, the substance by which foraging ants lay out linear paths to guide nestmates to newly found food sources. He also had devised some clever experiments that showed that individual ants, when assaulted, emit chemicals that alert nestmates nearby to the disturbance and enlist them to come help with the problem. He had shown further that in ant colonies corpses are recognized by certain fatty acids they contain. By tagging various inert objects with such fatty acids he could induce the ants to transport the objects to the graveyards of the colony, as they typically do with corpses. It was an impressive demonstration of the power of the bioassay.

I myself had had experiences that predisposed me for the study of insect chemistry. For one thing, my father was a chemist. He was one of the last graduate students of Fritz Haber, the Nobel laureate who first synthesized ammonia from its constitutive elements. My father would have loved it if I became a chemist, but he was reconciled to the notion that I was destined to study bugs. Yet there was a subtle influence my father was to have on me, and it relates to his having been an amateur perfumer. Wherever we lived before settling in the United States, whether in Spain or Uruguay, my father always had a basement lab in which he concocted, for friends and relatives, perfumes, skin lotions, sun tan oils, and colognes. The house was often mysteriously redolent as a result, which as a little boy I found wonderful. As I grew older, though, I became interested in the odors themselves and in the reasons why they might exist in nature. At the age of 13 in Uruguay, I hadn't read Darwin yet. In fact, evolution hadn't even been mentioned in the biology course I had taken. But quite instinctively I had begun to think in adaptationist terms. What does the fragrance of lavender do for the lavender plant itself? I don't know when the idea occurred to me that plant odors might be defensive, but I know I didn't get the idea from books.

In our summer house in Uruguay we had an icebox that was periodically invaded by ants. Following local custom, we immersed the legs of the icebox in tin cans filled to the height of a centimeter or so with kerosene or turpentine. Either kept the ants out, but turpentine worked better. So why should turpentine be a good insect repellent? I remember putting two and two together when I learned that turpentine is derived from pine resin. The resin must be the pine's defensive juice! I think I must have been 14 or 15 when I did an actual experiment in which I showed that a dab of pine resin placed in the path of ants would cause them to shy away.

I am sure that I owe it to my father that I became so conscious of odors. But I was apparently "nasal" right from the start. My parents recalled that when I was little I could tell from the scent that lingered in the coat closet in the morning that my grandmother had visited the night before. But now, as a teenager, I was coming to realize that I could really learn things from my nose. I was already collecting insects by then, but what had begun as a hobby at the age of 8, and had been directed almost exclusively to the capture of butterflies, was now developing into an interest in live insects, irrespective of kind. Quite casually at first, but with ever increasing fascination, I noted that insects, ever so often, have odors. Some were faintly scented all of the time. Others gave off odors when you handled them, from fluids they emitted when disturbed. In the latter case the odors were often pungent, and I learned to sniff insects carefully lest I end up sneezing and coughing. I also came to realize that I had a good memory for smells. Insect odors seemed to come in categories. Many ants, for instance, had the same acidic odor. I did not know then that this was well documented, and that as early as 1670 a British naturalist by the name of John Wray had published a paper on the acid "juyce" of ants.

I was struck also by one particular odor, very noxious, that I came to associate with millipedes, and with one particular arachnid, a daddy-long-legs, that I had collected in numbers in Atlántida, the seaside resort near Montevideo where we had our summer house. That odor was like none other that I had encountered and it came from juices that the millipedes discharged from pores along the sides of the body and the daddy-long-legs emitted from the edge of its carapace. There was something peculiar about these fluids. They stained the fingers brown, like iodine. The effect was not immediate, but it was invariable: handle any number of those Uruguayan millipedes or daddy-long-legs and within minutes you would end up with stained fingertips. I did not give much thought to these observations at the time, but filed them away in the memory bank. The bombardier beetle was to bring them back to mind.

■ **I KNEW THE MOMENT** I turned over that rock and caught sight of those beetles that they were members of the family Carabidae, the so-called ground beetles. I had picked up many a carabid in my time, so I knew a bit about them. They are quick on their feet, but like most beetles not quick to take flight. They tend to scurry for cover, so if you want to catch them you have to be quick yourself. And I knew that as carabids they belong to that category of insects that give off odors when disturbed.

The ones under that rock were unlike any carabids I knew. With a reddish-brown body and blue iridescent wing covers, they were a pretty sight. And there were several beneath that stone, huddling close together. I had a vial in hand and made my move at once, but they dispersed in all directions and I managed to catch only one. I grasped it in my fingers and was about to put it in the vial when it emitted a series of distinctly audible pops that so startled me I nearly let go of it. I held it closer for a better look, and found that by giving it a squeeze I could cause it to pop again. I also noted that every time it popped it discharged a visible cloud from the rear, at the very moment that I felt a hot sensation in my fingers. I took a sniff and thought I recognized the familiar unpleasant odor of the millipedes and daddy-long-legs from Uruguay. Sure enough, when I checked my fingers after putting the beetle in the vial, there were brown spots on them. I decided then and there that this was a beetle I'd get to know.

I spent another hour or two in the meadow and managed to capture upward of a dozen of the beetles. I took them back to the Bio Labs, where I found I could maintain them in small plastic containers filled with soil, on a diet of freshly cut-up insect larvae supplemented with water. When I showed them to my friend and eventual Cornell colleague, William L. Brown Jr., who often set me straight on matters entomological, he said, "Oh, you've got yourself some bombardier beetles. They go pop when you pick them up and shoot out some real nasty stuff." Shoot they did indeed, and as I was to find out in the months ahead, they even aimed their discharges. The irony that it should have been in Lexington, Massachusetts, of all places, that I first heard those shots didn't strike me until later.

At about that time I had a visit from a young Uruguayan scientist who was working in the Chemistry Laboratories at Harvard only yards away from the Bio Labs. María Isabel Ardao had known my father in Uruguay and having heard that I was at Harvard stopped by to say hello. She had a fellowship and was working in the laboratories of Louis Fieser, the eminent organic chemist. It seemed she was studying an arachnid, an Uruguayan species that apparently produced an antibiotic. As she described the animal it became clear that she had been working on the very daddy-long-legs I remembered from Atlántida. I perked up because it was evidently the "juice" from that animal she and Fieser were trying to identify. When I asked whether they had succeeded she said that yes indeed, they had isolated two compounds, and that these turned out to be benzoquinones. Nasty stuff, she said. The work was being published, and they were calling the chemical mixture gonyleptidine, after the generic name of the animal.

So it was benzoquinones. Finally I had an idea of what that pungent odor was all about. My Uruguayan millipedes and my newly found bombardier beetles probably produced benzoquinones as well.

A second person I met coincidentally at the time was Louis M. Roth, an expert on cockroaches, who was working at the Army Quartermaster Research and Engineering Center outside Boston. He and Barbara Stay, a friend of mine who had recently obtained her Ph.D. at Harvard, had identified benzoquinones from the glands of a cockroach, *Diploptera punctata*. They thought the glands served a defensive purpose, but were not sure. Lou himself had earlier worked on some tiny beetles that also produced benzoquinones. They let me sniff *Diploptera* and the beetles, and there could be no doubt. It was the familiar stink again. I decided there had to be something very special about benzoquinones if so many insects and millipedes were making use of them.

BOMBARDIER

I told Lou that I'd like to work on *Diploptera* and he gave me a cage full of them. Back at Harvard the first thing I did was to obtain some crystalline benzoquinones, which were available commercially. I found that there were all kinds of warnings on the labels, so I decided to be careful. "Toxic," one label said, "harmful by inhalation or contact with skin." Too late for that, I thought, given the dousing I had been getting from all those benzoquinone producers.

There were two things I wanted to do. First I wanted to see whether *Diploptera* ejected its secretion in response to provocation, and second I wanted to find out whether the secretion was repellent to its enemies.

■ **THE GLANDS** of *Diploptera* were two small saclike structures opening about midway along the sides of the body. They connected to respiratory tubes in such a fashion that one could imagine the animal ejecting its benzoquinones by forcing air through the sacs. I thought I could smell benzoquinones when I handled the roaches but I had no visible evidence of emissions. Whatever *Diploptera* was ejecting, it was in too small a quantity or in too dispersed a form to see.

I decided I would develop a bioassay. I had obtained some cultures of protozoans, aquatic one-celled organisms, and found that the benzoquinones were hazardous to them as well. Addition of benzoquinone crystals to droplets of culture medium quickly killed the protozoans within. Crystals placed near a droplet caused the protozoans to shun the droplet surface. I thought that if I placed a microaquarium with a protozoan in a confined space with a *Diploptera* and monitored the behavior of the protozoan when I subjected the cockroach to a simulated attack, I might have devised an indirect way of telling whether benzoquinone discharge took place.

I built the necessary apparatus and the assay worked. I had rigged things so I could stimulate the roach with a warm probe while at the same time observing the microaquarium with a microscope. I had chosen a large protozoan, *Spirostomum,* as the target organism, and used only one per microaquarium. It became clear that for as long as the cockroach remained undisturbed, *Spirostomum* swam about actively in its "pool." But no sooner had the cockroach been prodded than the protozoan began showing surface avoidance. It gradually moved to increasingly greater

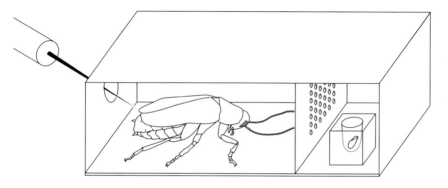

The bioassay used for demonstrating that the cockroach *Diploptera punctata* discharges its benzoquinones in response to disturbance. Vapor from the secretion diffuses through the perforated partition in the center, killing the protozoan in the microaquarium on the right.

depths within its confines, until the diffusing benzoquinones forced it to the very bottom and to its demise. The evidence was compelling. Control tests with cockroaches from which I had surgically removed the glands containing benzoquinones brought about no response in the *Spirostomum* when the roaches were stimulated. I was happy, but not satisfied. There had to be a more direct way of demonstrating *Diploptera's* discharges.

What I thought would do the trick was some kind of indicator paper that changed color in response to benzoquinones—in other words, paper that I could place under the roach when I disturbed it and on which the discharges would register visibly when they occurred. Benzoquinones are powerful oxidizing agents and there were commercially available indicator papers specifically designed to test for such agents. I looked up information on such papers and found that I could easily devise a kind of my own. All I would need to do is mix up some powdered starch with crystalline potassium iodide, stir the mixture with water, add a bit of hydrochloric acid, and soak a sheet of filter paper in it. The paper could then be blotted and would be ready for testing as soon as it dried. I took some quinone crystals and put them on the paper and found that the paper immediately turned dark brown. The same thing happened when I squeezed a freshly dissected *Diploptera* gland on the paper.

I next had to find a way to get *Diploptera* to stay put on the paper so I could subject it to "assault." I devised a technique that I still find handy nowadays when I experiment with insects. I took individual *Diploptera*, attached a small aluminum hook to each one's back, then connected them by way of the hook to the end of a metal rod, and used the rod to position them on the paper. The hooks were attached to the roaches with dabs of wax. A small piece of plastic tubing provided the link by which the hook was connected to the rod. The hooks seemed not unduly to encumber the

Spray patterns on indicator paper generated by the discharged secretion of the cockroach *Diploptera punctata*. In the top photos, the roach has been stimulated by pinching, first a right leg and then a left leg. In each case the roach sprayed from the gland of the body side stimulated. In the bottom photos, the roach on the right was attacked by ants, but repelled these by spraying; the roach on the left has had its glands removed and is under persistent attack.

roaches. They were easy to pry off, so it was never difficult to restore a cockroach to its normal hookless status after testing.

The experiments worked like a charm. I wondered at first how to stimulate the roaches but soon decided what to do. I had to fake being an ant. Ants loom large in the world of ground-dwellers and there was no question that they were a real threat to *Diploptera*. So I placed my first roach on freshly made paper and "bit" it in the legs with forceps. Instantly, as if by magic, a broad pattern of spots materialized on the paper. When I pinched a right leg, the pattern appeared on the right side. When I then pinched a left leg, a second pattern appeared on the left. Not only was *Diploptera* using its glands defensively, it was using them unilaterally in a way that made sense.

I went on to stimulate the antennae and parts of the body. Consistently the roach responded by discharging from the side stimulated. I was also able to show that the roach blew the secretion out of the glands with respiratory air. If I severed the tiny tubes connecting the glands to the respiratory system, a procedure that I found easy to carry out surgically on anesthetized roaches, I ended up with *Diploptera* that couldn't spray.

The question of course was how would *Diploptera* fare under real attack. Ed Wilson had colonies of *Pogonomyrmex* ants in his lab and he was kind enough to let me use these. So I affixed some *Diploptera* to rods and positioned them in the foraging arena beside one of the ant colonies. The ants attacked at once, but no sooner had an individual ant succeeded in grasping a roach than it was forced to let go again. In some of the encounters I had placed indicator paper under the roaches. It was dramatic to see how the attacks triggered the appearance of spray patterns, and how the ants were thrown into flight at the same time.

Diploptera from which I had surgically removed the glands proved vulnerable, and so did newly molted individuals. When shedding its skin *Diploptera* casts away the saclike lining of the glands, thereby losing the gland contents. It takes the roach the better part of a day after molting to replenish its secretory supply, and until it does it is helpless. The adult female, when she undergoes the molt that transforms her into an adult, is therefore at first quite vulnerable. She is, however, sexually receptive right after the molt, and by mating with a mature, secretion-endowed male is able to take advantage of his defended status, at least for the hour or so that the two remain coupled.

But it was really the bombardiers that taught me how uncannily precise insect marksmanship can be. I took some beetles from their cages, and after fastening them to rods and positioning them on indicator paper, proceeded to assault them in ant fashion. I had taken special precautions to keep them from discharging when I put on their hooks. I used a brush to coax them from their containers into a saucer full of ice water, and then stuck the little "handles" on them when they became immobilized by the cold. They would recover within minutes and be warmed up and ready to go.

The bombardier beetles in the United States belong to the genus *Brachinus,* a term that means short-winged. Indeed, the tip of the abdomen in bombardiers projects beyond the wings and is therefore free to revolve in all directions. As I was to find out, the beetle takes full advantage of this capacity by using its abdominal tip as a gun emplacement. I had dissected some bombardiers and confirmed what others had already shown, that the two large defensive glands of the beetles open close together on the abdominal tip.

It was marvelous to watch the beetles use their gun turret. Within their little nervous system they seemed to have their entire periphery mapped out. No matter where I pinched them, they took aim, went pop, and hit the target. Unlike *Diploptera,* which seemed to emit their secretion as an atomized cloud, *Brachinus* discharged their spray forcibly as jets. And there was one more point. The bombardiers seemed never to shoot preemptively. They appeared to fire only when directly contacted.

As expected, the ants proved vulnerable. They were hit full blast no matter where they bit the beetles, and were instantly repelled. The beetles survived unscathed. As they fled, the ants rubbed their bodies in the substrate, and in an obvious effort to cleanse themselves, paused frequently to lick individual legs or antennae. I found that the beetles could spray more than 20 times before they exhausted their secretory reserves. If presented to the ants with depleted glands, they were overwhelmed.

I was eager to publish some of these results and I remember drafting an outline for a bombardier paper in the waiting room of the Boston Lying-in Hospital while Maria was delivering our second daughter, Vivian. Fathers were excluded from the delivery room in those days. Having the bombardiers to think about helped immensely in keeping me calm during the wait. I also wonder whether by such preoccupation, at the moment of Vivi's birth, I might have infused the little girl with the seeds of entomophilia, a condition I am happy to say she now displays in full. Like her older sister Yvonne, Vivi was utterly enchanting.

A few weeks after Vivi's birth, in September 1957, we packed up and moved to Ithaca, New York. I had accepted a faculty post at Cornell University and the time had come to bid farewell to Harvard. In the front seat of the car, where I could keep them under observation during the long drive, were my captive bombardiers. They were part of the family and were treated as such after our arrival in Ithaca. Teaching the introductory biology course was time consuming but I was determined to do research no matter what the circumstances. I found out that bombardier beetles were relatively easy to come by in the environs of Ithaca, so I knew I would be able to have them in adequate numbers in the laboratory at all times. It took longer than I had hoped to get my laboratory up and running, but even so within a few weeks of arrival I was doing experiments. The paper I

Spray patterns of four consecutive discharges of a bombardier beetle, elicited by pinching, first the front and rear legs of the right side (top), then the rear and front legs of the left side (bottom).

had begun writing while Vivi was being born had been accepted for publication and so had a second paper I had written on the defense of *Diploptera*. And there was exciting news from Germany that the components of the bombardier's secretion had been identified. The analytical work had been done by Hermann Schildknecht and his group in Erlangen, and what they had found was indeed benzoquinones.

■ **AMONG THE FIRST EXPERIMENTS** I did at Cornell were predation tests in which I offered bombardiers to *Hyla versicolor,* a species of tree frogs. The frogs proved willing initially to attack the beetles, but with experience tended to develop an aversion to them. In only a few cases did they actually swallow a beetle. They usually got sprayed the moment they flipped their tongue at the bombardier and were then left, quite often, frantically having to wipe the beetle from the tongue. The spectacle was clearly worth photographing, but where to get a camera?

I had been interested in photography since my teens, but it wasn't until graduate school that I came to realize how absolutely essential the camera was to anyone who wanted to study live insects. I was inspired by my mother, no doubt, who was an artist, but I also loved things visual just for their own sake.

I had been envious as a student of fellow students who could afford the small-format, single-lens reflex cameras that were coming into vogue at the time—the Exacta being a particularly fashionable model—and I vowed that if I ever got a grant, such a camera would be the first item I'd buy. In the meantime, though, I'd have to be patient and improvise.

I had a memorable experience in my first year of graduate studies. I was tinkering with an ancient 4 × 5 bellows camera in the basement photo lab of the Biological Laboratories, trying to photograph a green lacewing larva feeding on aphids. Because the camera had no provision for a flash attachment, I was taking long exposures in ambient light and my pictures were coming out blurred.

An elderly man working in the lab at the time noted my frustration. "Use hand-held flash," he advised, in a heavy accent. "Darken room, close lens down, open shutter, and trigger flash manually." Needless to say, that worked. We conversed about photography at length that evening and on subsequent days. He introduced himself as Roman Vishniac. He was on assignment for *Life* magazine, doing a story on Carrol Williams's work on insect metamorphosis. He let me watch as he staged his subjects, balanced the lighting, and composed the image. Although I was not to learn of his fame until much later, that experience sufficed to get me hooked.

At Cornell, grant money was not at hand during that first year, but two graduate students, Benjamin Dane and Charles Walcott, close friends both to this day, were willing to lend me their own cameras. I had known Charlie from his undergraduate days at Harvard and from one memorable occasion when the two of us managed to present some protozoans live to a television audience. The program was being broadcast from MIT by the fledgling station WGBH, in Boston. The transmission was live and everything had to work on cue. The protozoans in question were the wood-

Attack of a frog *(Hyla versicolor)* on a bombardier beetle.

digesting species from the gut of termites. We had dissected them onto a slide, and the slide was on the stage of the microscope when we noticed that light was leaking into the improvised connection we had rigged between the microscope and the television camera. We had to find a tube with which to shield the junction and we had to find it at once. Mary Lela Grimes, the program's presenter, was closing in on the segment where she would have to introduce our "beasties." There was only one thing to do. We ran to the nearest men's room, grabbed a roll of toilet paper, unwound it, and once we had the tubular core ran back to the studio and used the tube to shield the junction. The protozoans made their appearance on schedule, in what may have been a television first. What I remember best though, is the bewildered look on the men in the men's room, when we burst in and unraveled the roll.

With Charlie's Hasselbladt camera, and Charlie's patient help, we got some really good pictures. Three photos in particular, which spell out the sequence of events in an attack of *Hyla versicolor* upon a bombardier, convinced me that I could no longer exist without a camera. Those three pictures were reproduced in a Time-Life book on insects and made me feel that if I ever got the proper equipment I might be able to succeed as an insect photographer. There was certainly no doubt that I loved to capture events on film.

■ **IN 1961** a second seminal paper on bombardiers was published by Schildknecht's group. I had met Hermann Schildknecht in 1960, in Vienna, at a symposium on insect chemistry that I had organized in conjunction with the Tenth International Congress of Entomology. I had corresponded with him and with a number of others who were studying various chemical aspects of insect defense and communication and I thought it would be good to get us all together. It was a small group and it served to bring our shared interests into focus. Among the participants were Mario Pavan, who had been doing pioneer work on the chemistry of ants; Murray Blum, who was looking into various glandular defensive systems; John Edwards, who had just completed an interesting study of an African insect that sprayed its poisonous saliva in spitting-cobra fashion; and Miriam Rothschild, who introduced the concept of chemical mimicry and probably brought more ideas to the meeting than the rest of us put together. Schildknecht and I talked at length and it was clear that we were both intensely committed to bombardiers. The paper he was about to publish dealt with the explosive nature of the beetle's discharge mechanism and it was a pioneering piece of work.

To understand what goes on chemically when a bombardier shoots requires knowledge of the anatomy of the glands. When one dissects a bombardier the glands are easily exposed. They are an identical pair of organs, lying side by side in the posterior half of the abdomen. The most important thing about them is that each consists of two confluent chambers, an inner large, compressible reservoir, enveloped by muscles, and a smaller,

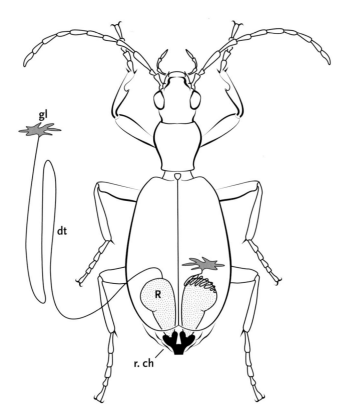

Diagram of a bombardier beetle with its two glands in place. R, reservoir; r. ch, reaction chamber; gl, glandular tissue; dt, duct.

rigid reaction chamber, interposed between the reservoir and the outer opening of the gland. The relationship of the two chambers is such that the contents of the reservoir must pass through the reaction chamber on their way to being expelled from the glands. The reservoir is ordinarily filled with fluid. The source of this secretion is a lobed mass of tissue connected to the reservoir by a highly coiled duct, longer than the beetle itself.

What Schildknecht's group had found is that the beetle does not store the benzoquinones as such in the glands, but instead stores chemical precursors called hydroquinones. It stores the hydroquinones in the reservoir together with hydrogen peroxide, an oxidizing agent. Ordinarily the hydroquinones and the hydrogen peroxide don't interact. But if appropriate enzymes are added to the mixture the two substances do interact and they can do so explosively. The bombardier has the right kind of enzymes and stores these in the reaction chambers. In a glandular ejection, which is initiated by compression of the reservoir, the reservoir's contents are forced into the reaction chamber and as a consequence are brought into contact with the enzymes.

BOMBARDIER

The mechanism of operation of the bombardier beetle glands. E, enzymes in the reaction chamber; R indicates that either a hydrogen atom (H) or a methyl group (CH₃) can occur at that site on the hydroquinone or quinone molecule.

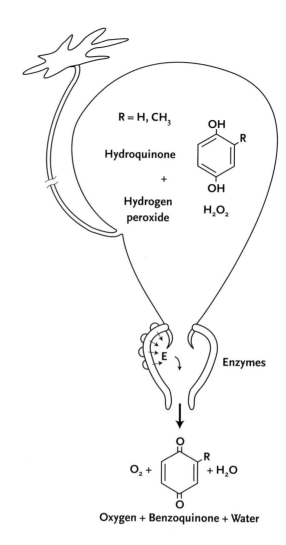

Oxygen + Benzoquinone + Water

As Schildknecht's group was to show some years later, the enzymes are of two kinds, catalases and peroxidases. In combination, the two do the trick. The moment reservoir fluid enters the reaction chamber, the catalases promote the breakdown of hydrogen peroxide into oxygen and water, while the peroxidases promote the oxidation of the hydroquinones to benzoquinones. In other words, oxygen is freed from the hydrogen peroxide, and that oxygen is used to convert the hydroquinones to benzoquinones. The liberated oxygen appears also to provide the force that expels the reacting mixture from the reaction chamber. The explosive nature of the expulsion supposedly accounts for the popping sound of the ejections.

And why so long a duct for conveying secretion into the reservoir? One possibility is that the great length of the duct protects the secretory tissue from fluid reflux when the reservoir is compressed during ejection.

It was a fine piece of work by Schildknecht's group. And it could explain something about the secretion that I thought was true but was being ignored. Bombardiers, I was convinced, discharged their secretions hot.

■ **I HAD FELT THE HEAT** when I first got "blasted" in Lexington, and I'd gotten used to feeling it every time I handled the beetles. I had even popped a bombardier in my mouth to see what a predator might experience. The "pops" felt hot, and I was quick to spit the beetle out. The foul taste lingered. Later I found out that Darwin himself had had such an experience. I quote from his *Life and Letters* because his words say something about the spirit that drives a naturalist.

> No pursuit at Cambridge [University] was followed with nearly so much eagerness or gave me so much pleasure as collecting beetles. It was the mere passion for collecting, for I did not dissect them, and rarely compared their external characters with published descriptions . . . I will give proof of my zeal: one day, on tearing off some old bark, I saw two rare beetles, and seized one in each hand; then I saw a third and new kind, which I could not bear to lose, so I popped the one which I held in my right hand into my mouth. Alas! It ejected some intensely acrid fluid, which burnt my tongue so that I was forced to spit the beetle out, which was lost, as was the third one.

I can't be sure that it was a bombardier that Darwin tasted, but I'd like to believe it was. It is certain that Darwin knew bombardiers. When in 1982 I visited Down House, Darwin's residence in Kent, I took time to check his beetle collection, which is on deposit there, and found that it included a pinned specimen of a bombardier.

Eventually, thanks to some exciting experiments, I learned that the bombardier's spray is indeed hot. The person who was instrumental in getting that proof was Daniel Aneshansley.

Dan was a first-year graduate student when he appeared in my office

Specimens from Charles Darwin's personal beetle collection. The arrow points to a bombardier beetle.

one sunny day in the fall of 1966 in search of a doctoral project. His field was electrical engineering so I didn't know whether we could find any common ground. As soon as we started chatting, though, it became clear that we would hit it off, so finding a problem of mutual interest became a priority.

I told Dan about bombardiers, and about the dramatic way in which they engender their spray, and added that I thought the spray was hot. Why not look into the thermal properties of the beetle's spray? Dan perked up, and within days joined our group.

For a start we got some help from two friends in the Chemistry Department, Joanne Widom and Benjamin Widom, who as physical chemists, we thought, should be able to tell us whether it made sense to expect the secretion to be hot. We knew the concentration of the two reactants, hydrogen peroxide and hydroquinones. Both were present at high concentration in the reservoir of the glands. Would enough heat be liberated in the process of interaction of these chemicals—in other words, in the process of benzoquinone formation—to heat the spray?

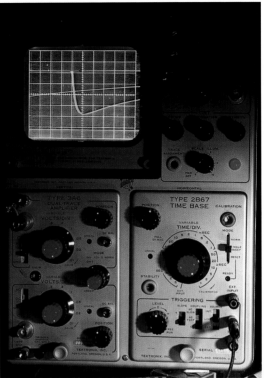

The answer, according to the Widoms, was a definitive yes. About a fifth of a calorie of heat should be produced—0.19 calorie to be exact—for every milligram of spray ejected. This was enough heat, they said, to bring the spray to the boiling point and to vaporize about a fifth of it. And what was the boiling point? About 100°C, they said, since the spray contains water as a solvent.

I was delighted with their prediction. But scalding hot? Was the spray temperature really that high? Dan was a fabulous gadgeteer and he was quick to create an electronic device for gauging spray temperature. In essence, what he did was to use a tiny thermistor bead as a thermometer. Thermistors are resistors commonly used in electrical circuits. The amount of current they conduct depends on their temperature. This means that if you calibrate a thermistor, so you know how much electricity it conducts at various temperatures, you can use it as a thermometer. Dan rigged up a thermistor in such a way that we could cause the beetle to spray directly upon it. By measuring the amount of current flowing through the thermistor at the moment it was hit, we could determine the

Measurement of the temperature of bombardier beetle spray. The beetle has been tethered so as to "fire" upon a tiny thermistor bead positioned just behind its abdominal tip. The beetle's discharge is registered on the oscilloscope screen. The downward deflection of the green line on the screen provides a measure of the spray's temperature.

A bombardier beetle with its rear stuck into the opening of the microcalorimeter. An electronic thermocouple within the chamber of the instrument registered the amount of heat released when the beetle was made to spray.

temperature of the spray. The value we obtained was quite consistently very close to 100°C.

That night neither of us could sleep. We lived for moments like these, and the fact that they came infrequently didn't matter. Maria and the kids shared in the excitement. We had three daughters by then, and "How are the bowbadeers" was the greeting I often got from Christina, our youngest.

Dan next measured the heat content of the spray. He built a microcalorimeter, in the form of a cylindrical chamber into which the beetle could be slid rear-end first. The cylinder bore a probe on the inside that poked the beetle when it was backed in, thereby causing it to spray. Again, an electronic means was used to measure the heat delivered with the discharges. The value obtained—0.22 calorie per milligram of secretion expelled—was astoundingly close to the 0.19 calorie predicted. We lost some more sleep.

We decided we'd publish in the journal *Science*. The paper was accepted in 1969, complete with cover photograph of spray droplets bursting forth in midair. The editors saved the paper for the July Fourth issue.

■ **DAN AND I** next set ourselves the goal of photographing the beetle in the act of spraying. We knew we had to resort to a special technique, since the discharges were of very short duration. On average they lasted only a few thousandths of a second. We knew this from sound recordings we had made of the ejections and from having measured the duration of the response of the thermistor to spray impact.

SCIENCE

4 July 1969
Vol. 165, No. 3888

AMERICAN ASSOCIATION FOR THE ADVANCEMENT OF SCIENCE

The cover of the 1969 *Science* magazine issue in which we reported our finding that the bombardier beetle spray is hot. The photo shows a close-up of the beetle's "fireworks," appropriate for the July Fourth issue.

We tried conventional photography but kept missing the discharges. It was simply impossible to release the shutter by hand in synchrony with the brief moment of spray emission. We would have to design a system whereby the spray itself acted to trigger the camera. In other words, we had to get the beetle to take its own picture.

We found that there was a relatively simple way of doing this. First of all, since the beetle was unwilling to pose for us, we had to tether it, so we made use of our standard technique of attaching it to a rod with a droplet of wax. We knew also that we would have to use a very brief and bright pulse of illumination, which meant using an electronic flash unit. The trick was to cause the unit to be triggered by the popping sound of the beetle's ejection. To do this we adopted a standard procedure. After focusing on the beetle in the camera's viewfinder, we dimmed the room lights to an absolute minimum and opened the shutter. Using forceps we then pinched one of the beetle's legs, causing it to spray. A microphone placed directly above the beetle picked up the accompanying sound, which was relayed instantly by electronic circuitry to the flash unit, causing it to discharge. The moment we saw the flash we closed the shutter again and got ready for the next picture. There were some skills we had to master to ensure that the procedure worked. For instance, after dimming the lights and opening the shutter, we had to work fast, to prevent the background illumination from giving us a ghost image on the film. But with experience we got to the point where pinching the beetle's leg was something we could accomplish in a second or two.

We took several rolls of film in our very first portrait session. Waiting for the pictures to be developed was sheer agony. But the reward was worth the wait. The beetles had taken fabulous pictures of themselves. The best one, from that session, was the very first picture of the first roll we shot. It shows the beetle, with abdomen bent forward, firing precisely toward the leg being pinched.

We eventually took many pictures. Most of them were of a large flightless African bombardier beetle that was being sent to us live from Kenya by Glen Prestwich, a former postdoctoral associate, now a prominent insect biochemist at the University of Utah. It is amazing how long bombardiers can live in captivity. Our African ones, *Stenaptinus insignis,* survived for over a year and we became very attached to them. To many we gave individual names and we came to think of them as partners rather than subjects. The pinchings we inflicted on so many of them to make them spray seemed to cause no injury. Their tough armor must in itself be helpful to them in nature.

The precision with which the bombardiers eject their spray is remarkable. When pinched in a leg, they do not simply fire in the direction of that leg, but take into account the positional orientation of the leg. If the leg happens to be lifted, they fire upward toward the leg, and if the leg is in a downward position they fire downward. They take into account, further, whether the stimulus is applied to the tip of the leg or to somewhere along the leg's length. In real life, therefore, no matter where a leg is bitten, or how that leg is positioned when being bitten, the predator cannot avoid being hit.

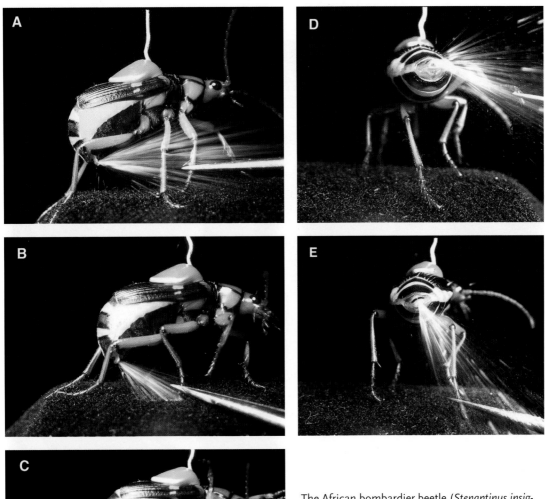

The African bombardier beetle (*Stenaptinus insignis*). Pinching of the right foreleg (A), midleg (B), and hindleg (C), and pinching of the tarsus of the right hindleg (beetle in rear view), with the leg in raised position (D) and in downward position (E).

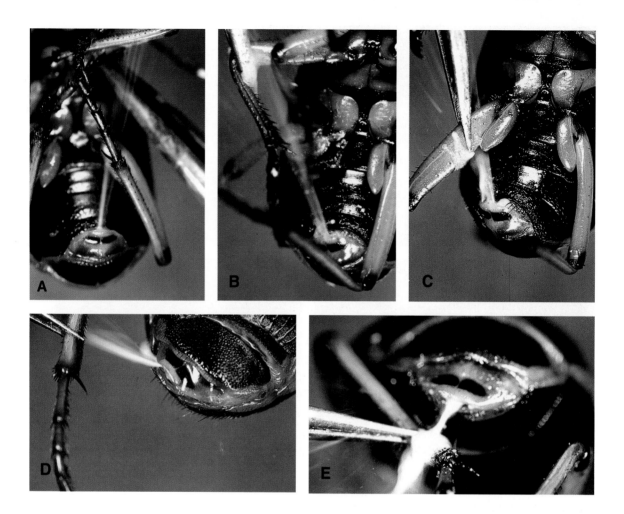

More pinching of the African bombardier beetle. Pinching of the base of the left midleg (A), of the distal portion of the femur of the right midleg (B), and of the base of the femur of the right hindleg (C), and pinching of the distal portion of the left femur, with the femur positioned beside the body (D) and positioned behind the abdominal tip (E). A-C show the ventral view of the abdomen, D shows the dorsal view of the abdominal tip, and E shows the end-on view of the abdominal tip.

The bombardier is even able to fire forward over its back. It does this by bouncing the spray off a pair of skeletal reflectors that it manages to stick out from the tip of the abdomen at the moment of ejection. Ants are therefore at risk even if they scale a beetle's back.

We also took pictures of the moment of an ant attack. For these we triggered the flash not by sound but by the output of an electronic sensor, a thermocouple, that responded to the heat of the spray. We tethered the beetle directly beside the sensor, so when the ant attacked the sensor would be hit at the same time as the ant. The technique worked. We only wish we had taken the pictures in color.

At that point we thought, rather smugly, that we had figured out everything there was to know about bombardier beetles. Little did we realize that the best was yet to come.

Dan Aneshansley (with pipe) and the author in 1978, photographing bombardier beetles in the process of discharging.

■ **A SECOND GRADUATE STUDENT** had joined us in the late sixties to work on bombardiers. Jeffrey Dean was a quiet, gifted, independent scholar who liked nothing better than to work on his own. He eventually went on to accept a professorship in Germany and to turn his attention to the neurobiology of muscle control. He is now back in the United States, on the faculty of Cleveland State University in Ohio.

Jeff's doctoral thesis had dealt with the predator-prey relations of toads and bombardier beetles. In an elegant series of experiments he had shown that the beetle's spray is improved by being hot. He had built an artificial bombardier consisting of a benzoquinone gun, with which he could stimulate the toad's tongue. He found that when the benzoquinones were delivered hot, the sensory nerve in the tongue was quicker to respond by sending signals to the brain. He also found that one large species of toad, *Bufo marinus,* gulps bombardiers down in one fell swoop, so that the beetle does not really get the chance to fire in time. It does discharge once it is in the stomach of the toad—you can actually hear the popping sounds through the body wall—but by then it is too late. The toad as a rule keeps the beetle down.

Jeff and Dan had independently made an observation. They had recorded the popping sound made by bombardiers when they fired and found the sound to be pulsed. They had taken their tape recordings, slowed them down on playback, and found the sound to have a distinct

An ant attack on a tethered bombardier beetle. The thermocouple that sensed the discharge and triggered the flash is seen projecting from the substrate just behind the beetle.

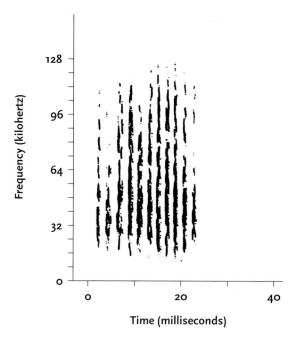

A sound spectrogram of a bombardier beetle's discharge. The pulsed nature of the sound is denoted by the parallel banding. The pulse duration in this particular discharge is approximately 2 milliseconds.

rat-tat-tat quality to it. The pulse repetition rate was very high, meaning that the tat sounds were following one another in very quick sequence. You can visualize this clearly in sound spectrograms, where the sound is plotted out in two dimensions on graph paper. The pulse repetition rate was in the order of 500 to 1,000 pulses per second. There could be several explanations for this, but the one that appealed to us was that the delivery of the beetle's spray during an ejection was discontinuous. The rat-tat-tat could be that of a rapid-fire machine gun. The image of a dental waterpick also came to mind, delivering its fluid at an exaggerated pulsation rate.

A simple experiment proved that we were right. We tethered a bombardier and set it up in such a way that it could be made to discharge on a piezoelectric crystal. Such a crystal has a very special property. If you deform it, even slightly, it produces a burst of electricity. Deform it intermittently, and it produces intermittent electric bursts. The prediction was that if the spray was pulsed it would hit the crystal discontinuously and elicit from it a discontinuous electric response. That is exactly what happened. We made simultaneous recordings of the sound of the ejections and found that the discontinuity in the sound matched the discontinuity in the crystal's electric output. The spray, in other words, was emitted in bursts, and for every burst there was a pulse of sound. We were indeed dealing with a machine gun, one that delivered fluid at an unprecedented pulsation rate,

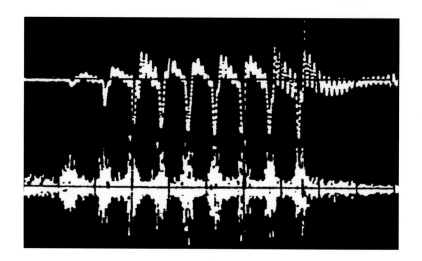

Simultaneous recording of the sound (lower trace) of a bombardier beetle's discharge, and of the electronic output of a piezoelectric crystal (upper trace) on which the beetle directed that discharge. Note the precise peak concordance of the two traces, indicating that for each pulse of sound there is a pulse of secretory impact on the crystal.

and with acoustical accompaniment. Could we somehow get to see that gun in action?

■ **THE IDEA, OF COURSE,** was to film the ejections. But we would need to use someone else's camera. Our own motion picture camera had a maximum filming speed of 400 frames per second, not fast enough to record an event with a repetition rate of 500 to 1,000 per second. But I knew that there were faster cameras in existence and I had heard a great deal about a man at MIT who had them. I always wanted to meet that man.

I forget how I got in touch with Harold Edgerton, whether by phone or by letter, but his response was as warm as I had come to expect, judging from his reputation. He gave me a choice of dates. "Try to be here before eleven, and I'll take you to lunch," he said. "Bring your beetle and we'll see if we can film it in the afternoon." And he added, wistfully, "I'm not sure I believe a word you said about the critter."

So on September 26, 1976, shortly before eleven, I showed up at "Doc" Edgerton's lab. "I've got my freshman seminar at eleven. Come on. You don't mind, do you? I told them about your beetle. Why don't you tell them about it yourself?" The class was immensely enjoyable, as was lunch. The man I had admired from a distance for so many years, whom I knew as the inventor of the electronic flash, the stroboscope, and high-speed photography, and who was by all accounts larger than life, was more than anything else fun to be with.

He showed me around his lab before we got started. There was equipment everywhere, cameras and flash units, stroboscopic apparatus, motors and generators, plus miscellaneous gadgetry of every imaginable kind. But everything had its place, and a neat corner had been prepared for the high-speed film we were to shoot. The camera to be used was a 16-millimeter Fastax, and we were to shoot at 2,680 frames per second. The film I had brought was TriX, the fastest I knew of, and I had brought it in 200-foot rolls. At the intended filming speed, a 200-foot roll gives you less than 4 seconds of filming time. I gulped. I would need to pinch the beetle's leg precisely within that period or we'd miss the action. I knew from measurements we had made that the beetle responds to pinching in a matter of 100 to 300 milliseconds. So the beetle was not going to be a problem as long as I pinched it soon enough after the camera started rolling. In fact, said Doc, I should listen to the camera's sound and delay the pinch until the sound reached an even high pitch—in other words, until the camera reached the desired speed. That would take about half a second. I memorized my task. I would first bring the forceps to within millimeters of the beetle's leg, and then, when I heard the camera, wait an instant and give the pinch. Oh yes, said Doc, and watch your eyes. The flash gets to be pretty bright. How I was supposed to watch my eyes and the beetle at the same time escaped me.

Doc's Fastax was a prismless camera, meaning essentially that it had no shutter. The film in such a camera runs by an open gate, and is exposed by the pulsation of the flash. Doc's flashes were special because they could be set at such high repetition rates. The film goes through the camera so quickly when you shoot at thousands of exposures per second that the last feet of the roll are usually shredded. The result is that the camera is always in need of a thorough cleaning between takes.

With the help of two of Doc's assistants, Bill MacRoberts and Charley Miller, we shot three sequences that afternoon—three shots, from three beetles. MIT being MIT meant that development could be done in-house, and within the hour we were holding the wet film against light to see whether we had captured the action. We had, and there could be no doubt. The spray delivery was pulsed. The films were truly spectacular. Each pulse was clearly resolved. The repetition rate of the pulses was well in line with what had been predicted from the acoustical and electronic data.

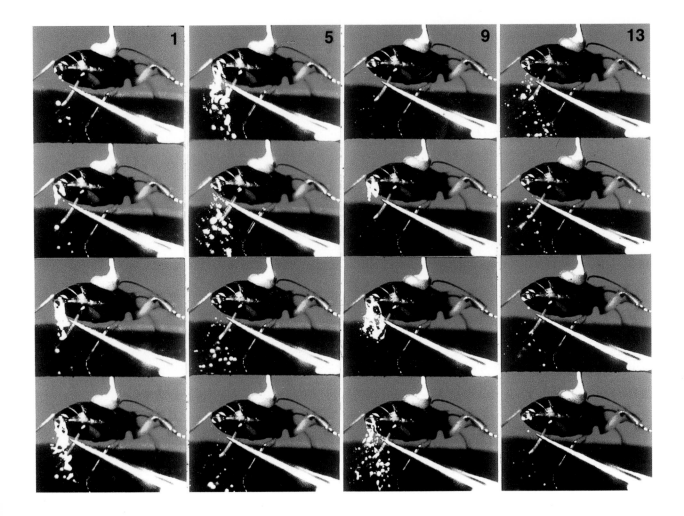

Excerpt from a motion picture film taken at 4,000 frames per second, showing two full pulsation cycles of a bombardier beetle's discharge. The four vertical sequences are read downward, beginning with the sequence on the left.

We stayed in touch with Doc, and went to see him again in November 1981, this time to film the beetle's ejections in color, at 4,000 frames per second. The results were even more dramatic.

■ **IT SEEMED TO US** that we were now in a position to explain the pulsation mechanism. We postulated that the individual pulsations represent individual microexplosions, repeated one after the other as the beetle delivers its spray. Critical to the operation of such a cyclic mechanism is the maintenance of sustained pressure on the reservoir through contraction of the muscles in its wall, and the oscillatory opening and closing of the

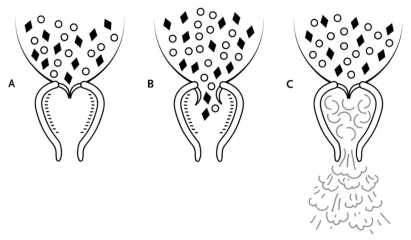

◆ Hydroquinone
○ Hydrogen peroxide
⁗ Enzymes

A   B   C

The pulsation mechanism. In A, at rest, the valve leading from the reservoir into the reaction chamber is closed by its own springlike elasticity. In B, the intake phase, the reservoir is under muscular compression and some of the reactants are forced into the reaction chamber. In C, the exhaust phase, the explosive interaction of reactants and enzymes causes a rise in pressure that closes the valve, with the result that the products of the explosion are ejected to the outside. After the reaction chamber has vented itself, the cycle recommences by returning to B, and it continues to repeat itself for as long as the reservoir remains under muscular pressure. The result is the pulsed emission that characterizes the discharges.

valve that controls access to the reaction chamber. We imagined that this valve oscillation occurred passively, without the help of muscles. To initiate an ejection all the beetle needs to do is compress the reservoir. When the resultant pressure overcomes the occlusory force of the valve, reservoir fluid passes into the reaction chamber, initiating the explosive chemical events. As oxygen is liberated in the chamber and temperature and pressure rise, the point is quickly reached where the valve is forced closed. This results in a further rise in pressure in the chamber until the chamber vents itself, causing its contents to shoot out. With the chamber emptied and its internal pressure restored to a level below that exerted on the reservoir, the cycle promptly recommences, and it will continue to repeat itself as long as the reservoir is compressed.

The beetle derives distinct advantages from the pulsation. For one thing, the pulsation enables it to dose the delivery of the spray. Quantity of output being set quite precisely by the volume of the reaction chamber means that the beetle probably expels fairly constant quantities of secretion per pulse. This means that it is able to exert control over the total amount of secretion that it expels per discharge simply by controlling the duration of compression of the reservoir. Spray delivery can thus be maintained constant over time, with the length of the pulse train adjusted to the magnitude and duration of an attack. The beetle, in other words, ejects liquid bullets of constant caliber, at a rate that is automatically set, in a quantity that it can control.

The pulsation also provides the beetle with an alternative to muscular force for ejecting the spray at a high velocity. The pressure that the beetle needs to exert on the reservoir to get the discharge going and to sustain the ejection is probably relatively modest. The explosive force that empties the reaction chamber is doubtless higher and is engendered by chemical rather then muscular means.

The pulsation may also keep the beetle from getting an overheated rear during ejections. You can envision the periodic infusion of cold reservoir fluid into the reaction chamber, such as occurs in the course of the ejections, acting to prevent overheating. The beetle's machine gun, it seems, may be cooled by its own reloading. Keeping the reaction chamber from getting too hot may be particularly helpful in preventing the enzymes from undergoing thermal denaturation.

◼ **THERE IS A SAD SEQUEL** to the story. When in December 1989 I wrote up these results, I phoned Doc to ask whether he might be willing to be a coauthor of our paper. He initially resisted, claiming that he had not contributed sufficiently to the work. Upon my insistence that we could never have done the study without his camera, his stroboscopic flashes, and his know-how, he relented but said he wanted to think it over and would let me know. He eventually did, in a letter dated December 19, which I treasure. "I offer you full cooperation with whatever you have in mind," he said, concisely and to the point. We proceeded to write the paper and I added the final touches shortly after New Year's, on Thursday, January 4. I tried to call Doc the next day to alert him that I was sending the manu-script to get his input, but it was too late. Doc had died on that very Thursday, from a heart attack suffered while he was having lunch at the MIT faculty club. He was eighty-six. He never saw the *Science* paper he co-authored, but the three of us, Dan Aneshansley, Jeff Dean, and I, think of it as having been written in his honor.

◼ **WHEN IN LONDON** some years back, in an antiquarian store, I came upon a charming little book that I could not resist buying. Published in 1819 under the title *Dialogues on Entomology,* and intended for the "little pupil," it takes the form of conversations between a mother and daughter. It contains the following passage:

> *Mother.* . . . I shall mention only one [species of carabid beetle], the
>   *crepitans,* or, more commonly, the *bombardier.*
> *Lucy.* What a very droll name mamma; what does it mean?
> *Mother.* Bombs are hollow iron balls, filled with gunpowder: they burst
>   at a distance from the place whence they are thrown, with a very loud
>   report, and do great execution: they are generally used in the attack
>   on fortified towns, and the men, whose business it is to fire them
>   from mortars, or short, wide cannons, are called bombardiers.
>   Our beetle bombardier has been thus named from a singular noise
>   that it is enabled to make, as a means of defence. It is seldom known
>   to use its wings . . . and when pursued, or touched, its assailant is sur-
>   prised by a sudden noise, like the discharge of a diminutive mortar or
>   bomb.
> *Lucy.* A most excellent name for it, indeed.

Bombardier beetles have not always been called bombardiers. The earliest account I know of, in the *Proceedings of the Royal Swedish Academy of Sciences* of 1750 (I have seen the German translation) speaks of a *Schuss-fliege* (shooting fly) encountered by the student author, Daniel Rolander, under a rock. Rolander noted that the animal gave off both sound and smoke *(Rauch)* when first captured, and that it could be made to fire some twenty times when tickled *(gekitzelt)* with a needle. He dissected the bee-

The bombardier beetle, recognized at last.

USA 33

Bombardier Beetle

tle and thought he could detect a gland. *"So ist die Natur in ihrem Werke wunderbar und manichfaltig,"* he concluded—"So is nature full of wonder in its diversity". He told Linnaeus about his find.

■ **THE BOMBARDIER BEETLE** even won praise from the Pulitzer Prize–winning humorist Dave Barry. The Entomological Society of America had been lobbying Congress to name the monarch butterfly the official national insect. Barry had come upon the results of a survey that indicated there was support for other insects as well, including the bombardier beetle.

Pointing out that he was not "making this up," Barry described the bombardier as having "an internal reaction chamber where it mixes chemicals that actually explode, enabling the beetle to shoot a foul-smelling, high-temperature jet of gas out its rear end with a distinct 'crack.'"

The Time-Life insect book "has a series of photographs in which what is described as a 'self-assured bombardier beetle' defeats a frog. In the first picture, the frog is about to chomp the beetle; in the second, the beetle blasts it; and in the third, the frog is staggering away, gagging, clearly wondering how come it never learned about this in Frog School." "I would be darned proud," added Barry, "as an American, to be represented by this insect. An engraving of a bombardier beetle emitting a defiant blast from his butt would look great on a coin."

I wrote Barry, congratulating him on his remarks. He was running for President at the time. In his reply he graciously asked whether I'd be willing take the post of "Biologist General" in his administration.

I fear it'll be a while before we see a bombardier on a coin. The beetle did, however, quite recently make it in philately. When the insect stamps issued on October 1, 1999, by the United States were in the planning stage, I was asked to write the descriptive text for the back of the stamps. When I received the preliminary sketches of the insects that had been selected, I noted that Barry's favorite beetle was not among them. I suggested that the oversight be corrected, and it was. The United States now has a bombardier beetle stamp.

2

Vinegaroons and
Other Wizards

By the end of my first year at Cornell I was starved for fieldwork and somewhat at loose ends. I had made up my mind that I would take a broader look at the chemical defenses of insects, but I wasn't quite sure that I could handle the chemistry myself. I thought I'd enlist a collaborator, but was uncertain whether chemists would be sufficiently intrigued by the topic. At any rate, it seemed that I should first do a bit more of what I was good at: going into the field and seeing if I could come up with some chemical mysteries worth solving. And I thought that I would not restrict myself to insects. There were other arthropods worth studying, other members of that large group of animals that share with insects the possession of a segmented body, an exoskeleton, and jointed appendages. Millipedes, centipedes, and arachnids were high on my list of arthropods to watch. So off I went to Florida in the summer of 1958 and postponed my search for a chemist until the fall.

The next year I met Jerrold Meinwald, not because I was looking for a chemist but because I was looking for musicians. As a keyboard player I was eager to meet instrumentalists in need of an accompanist, and a mutual friend thought I might enjoy meeting a flutist. The flutist turned out to be not only a marvelous musician but a natural products chemist of precisely the right inclination.

I don't have exact recollections of my first meeting with Jerry, except that I know we clicked right away. Here was someone who could talk about a molecule as if it were a jewel, someone who was as familiar with Carl Philip Emmanuel's sonatas as with those of "Papa" Bach himself, someone with an unexaggerated sense of self, a quick smile, and a true passion for science. I was to learn later that he was also a man of immense generosity, unbending in his liberal convictions, and genuinely interested in biology. My experience with chemists had been that although they could be interested in functional biological explanations, they were closed to adaptive reasoning. They could show interest in how it works, but not in how it got to be what it is. The mere thought of evolution caused their eyes to glaze over. Evolution was one of my passions and realizing that Jerry was open to evolutionary reasoning made the prospects of our collaboration even more attractive. As I recall, the outcome of our first lunch was an agreement that we would team up. I had a field trip to Arizona planned for the forthcoming summer. "Let me see what I can bring back for us," I said.

In the weeks ahead we had our first music session. We were both good sight-readers, so Jerry had simply brought a pile of music, from which we picked and chose. From the very first measures it became clear that we "felt" music the same way. We agreed on tempos, on retards and ornamentation, and even on the music we liked best. I couldn't believe my luck. Here we were, at the mere threshold of our joint scientific venture, and we had already found splendid common ground. As I recall, it was Bach, Handel, and Telemann that day. We shared a love of the baroque. Coincidentally, we had in some respects a shared musical background. In Uruguay I had briefly studied piano with Fritz Busch, the noted conductor who as a refugee from Hitler's Germany had taken up residence in South America. Jerry had studied flute with Marcel Moyse, of the famed Marlboro Music School in Vermont, which had been founded by Fritz Busch's brother, the violinist Adolf Busch.

Jerry and I were to have many joint musical sessions over the years. We went so far as to give a number of recitals, usually in connection with joint lecture engagements or scientific meetings. We even belonged to the Union together in the late seventies, when as part of a trio, we played baroque music at brunch time on Sunday mornings in various eating establishments around Ithaca. In the late seventies I conducted an amateur orchestra at Cornell. Its name was BRAHMS (Biweekly Rehearsal Association of Honorary Musical Scientists) and we played under the motto "we're not as bad as we sound." A high point in the history of that orchestra was the evening when Jerry was one of the soloists in the Vivaldi concerto for two flutes.

Jerry and I continue to collaborate to this day. We have published over 200 joint papers, of which, as of this writing, 183 are in the series Defense Mechanisms of Arthropods. Dozens of our doctoral students, undergraduate research students, and postdoctoral fellows are coauthors of these publications.

■ **MY DESTINATION** in Arizona in the summer of 1959 was the Southwestern Research Station of the American Museum of Natural History in Portal. I had been to the Arizona desert before, on my field trip with Ed

Wilson 7 years earlier, and had promised myself at the time that I would make every effort to return at the earliest opportunity. What I found so extraordinary about a desert is that it is in so many ways an open stage. With plants spaced out and the line of sight unobstructed, life's events in the desert tend to unfurl in full view. The desert is also a highly competitive place, where resources are at a premium, and where survival is often a matter of defense. Arthropods in the desert, I was to learn, are indeed well defended, and their weaponry is mostly chemical. I had come to the right place.

What I had not previously experienced is a desert at night. It is hard to convey in words what it is like to take a first stroll through an unfamiliar desert at dusk, and to continue through the night into the dawn, unaware of the passage of time. Every observation is new, every fragrance, every sound, unfamiliar. You gradually come to make associations, a certain chirp with a cricket, a terpenoid whiff with a bush, a track in the sand with a millipede. You are driven by anticipation, by the expectation of discovery. You walk slowly, casting around with your headlamp, hoping to reveal rather than to startle. There is life everywhere, vertebrate and invertebrate, on the ground, on vegetation, and in the air. And there is predation, death, copulation, egg laying, and birth. The experience is overwhelming at first, but the observations eventually cross-connect and lead to queries. Why are there so many black beetles in the desert, active mostly at dusk? Is black to be viewed as bleakness, or is it a form of advertisement? So many black desert beetles stink when handled. Does it not make sense that they should proclaim their noxiousness by being conspicuous? And what better way to achieve conspicuousness against sandy soil in twilight than by being black? I decided that it might be worthwhile to study these beetles.

Early one evening I came upon what can only be described as the most beautiful of monsters. Dark brown and scorpion-like but with a slender filamentous process at the rear in lieu of the stinging tail, it was obviously an arachnid, but an unfamiliar one. There was an attitude about the animal that suggested it was defended. It walked slowly and deliberately and made no effort to flee when touched. It simply halted in its tracks, raised its front, and, with its armlike front appendages spread wide, assumed a threatening posture. The animal was big enough to fill my palm, but I didn't risk picking it up. Instead I coaxed it into a plastic container—

A fully grown whipscorpion. The "whip" at its rear is a feeler. The knob at the base of the whip (shown enlarged in the bottom picture) is the revolvable "gun emplacement." The arrow denotes the site of spray emission.

of which I always have a supply in my collecting bag—and took it back to the station. I had found my first whipscorpion. Mont Cazier, the director of the station, set me straight on the taxonomy. Its name was *Mastigoproctus giganteus* and it was the only species of its kind in the United States. "And watch this," Mont said, as he gave the animal a vigorous tap. "It's got one hell of a spray." He let me have a whiff of his hand. "What does it smell like?" he asked. "Vinegar," I said, suppressing a cough. "Strong vinegar." "You got it," said Mont. "Its common name is vinegaroon."

The odor was unmistakably that of acetic acid, the chief ingredient of vinegar, but that compound could not be the only one in the secretion. After dissipation of the initial pungent odor, there was always a rancid residual stink that persisted for hours. There had to be at least one other component in the fluid.

I collected several additional whipscorpions and put them in the same cage, only to find that they ate each other. The secretion evidently did not protect against cannibalism. I then confined them singly, and presented them to predators. I had young whipscorpions, little enough to offer to harvester ants *(Pogonomyrmex occidentalis),* and big whipscorpions that I could offer to grasshopper mice *(Onychomys torridus).* I did the tests with the ants directly beside the nest entrance of a natural colony. I tethered the whipscorpion to a rod, as I had done with bombardier beetles, and set the animal down among the ants. The ants attacked instantly and caused the whipscorpion to spray, but they were repelled at once. As they fled, they flailed their legs spastically and dragged their bodies in the soil, as if they were attempting to clean themselves. Each discharge was followed by a period of invulnerability during which vapor, or residual secretion remaining on the whipscorpion's body, appeared to deter further attack. The effect was dramatic because next to the nest entrance the ants, typically, were present in swarms. Following a discharge, the whipscorpion would be seen amidst the swarm, but with not a single ant directly on it or immediately beside it. Gradually the swarm would close in again, only to be dispersed by the next ejection.

The grasshopper mice, which my two students and I had trapped locally, were also thwarted. To quote from the account that a group of us eventually published, "The moment the whip scorpion was introduced, the mouse pounced upon it, but on contact was instantly repelled by a discharge. It jumped back and scurried about frantically, pausing occasionally to paw its muzzle . . . or to dig . . . itself into the sand . . . When the reaction subsided, after about 30 sec, the mouse always attempted to attack again, only to be (once more) repelled . . . After the fourth or fifth attack it finally desisted, settling in a corner, . . . its eyes closed and fur ruffled, breathing deeply and quickly." I might add that the mice recovered fully. Grasshopper mice are ferociously aggressive, and—as I was to learn from feeding tests—remarkably tolerant of insect defenses. The results with the whipscorpions were exceptional.

Eventually I did tests with solpugids (sunscorpions), lizards, birds, and an armadillo, and found that the whipscorpion could deal with them all. Small wonder whipscorpions have been around for over 300 million years. The spray could have been the very key to their endurance. Interestingly, the vernacular names given whipscorpions elsewhere are also suggestive of acetic acid. In Martinique they are known as *vinaigriers,* in Mexico as *vinagrillos,* and in Brazil as *escorpioes vinagre.* I collected a number of whipscorpions to take back to Ithaca, and on the last days at the station attempted to learn a bit more about their natural history. They appeared to be strictly nocturnal and territorial. Individuals seemed to return to the same shelter every day. At the station they could be found in cracks between the foundation stones of the main buildings. Next to their hiding places were piles of sowbug remnants. These little crustaceans can be assumed to have been around for about as long as whipscorpions and could have been part of the whipscorpion's diet since before the advent of insects.

The 3 weeks in Arizona went by all too quickly, but the experience was exhilarating. I had uncovered literally dozens of problems and in the process fallen in love with the desert. The late Frances Ann ("Andy") McKittrick and George Happ, the two graduate students who had joined in the trip, were also profoundly affected by the adventure. With a car full of live animals, and lacking air conditioning, we were forced to drive mostly at night on our way back to Ithaca, to avoid the heat of the day. We made it in less than 4 days, but not every motel took us in along the way. We had difficulty in some cases convincing the managers that our diverse caged denizens were harmless and qualified as pets.

I called Jerry the moment I got back. "How would you like to work on a prehistoric monster?" I asked. "Literally on a 300-million-year-old piece of artillery?"

He was enthusiastic. "But the chemistry is likely to be boring," I cautioned. "I think the beast sprays acetic acid." "Let's have a look," he said, and we set up a time to meet in the instrument room of the Chemistry Department.

Before we met, I did some experiments that proved the whipscorpion to be a great marksman. We had come to expect this. Arthropods that spray appeared quite consistently to have control over the directionality of their ejections. Again I used indicator paper for demonstration of the dis-

VINEGAROONS AND OTHER WIZARDS

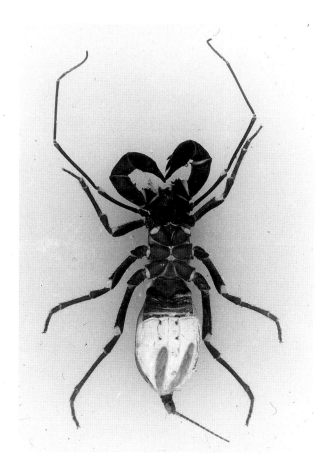

A whipscorpion in ventral view, dissected to show the two glands that release the spray.

charges, and I tethered the whipscorpions to rods. The indicator fluid I used was an alcoholic phenolphthalein solution, to which I added a bit of potassium hydroxide to turn the fluid red. I soaked sheets of filter paper in the solution, laid the papers out on glass, blotted them off, placed the tethered whipscorpions on them, and executed my "attack." Actually, it took no more than a mild pinch to get the vinegaroons to spray. They discharged accurately toward the appendage stimulated and proved capable of ejecting in all directions. The pattern of the spray registered beautifully in white against the red of the paper. I had dissected a whipscorpion and found it to have two large glands in the opisthosoma, the posterior body part that arachnids have in lieu of an abdomen. The glands open side by side, at the end of the small knob that forms the base of the animal's "whip." To aim its spray, the whipscorpion adjusts the posture of the "abdomen," revolves the knob, and fires. I managed to get as many as 19 discharges from a single individual.

A whipscorpion discharging four times in succession, in response to the pinching of individual appendages. On phenolphthalein-impregnated paper, the spray registers in white.

I will always remember our first analytical attempt because it was quite funny. I had told Jerry that I thought the secretion was pretty concentrated, to judge from the intensity of its odor. "If that's the case," he said, "we should be able to take its infrared spectrum without using a solvent."

VINEGAROONS AND OTHER WIZARDS

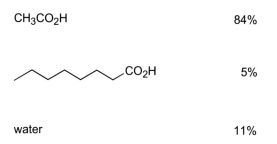

CH$_3$CO$_2$H	84%
⌁CO$_2$H	5%
water	11%

The composition of whipscorpion spray: acetic acid (top), caprylic acid (middle), and water (bottom).

He explained to me that we would need to squeeze a drop of secretion between two disks of potassium bromide, which would then be inserted, like a little sandwich, in the path of the infrared beam in the spectrophotometer. "Can you get some secretion on one of the disks?" he asked. "Sure," I said, "and what's more I can get the whipscorpion to do it." So I took the disk and instructed Jerry to hold it precisely above one of the whipscorpion's legs. I pinched the leg and, presto, we had secretion on the disk. Jerry took the second disk, pressed it against the first, placed the pair in the instrument, and minutes later we had the spectrum. It *was* acetic acid! The whole operation had taken only minutes. What a breeze, I thought. This collaboration was going to be fun.

Eventually the presence of acetic acid was verified by additional analytical procedures, and a value was obtained for its concentration in the secretion. It was 84 percent. Vinegar indeed, I thought.

Jerry and his assistants also identified a second compound in the fluid, the chemical responsible for the persistent rancid odor of the spray. It was caprylic acid, and it was present in a concentration of 5 percent. The remaining 11 percent of the spray was water.

Knowing exactly what the secretion was made of raised some questions. It occurred to me that the caprylic acid, despite its low concentration, might have a double function. It could serve at once as a wetting agent, to promote spread of the spray droplets on the enemy surface, and as a penetration-promoting agent, to increase the permeability of that surface to the spray. It could thus be very important in increasing the effectiveness of the secretion.

Acetic acid, while potent in its own right, particularly at a high concentration, suffers from a shortcoming. It is a water-soluble compound, without affinity for a lipid surface. On the wax-impregnated integumental surface of an insect (waxes are lipids), a droplet of acetic acid would tend not to stick but to roll off. Mixed with a small amount of caprylic acid it could

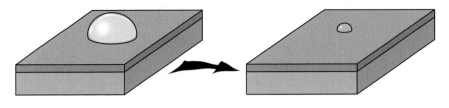

Acetic acid

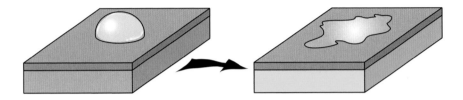

Acetic acid + caprylic acid

When a droplet of acetic acid is placed on a piece of insect cuticle it merely shrinks by evaporation, without spreading. Mixed with a small amount of caprylic acid, the droplet spreads broadly. The caprylic acid also causes the fluid to penetrate through the cuticle, as shown here by the color change in the layer of gelatin beneath the cuticle (the gelatin has been impregnated with bromthymol blue, which changes from blue to yellow in the presence of acid).

not only stick but spread, and might penetrate the surface much more readily. Caprylic acid could have these effects because as an oil it has an intrinsic affinity for lipids. Its properties are predicted by its molecular structure. Unlike acetic acid, which contains only two carbon atoms, caprylic acid contains eight carbon atoms in a row. It is because it consists of a relatively long carbon chain that caprylic acid is lipophilic (literally lipid-loving). The short-chained acetic acid is hydrophilic (water-loving). The chief enemies of whipscorpions are likely to be insects and other arthropods. It is therefore most often the waxy arthropod integument that the whipscorpion's spray must traverse, and caprylic acid could well be the "additive" that makes that penetration possible.

I did a simple experiment that showed caprylic acid was a spreading agent. I made up mixtures of acetic acid and caprylic acid in various proportions and applied droplets of the fluids to the surface of pieces of insect integument, pieces of the insect cuticle, or exoskeleton. I used cockroach cuticle because I had cockroaches in culture and because the smooth ab-

VINEGAROONS AND OTHER WIZARDS

domen of a cockroach provided the sort of cuticle pieces best suited for the test. I found that droplets of acetic acid retained their droplike shape on application and simply shriveled away as they evaporated, while droplets containing caprylic acid spread broadly.

I was also able to demonstrate the penetration-promoting effect of caprylic acid. Trying to come up with an appropriate experimental set-up, I had found that if I mounted a piece of cockroach cuticle on a layer of gelatin impregnated with an indicator dye, I could apply droplets of acidic fluid to the surface of the cuticle and check on their penetration by watching for a color change in the gel. As an indicator I used bromthymol blue, which turns from blue to yellow in the presence of acid. Again, droplets of acetic acid alone were relatively ineffective. They penetrated slowly. But with caprylic acid as an add-on, the droplets penetrated more quickly.

I used yet a different cockroach preparation to show that caprylic acid acts to shorten the response time of the living animal to surface application of acidic mixtures. I had developed that preparation at Harvard, while still a graduate student, for an entirely different purpose. I had been assisting in the introductory biology course one summer, and was present on that most horrible occasion when, in the name of neurobiology, the entire class had to take part in the frog-pithing ritual that would yield the spinal preparations needed to demonstrate reflex behavior. The idea was to destroy the brain of these poor frogs—to pith them as the expression goes, meaning to scrape out the cranial cavity with a pin—so that they could then be used to show how they scratched themselves when a droplet of acid is placed on the back. The purpose, I suppose, was to show that some behavior can run its normal course without use of the brain. I hated that lab, and so did many students and most of the professors I knew. So I decided to concoct an alternative preparation.

I had been working on a little research project on roaches—on how they absorb fat from food—which required my killing some individuals by quick decapitation so I could extricate the gut. I had noted that the resulting headless roaches remained viable for days and responsive to stimuli, including chemical stimuli. In fact, the headless roach was a vastly more instructive preparation than the pithed frog. The entire body surface appeared to be "mapped out" in the central nervous system of the roach. Wherever I stimulated the animal—I used formic acid, applied with a brush—whether on an appendage or on the body, there always followed a

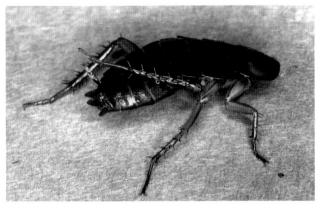

A decapitated cockroach *(Periplaneta americana),* responding to topical application of a mixture of acetic acid and caprylic acid. In the top left photo, I applied a droplet of the mixture to the right abdominal flank of the animal, causing it to scratch the site with the right hindleg. In the top right picture, I applied a droplet to the tip of the right hindleg, causing the animal to flex that leg forward in such a fashion that it would ordinarily be accessible to the licking action of the mouthparts. In the bottom photo, I applied a droplet mid-dorsally to the base of the abdomen, causing the animal to arch its back, in an attempt to bring the stimulated site into contact with an over-hanging surface. Had I provided such a surface, say by holding a pencil against its back, it would have moved forward and wiped its back against the pencil.

cleansing reflex accurately directed at the stimulated site. When I stimulated an abdominal segment, the roach scratched that segment. When I stimulated a leg, it brought that leg forward to a position that ordinarily would make it accessible to the licking action of the mouthparts. When I stimulated an area of its back that it couldn't reach with a leg, it arched its back upward in an obvious attempt to rub it against an overhanging structure. I eventually published the technique in *Turtox News,* a journal for teachers, in hopes that it would be adopted and would liberate frogs forever from the experience of being pithed. But no such luck. Pithed frogs are still used in courses to demonstrate reflex behavior, and cockroaches are being avoided, as they are in general.

In addition to lending itself to the demonstration of reflex behavior, my headless roach preparation illustrates another fact: that animals can sense chemicals with the body surface. One does not deal with olfaction or taste, in the traditional sense, when one applies acid to the cockroach's body and the animal responds by scratching, or when one stimulates the frog with acid and it scratches. One is then dealing with a special sense, sometimes called the common chemical sense, which animals have over greater or lesser parts of their body, and by which they are able to sense noxious substances such as irritants. The irritation I felt in my eyes when I got too close to a discharging whipscorpion was indicative of the common chemical sense at work. And so was the irritation I felt from the vapor of the bombardier beetles' benzoquinones. In fact, it seemed to me that pungent chemicals in general, sprayed by arthropods, might repel not so much because of odor or taste but because of irritancy, an irritancy detected by the prey through the common chemical sense. Could the ants that fled from the whipscorpion after a discharge, and that dragged their bodies and flailed their legs as they escaped, be "itching all over"? And was that not precisely the effect that the whipscorpion spray was designed to elicit? I suddenly envisioned survival in the arthropod world being a matter of "letting the enemy have it" by stimulating its common chemical sense.

My cockroach preparation was also nicely suited for assessment of the irritant effectiveness of acetic and caprylic acid. I took headless roaches—the reason headless roaches are needed for the test is that they don't run away when stimulated—and applied droplets of test fluid to their abdomen. My criterion for irritant effectiveness was the length of the delay between application of the droplet and the onset of scratching. The shorter the delay, the more powerful the irritant. As one might have predicted, acetic acid alone was less effective than when it was mixed with caprylic acid. Through its double action—the promotion of spreading and penetration—caprylic acid had the net effect of hastening the onset of the scratch response. The tests also showed that caprylic acid itself has the capacity to irritate. The roaches scratched quite vigorously when stimulated by caprylic acid alone.

I had learned something about how the secretion affects the predator, but had left one fundamental question unanswered. How does the whipscorpion produce acetic acid at such a formidable concentration without poisoning itself? Acetic acid is a general toxicant, the kind of compound which on account of general chemical reactivity can be a hazard to any living system. It is inconceivable that there should exist cells capable of

A mound of the European ant *Formica rufa* being tapped by hand. Hundreds of resident ants are responding by ejecting their formic acid–containing spray into the air.

tolerating exposure to concentrated acetic acid. In concentrated form the compound should be a universal poison. Yet the whipscorpion is able somehow to produce and store a solution that is 84 percent acetic acid. It appears also to tolerate the incidental dousing it inevitably receives when it discharges its spray.

The vinegaroon is by no means the only arthropod to have a specially formulated defensive secretion. Presence of surfactants—additives that promote spread and penetration—is now known to be common in arthropod secretions. The whipscorpion is also not alone in producing an acid at high concentration. Ants of the subfamily Formicinae, for instance, may spray formic acid at concentrations in excess of 50 percent. There is also a carabid beetle, *Galerita lecontei,* that ejects a spray containing 80 percent formic acid. The vexing question of tolerance of toxicants was evidently a more general one. Secretions containing 80 percent formic acid should be no easier to produce and store than those containing 84 percent acetic acid.

MY INTEREST in millipedes dates back to my boyhood in Uruguay and it was revived when I explored the outskirts of Ithaca. There is a particularly beautiful state park north of the town, Taughannock State Park, with a spectacular waterfall, where I used to take the family to picnic and where we would sometimes as a group search through the leaf litter for insects

A carabid beetle, *Galerita lecontei*, spraying in response to the pinching of a foreleg and a hindleg. The secretion consists of 80 percent formic acid. The indicator paper is the same as that used in the whipscorpion experiments.

and other arthropods. The gorge leading to the waterfall was a particularly good site for millipedes, including some of a group I was especially interested in, because its members were reported to produce hydrogen cyanide. Some of the reports were anecdotal, but others, including one dating back to 1882 by a Dutch investigator named C. Guldensteeden-Egeling, were credible. I assumed that millipedes produced hydrogen cyanide for defense, and that the animals exercised some level of control over release of the compound, but it was by no means clear how they might be able to do this. I was also intrigued by the fact that there should even exist arthropods capable of producing hydrogen cyanide, although after my work with the whipscorpions I had begun to believe arthropods are capable of virtually anything. In a way, though, hydrogen cyanide production seemed even more impressive than production of a spray containing 84 percent acetic acid.

So in what became a family effort we collected those particular millipedes we thought might be hydrogen cyanide producers and brought them live into the laboratory for study. Maintaining them proved relatively easy. All they needed for food was leaf litter. The project was to profit from the help of yet another family member. My parents, in 1958, decided to retire in Ithaca, and my father, eager for involvement in chemical work, was quick to join in the millipede venture. At the age of sixty-eight he was still willing and able to help us in our collecting efforts. He was a welcome addition to the lab. Good natured in the extreme, he bonded quickly with everyone, and worked with us for nearly 20 years.

Hydrogen cyanide (or HCN) is of unmatched notoriety in the annals of crime. Known also as prussic acid or simply as cyanide, and said to have a characteristic odor of bitter almonds, it is quickly fatal in small quantities. It is as nearly universal a poison as one can find, since it effects its toxic action by interfering with certain molecular respiratory processes upon which most living cells depend.

HCN production has been well worked out in plants. The compound is a gas at ordinary temperatures and therefore rather difficult to store. Instead plants store HCN in combination with an aldehyde, to form a cyanohydrin, which in turn is bound to one or more sugar molecules. The HCN-aldehyde-sugar combination is called a cyanogenic glycoside, literally a cyanide-generating, sugar-containing molecule. To liberate HCN from the glycoside, two enzymes are required, a glycosidase to remove the sugar and a nitrilase to split off the aldehyde. Ordinarily the plant stores the glycoside and the enzymes in separate cells or parts of cells, so that under normal conditions no HCN is generated. But when there is injury, as when an herbivore bites into the tissue, the glycoside and enzymes are brought into contact and cyanogenesis is triggered. This deters the herbivore. The classic example of a plant cyanogenic glycoside is amygdalin, from bitter almonds. In amygdalin the sugar is glucose and the aldehyde is benzaldehyde.

Millipedes belong to the class Diplopoda. The species purported to produce HCN all belong to one order, the Polydesmida. Several species of polydesmoid millipedes occur in Ithaca, and one, *Apheloria corrugata,* appealed to us the most. *Apheloria* was not only beautiful, with its yellow and pink markings, but relatively easy to come by. With an adult body length of 5 centimeters it was also of a nice size for experimental study.

Most millipedes have their defensive glands arranged in rows along the sides of the body. This holds true for Polydesmida, as it does for millipedes of several other orders that produce benzoquinones and phenols. In these other orders the glands consist of individual sacs, each drained by an exit duct that leads to a pore on the body wall. The exit duct, near the pore, is ordinarily inflected, forming what is essentially a valve that keeps secretion from leaking out of the sac. A muscle that inserts on the inflection and originates on the body wall serves to pry open the valve. Contraction of this muscle clears the duct for secretory discharge. The sac compression that presumably effects the discharge cannot occur through action of "squeezer" muscles because the sacs lack such muscles. Instead, sac compression is probably brought about by a momentary telescoping of the body segments, an action that would exert pressure on the internal body fluids and indirectly on the sacs. I had become familiar with this kind of gland, because I had begun studying some of the benzoquinone-producing millipedes. I was therefore quite surprised to find that the glands of *Apheloria* were somewhat different. Each consisted of the familiar structures—the sac, the exit duct, and the muscle-operated valve—but interposed between the valve and the outer opening was a second chamber. It was as if the body wall had become invaginated at the level of the original gland opening, to form the new chamber and a new outer opening. I was immediately struck by the similarity of this two-chambered arrangement to that prevailing in bombardier beetles. The parallel was striking and raised the question whether it might extend to glandular operation as well. Was the second chamber in the millipede a reaction chamber, a housing for the enzyme or enzymes that promoted the chemical "ac-

The breakdown of amygdalin (left) into its constituents: sugar, benzaldehyde, and hydrogen cyanide.

Top: The millipede *Apheloria corrugata* in its typical habitat. Bottom: A close-up view of the right flank of the same millipede, showing two droplets of cyanogenic secretion being emitted from the glandular pores.

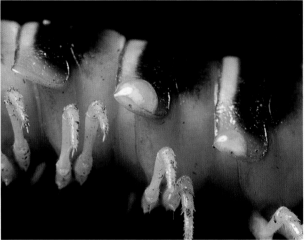

tivation" of the secretion? And could the first chamber be analogous to the bombardier's reservoir in that it stored the substrate or substrates upon which these enzymes act? My initial hypothesis was that the reservoir in *Apheloria* contained a cyanogenic glycoside such as amygdalin, and the reaction chamber a mixture of glucosidase and nitrilase. If I was correct, then the *Apheloria* secretion should consist of HCN, benzaldehyde, and glucose.

Working together with Jerry's group, my students and I looked into the composition of the secretion. The *Apheloria* gave off the typical odor of bit-

VINEGAROONS AND OTHER WIZARDS

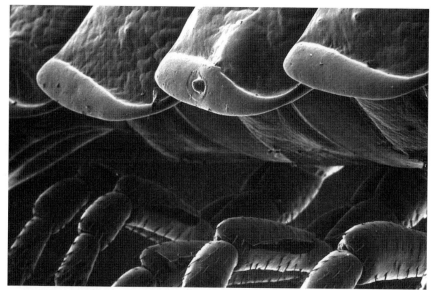

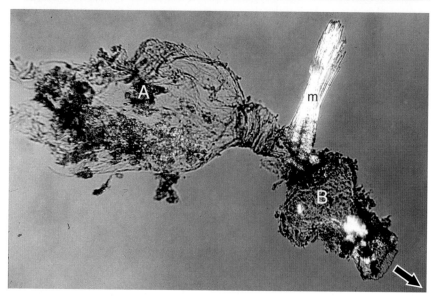

Top: A close-up view of the right flank of *Apheloria corrugata*, showing a glandular pore. Bottom: A freshly excised gland, showing its bi-compartmented structure. The inner reservoir (A) connects to the reaction chamber (B) by way of a valve operated by an opener muscle (m). The reaction chamber opens directly to the outside by way of a glandular pore (arrow).

ter almonds when we first handled them, and we were impressed by the potency of the aroma. Ordinarily, when kept unmolested in their cages, they were quite odorless, but they were quick to emit their stink whenever we disturbed them. We noticed that the onset of the stink was coincident with the appearance of secretory droplets at the gland openings, so there was little doubt that the odor came from the secretion. Oddly, in *Apheloria*, as in most polydesmoid millipedes, the glands are present only in segments 5, 7, 9, 10, 12, 13, and 15 to 19. Discharges did not necessarily involve emission from all glands at once. Typically, in response to localized stimulation, such as the pinching of a leg with forceps, the animal responded by discharging from the nearest gland or glands only. The discharges were never copious. A single droplet is all that a given gland ever produced. The droplet was not forcibly expelled but simply oozed from the gland and remained clinging to the gland opening unless it was somehow wiped away.

A simple experiment showed that disturbed *Apheloria* do indeed emit HCN. Using a circulating pump, we bubbled the vapor from a container of agitated millipedes through a silver nitrate solution, thereby forming a precipitate of silver cyanide. The precipitate was shown to liberate HCN on treatment with hydrochloric acid. The presence of HCN was also demonstrated in the secretion itself on the basis of several other diagnostic criteria.

Additional chemical tests revealed the presence of benzaldehyde in the secretion, but not of sugar. We searched carefully for sugar, using a number of analytical procedures, but all tests proved negative. This result forced us to modify our hypothesis. Our assumption now was that the reservoir of the gland stored a cyanohydrin rather than a glycoside, and specifically the cyanohydrin combining HCN and benzaldehyde. That specific cyanohydrin is called mandelonitrile, and it is an oil.

We analyzed fresh secretion and found that it did indeed contain mandelonitrile. Also, as was to be expected, the secretion, on standing, as it gave off HCN, underwent a change in composition. While the content of mandelonitrile gradually decreased, the benzaldehyde content increased.

We next did some tests with the glands themselves, and specifically with the isolated reservoir and reaction chamber. It proved possible to dissect the glands intact, and to separate the reservoir from the reaction chamber, without loss of contents of either glandular component. The reservoir had to be handled with care because it is membranous and easily punctured. The reaction chamber, just as in the bombardier beetle, is rigid and more resistant to breakage. In subjecting these chambers to microchemical testing, we made use of a simple color test for HCN, based on an indicator solution (a benzidine acetate–copper acetate reagent) that turns from colorless to blue in the presence of the compound. We also used a commercially available enzymatic preparation (emulsin) that strongly catalyzed the liberation of HCN from mandelonitrile.

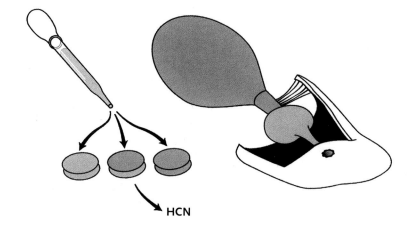

Experimental demonstration of the mechanism of cyanogenesis in *Apheloria corrugata*. HCN production (demonstrated by the addition of a droplet of indicator solution) occurred only when the contents of a storage chamber were mixed with those of a reaction chamber. The pairing of two storage chambers or two reaction chambers failed to elicit HCN liberation.

We found that if we drained the contents of a gland reservoir and reaction chamber separately on a piece of filter paper, and applied emulsin plus indicator solution to the drainage sites, a blue color would appear only where the reservoir had been drained. This established that the cyanogenic material is restricted to the reservoir.

We next excised reservoirs and reaction chambers, drained these on filter paper, and cut the individual spotted sites from the paper. The resulting paper disks were then pressed together in pairs one over the other, and each double disk was tested with indicator solution. The test was negative when the reservoir contents or reaction chamber contents were paired with their own kind, but intensely positive where the match was between a reservoir and a reaction chamber. This experiment proved that it is through contact of the contents of the two chambers that HCN is liberated. Since the reservoir was the proven source of the undissociated cyanogenic compound, the factor that causes the emission of HCN had to reside in the reaction chamber. We presumed that factor to be enzymatic, and therefore heat-labile. Indeed, we found that if we heated the disks with drained reaction chamber contents to 130°C for 20 minutes, we destroyed their HCN-liberating capacity.

The defensive gland of *Apheloria* is an admirably refined weapon. The storage of mandelonitrile provides the animal with a convenient means of retaining relatively large quantities of HCN in stable form. Release of HCN is timed to coincide with moments of emergency, when mandelonitrile is forced through the reaction chamber and thereby exposed to the dissociating catalyst. The mechanism may be no different in its essentials in other polydesmoid millipedes, although we know from more recent work by a number of investigators that the secretion in some spe-

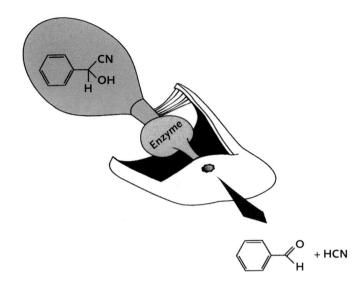

The cyanogenic mechanism in *Apheloria corrugata*. The major compound in the reservoir is mandelonitrile. When, on being discharged, mandelonitrile is forced through the reaction chamber, it is enzymatically broken down into benzaldehyde and HCN.

cies contains additional compounds. In fact, as we ourselves determined, *Apheloria* itself contains a second cyanogenic compound in the reservoir, benzoyl cyanide, which on passage through the reaction chamber can be expected to be broken down into benzoic acid and HCN.

Tests that we did with predators showed the secretion to be effectively protective. We found that *Apheloria* is capable of fending off ants and of being rejected by toads. The rejection from toads was not consistent. It seemed that the millipedes were helpless if gulped down in one fell swoop. Under such conditions they may literally lack the time to discharge. A millipede seemed to be rejected mostly when the toad had taken it crosswise in the mouth and needed to reposition it prior to swallowing. It was in the course of such repositioning that the millipede was usually spat out. The toad would then often be left pawing its tongue, in obvious discomfort. In no case was a millipede regurgitated by a toad once it had been swallowed, nor did the toads show delayed ill effects from the meal.

It is worth noting that the benzaldehyde may itself contribute to the defensive action of the secretion in the case of ants. Benzaldehyde is intrinsically repellent to ants, and because of its relatively low volatility may actually prolong the period of effectiveness of the discharged fluid.

The amount of HCN produced by a millipede may be considerable. We found adult *Apheloria* (body mass about 1 gram) to have a maximal HCN output of 0.6 milligrams. This is equivalent to 18 times the lethal dose (LD)of a 300-gram pigeon, 6 times the LD of a 25-gram mouse, 0.4 times the LD of a 25-gram frog, and 0.01 times the LD of a human.

VINEGAROONS AND OTHER WIZARDS

An American toad *(Bufo americanus)* rejecting an *Apheloria corrugata* that it had just taken into the mouth.

Production of HCN among arthropods is not restricted to millipedes. It occurs also in certain centipedes, beetle larvae, and moths.

■ **THE BLACK BEETLES** that my assistants and I had found to be so noticeable and widely distributed at our Arizona site were easy to maintain and we took numbers of them back to Ithaca. They had modest dietary requirements and got along well with one another in cages. Cereal and water appeared to be all they needed, although we spiked their fare occasionally with some lettuce, banana peel, and pieces of carrot. Like bombardier beetles, they were surprisingly long-lived as adults. Some survived for almost 2 years, making us wonder whether a long adult lifespan was a characteristic of insects that invest heavily in chemical defense.

We learned eventually that the beetles belonged to the family Tenebrionidae, and that we had several species on our hands. Most discharged secretion when anyone picked them up, and there seemed little question that the fluid contained quinones. Unlike bombardiers, though, they sprayed their secretion cold. While still at our desert site we had done some studies of their behavior and of their vulnerability to predation (see Chapter 6), but the reason we wanted them back in Ithaca was so that we could look into the composition of the secretion. The chemical work panned out, but more important, we got clues from these beetles to what

Left: A European HCN-producing moth of the genus *Zygaena*. Right: An Australian beetle larva of the genus *Paropsis*. When disturbed, such larvae commonly defecate, which is in itself a defensive response. They also evert a pair of glandular pouches from just anterior to the anus, from which they emit hydrogen cyanide.

Left: A female centipede *(Orphnaeus brasilianus)* guarding her eggs. The female is here responding to disturbance by emitting a sticky gelatinous fluid. The secretion is cyanogenic, and potently defensive. Right: Two ants that were repelled by a close relative of such a millipede and became contaminated by the millipede's secretion are here shown stuck together by the fluid.

had been mystifying us: How is it that arthropods can produce the potent toxins they use for defense without poisoning themselves?

We chose to concentrate on the largest of these beetles, which we identified as *Eleodes longicollis*. Its glands—two large sacs opening side by side near the abdominal tip—provided secretion in ample amounts. Jerry's group was quick to confirm that the beetle did indeed produce benzoquinones. In addition the secretion turned out to contain three unsat-

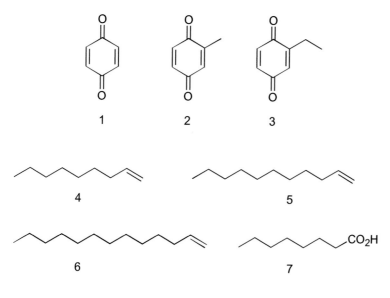

Components of the defensive secretion of *Eleodes longicollis:* 1, 1,4-benzoquinone; 2, 2,methyl-1,4-benzoquinone; 3, 2,ethyl-1,4-benzoquinone; 4, 1-nonene; 5, 1-undecene; 6, 1-tridecene; 7, caprylic acid.

Left: The beetle *Eleodes longicollis,* in dorsal view, dissected to reveal the large quinone-producing glands in the rear of the abdomen. Right: The abdominal tip of the beetle, with glands attached. Overnight immersion of the tip in potassium hydroxide solution caused all soft tissue components to be dissolved away, thus reducing the glands to their cuticular linings.

urated hydrocarbons and caprylic acid. The hydrocarbons, with chain lengths ranging from 9 to 11 carbon atoms, had a distinct odor that "came through" in the secretion, masking to some extent the odor of the benzoquinones. Finding caprylic acid brought pleasure. The compound in all probability played a surfactant role in *Eleodes,* just as it did in the whipscorpion. Independently , the two animals evolved a common solution to a shared problem.

Because of the size of the beetles, dissection and study of the glands

A: Freshly dissected defensive gland of *Eleodes longicollis*. White glandular tissue covers much of the surface of the gland. A muscle (m) operates the gland opening. B: An enlarged view of a portion of the glandular tissue. Note the clear vacuolar region in the center of the cells. C: An enlarged view of a single glandular cell showing the cuticular drainage apparatus within the vacuole. D: A pair of cuticular drainage apparatuses that were isolated by treating secretory cells with potassium hydroxide. The treatment also isolated the filamentous ducts that connect the apparatuses to the storage chamber of the gland.

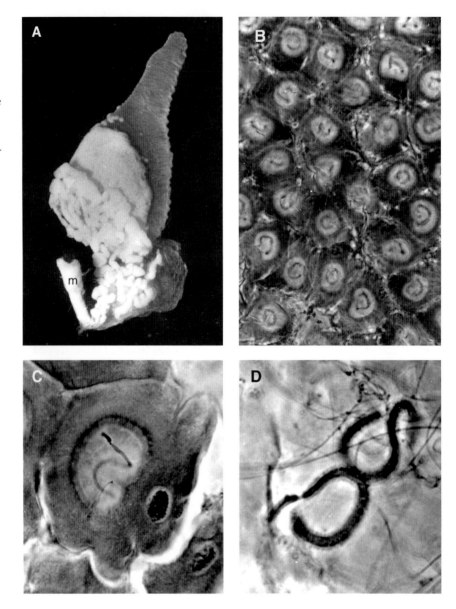

posed no problem. Notable, first of all, was that each gland consisted of a single sac, a reservoir, in which the secretion was stored as a finished product. There was no subdivision here into reservoir and reaction chamber, as was true of the bombardiers and *Apheloria*. The sac was thin-walled and ordinarily filled nearly to capacity with the brown secretion. Two types of glandular tissue were directly associated with the sacs. The tissues formed two layers, arranged one over the other, and covering part of the surface of each sac.

VINEGAROONS AND OTHER WIZARDS

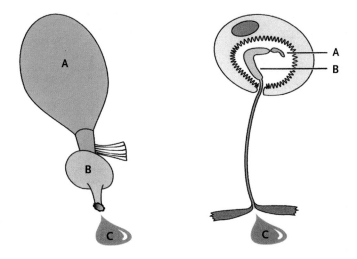

Mechanisms of poison production in arthropod defensive glands. In reactor glands such as those of bombardier beetles and *Apheloria* (left), the precursors (A and B) are secreted by different populations of cells into different subcompartments of the gland, and are not mixed to form the products (C) until the secretion is ejected. In the cellular model (right), the precursors (A and B) are produced by cells of one type, but kept from reacting until they are mixed within the cuticular drainage tubes that convey the precursors from the cells to the storage chamber of the gland.

Examination of the cells of these tissues at higher magnification showed right away that they were "special." Unlike cells generally, which usually are crammed full of material, these cells had large internal vacuoles, large inner spaces, mostly clear, that seemed to take up much of the volume of the cells. And, most interesting, inside these vacuoles were peculiar tubular structures that you could trace from the vacuoles to the wall of the sac itself. It seemed as though the cells were equipped with an inner chamber, the vacuole, into which they could secrete their products and also with a tubular drainage duct by which these products could be channeled to the sac.

More careful examination, particularly with the electron microscope, revealed that things were a bit more complicated. The vacuoles were, strictly speaking, not internal cellular "pools," but external spaces physically engulfed by the cells. By secreting into the vacuolar spaces, therefore, the cells were actually ridding themselves of material. They were literally excreting their products to the outside.

The ducts that drained the vacuoles did not simply arise open-ended within the vacuoles. Instead, where they lay within the vacuoles, they took the form of a more or less elaborate apparatus, tubular and subcompartmented, as if modified along their length for special uptake purposes. The vacuoles and their contained apparatuses imparted a very characteristic appearance upon the *Eleodes* cells.

There is a technique widely used by entomologists that involves dissolving away all soft tissues of an insect so that it is reduced to its skeletal remnants. All it entails is submerging the dead insect for some hours in warm dilute (10 percent) potassium hydroxide. The technique is particularly

VINEGAROONS AND OTHER WIZARDS

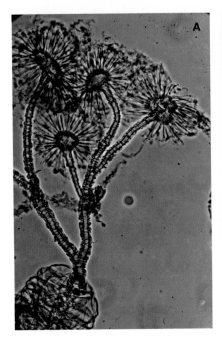

Cuticular drainage systems isolated (by potassium hydroxide treatment) from the secretory tissue of defensive glands of carabid beetles. B, C, and D are from quinone-producing glands of various bombardier beetles; A is from a formic acid–producing species *(Galerita lecontei)*. All pictures were taken by conventional microscopy except B, which was taken with a scanning electron microscope.

useful if you want to reveal internal structures that have cuticular—skeletal—components. Treating *Eleodes* by this procedure showed clearly that the glandular sacs were cuticular. This was to be expected, since the defensive glands of arthropods generally are infoldings of the body wall, and as such carry linings of cuticle. Treatment with potassium hydroxide reduced the *Eleodes* glands to two cuticular sacs. But what was most interesting is that the vacuolar drainage ducts themselves, complete with their intravacuolar apparatuses, survived the treatment. They too were made of cuticular material. It appeared to me that this had significant implications, for it could indicate that the beetle is in effect shielded from its own secretion from the moment the products are secreted into the vacuoles. If, as seems reasonable to assume, the cuticle of both the sac and the tubules is impermeable to the secretory products, then the only risk from exposure that the beetle incurs is at the level of the secretory cells themselves.

But suppose that these cells, instead of secreting actual toxins, produce harmless precursors that are kept from interacting until they are introduced into the drainage tubules? Could the secretory cells produce the precursors in separate parts of the cell body, and secrete them into different parts of the drainage tubules? Could different parts of the highly complex intravacuolar portion of the tubules be specialized for uptake of different chemicals?

VINEGAROONS AND OTHER WIZARDS

Such a cellular model for toxin production, if correct, would not be unlike the model operating on a macroscopic level in the glands of bombardiers and *Apheloria*. The difference would be that whereas in the glands of the latter two animals the reactants are secreted by separate populations of cells and stored in separate subcompartments of the gland for eventual mixture at the moment of discharge, in *Eleodes* they are secreted on a micro scale by different portions of the same cell and kept separate until introduced into the cuticular drainage ducts.

We now know that cuticular drainage ducts are a general feature of insect defensive glands. Many older European publications on the fine structure of insect glands attest to that. I have seen spectacularly beautiful examples of such duct systems and have isolated the ducts in some cases by potassium hydroxide treatment. My assumption is that it is within the insulated confines of these cuticle-lined drainage systems that the final synthetic steps take place by which arthropods produce their defensive toxins. The animals are therefore shielded from direct exposure to the toxins during the formative stages of the chemicals. They are also shielded against the final products, by the cuticular lining of the glandular sacs themselves.

3

Wonders from Wonderland

The introductory biology course, that first year at Cornell, had gone well. I organized the lectures around the unifying theme of evolution and made every effort to incorporate the new molecular information that was redefining the biological frontier at the time. DNA was not quite a household word yet, and it was wondrous to sense the intellectual excitement engendered in the students when they came to grips with the reality that the genetic material was now understood in chemical terms. Much of the credit for the success of the course belonged to the graduate students who assisted me that year. They were truly a remarkable lot. There was Jo Davis, a mammologist and talented artist, who was later to become associated with zoos. There was also Irwin Brodo, a born teacher and naturalist, destined to become the world's leading specialist on lichens (and author, in the year 2001, of a spectacularly beautiful book on these little-appreciated organisms). And there was Roger Payne, naturalist and visionary teacher, who was eventually to achieve fame as the world's authority on whales.

One of the great benefits of the academic year is that it leaves the summer months—the insect season—free for fieldwork. I knew all along that I would abscond the moment I finished grading the final exams and I had my mind set on Florida. I knew the panhandle but had never been to the peninsula itself, although the area had long been on my list of "musts." The further south the warmer, I reasoned, and the warmer, the better for bugs. I was raring to go.

I asked Roger whether he would like to come along and he said yes, with enthusiasm. Roger was really interested in birds. His doctoral research—on how owls locate prey by sound in darkness —had little to do with insects but he was ready for a diversion. We agreed we would explore together, each driven by his own interests. So we rented a car from Cornell's fleet of vehicles and took off. At the last minute, Joseph Nowosielski, a graduate student in entomology, decided that he too would join us, so we left as a trio. We had all the gear we needed, including vials of all sizes, some large insect cages, Coleman lanterns for night work, a borrowed camera, a Coleman stove, pup tents, and sleeping bags, but very little money. We were loaded to fender-scraping capacity, as we were reminded every time we hit a bump.

We went by way of the Smokey Mountains, where we camped and had occasion to observe daddy-long-legs as they came to life at night and carabid beetles as they fed on snails. I noticed that the daddy-long-legs emitted odorous secretions when disturbed, fluids that years later I was to investigate jointly with Jerry Meinwald. As we made our observations we were ourselves also observed at close range by a bear. We tried unsuccessfully to conceal from each other that we were quite terrified of such animals.

I recall also having long discussions on that trip about matters political. Both Roger and I were becoming convinced that being a biologist meant having to speak out on behalf of conservation. For me a decisive experience had been hearing as an undergraduate a lecture by Karl Sax, the author of *Standing Room Only*. The world was getting too crowded and nature was disappearing. Roger and I agreed that we would both need to become activists. This meant, for starters, including lectures on the human predicament in the introductory course when we taught it again in the fall. And so we did. I also recall Roger mentioning that he would want eventually to work on animals that could inspire concern. Take whales, said Roger. People really could be made to care about whales.

Roger followed through on those dreams. Years later, in 1968, when he was already famous as a whale expert, he was to provide me with one of the most memorable experiences of my life. I was to lecture at Rockefeller University, and Roger, who was on the faculty there, was my co-host. "I have something to show you," he said. "Let's have a glass of wine first, and then go to my lab." When we arrived, he sat me down in an easy chair and put a set of earphones over my head. "Now don't say anything. Just listen." I relaxed, and let myself be overwhelmed by the songs of humpback whales. Fresh out of the ocean, they had so far been heard by few. Roger knew he was on to something precious. Millions, including Judy Collins and Alan Hovannes, would eventually agree.

The trip was going well except that we hadn't counted on the Florida rains. We were drenched each day and were constantly running out of dry clothes. Not even the tents provided refuge. At Myakka State Park, inland by some 40 miles from Sarasota, Florida, we got caught in a torrential thunderstorm that soaked us to the bones. It had been raining off and on all day, weather that kept the insects in hiding and us pretty frustrated. Tired out, we had our hopes set on the dryness of the tents. Alas, we had not counted on the raccoons. They had broken into what we thought was our private domain, consumed our watermelons, and in the process made an utter mess of things. We didn't sleep much that night, and by the time we packed up and went on our way the next morning, we were feeling a bit dejected. Little did we know that we were about to stumble upon paradise.

From Myakka we followed Route 70 east, and then, about midway across the state, turned south on Route 8. Our intention was to head for the Everglades. Instead, after only a short distance following the turn, we came upon a sign that we found inviting: Archbold Biological Station. We turned right into the entranceway and found ourselves on a secluded path, flanked by pines, leading after a quarter of a mile or so to a cluster of what were clearly the main buildings. We rang the bell and were greeted by a gray-haired, middle-aged man who quietly and politely asked us where we were from and what we were doing in Florida. I was struck by his rather formidable looks and by his shy demeanor. He reminded me of someone but I couldn't make the connection. "We are here on a lark," I said. "We found you by chance."

"Do you want to stay at the station?" he asked. "We can put you up." It was a wonderful offer but I worried that our rag-tag appearance might be a problem. "The clothes we're in are the best we've got," I said. He ignored the comment. "I'll show you to your quarters," he said. "Dinner is at six. I am Richard Archbold."

We were ushered one flight up, shown our rooms and a private bath, and told we could stay as long as we liked. And yes, there was a washer and drier available to us. As soon as we were by ourselves we looked at each other in disbelief. Fresh beds and air conditioning. Was this for real?

We showered, got dressed, and went downstairs. It was late afternoon and the humidity hung heavily in the air. If we could get to some lights after dark, we might have the thrill of sampling the insect fauna. We had a quick look around and realized we were in unfamiliar surroundings. It was very much an open sort of habitat. Sandy terrain, lots of palmetto and scrubby oaks, plus much else that we didn't recognize. We knew the next days would be adventurous. The station owned some 1,500 acres.

It was an hour before dinnertime, and Richard Archbold was just outside the front door, in the parking area, scattering seeds and other edibles to an assemblage of the local fauna, including bunnies and a variety of

Richard Archbold.

birds. This was a daily ritual, we were told, in which Mr. Archbold took delight. Prominent among the animals that gathered for the feeding were birds called scrub jays, which were endemic to the area. They were tame and readily took food from Mr. Archbold's hand, and would even fly casually to his shoulder to beg for morsels they were sure to get. What a wonderful setting, I thought, for feeding insects to a bird, to check on which insects were edible and which were noxious.

Richard Archbold bid us to dinner, which meant joining him at his private table in the station's dining room. It also meant joining him in a drink beforehand, a double martini as I recall, which none of us were physiologically constituted to withstand. We hadn't had a drop of alcohol on the entire trip and the martinis hit us like dynamite. I still don't know how we managed to stay upright in our seats. Dinner was sumptuous, and in such sharp contrast to what we had been used to that we were quick to regain our senses. Mr. Archbold led the conversation. He reminisced about New Guinea and Madagascar. He had organized expeditions

Top: Archbold Biological Station, main buildings. Bottom: Florida scrub at dawn.

there before the war, in the thirties, and he had been a pilot. He had owned an amphibian aircraft, a PBY, which he had flown on these expeditions. As we listened spellbound a thought suddenly came to me: Albert Schweitzer. Richard Archbold had the same imposing looks.

I learned eventually of the importance of these expeditions, which had brought to light so much new information about Madagascar's animal and plant life, and led to the first contacts with some of New Guinea's indigenous tribes. I learned also about the volumes upon volumes of scientific documentation published under the heading of Archbold Expeditions. And I learned how Mr. Archbold came to possess the station, in realization of a lifetime dream to create a natural haven for the study of wildlife. The Archbold Station was to become my primary natural laboratory, and is to this day my favorite outdoor haunt. It is where I made most of my discoveries and where nowadays I feel most at home as a naturalist. I fell in love with the Florida scrub on that very first trip, and have remained in love with that unique habitat ever since, acutely aware of its threatened status.

We lit the Coleman lanterns after dinner that evening and, still a bit shaky from the martinis, set out on foot to explore the surroundings. The night was balmy. Among the first animals we encountered were two species of millipedes, *Narceus gordanus* and *Floridobolus penneri*. Both were giants of their kind and they were out in abundance, crawling on the sand. Both gave off secretions when we handled them, from glands along their flank, and the odor was instantly familiar. Could it be benzoquinones again? Jerry's group eventually confirmed that this was indeed the case. I had never seen millipedes that large. The size of index fingers, they left trails in the sand, which in the daytime betrayed the sites where they were commonly found. I learned from that first evening's experience that the time to collect millipedes is at night, when they are out in the open. *Floridobolus,* for instance, had so far been sought by biologists only in the daytime, by looking under logs and in other presumed hiding places. The species was known from a single specimen collected at Archbold's and was considered to be one of America's rarest millipedes. We must have seen more than a hundred on that night alone.

Narceus was interesting in another respect. Unlike most millipedes, which are dark in color, *Narceus* closely matched the color of sand. I found this puzzling since I couldn't see how an animal active in the dark would benefit from blending with the background. I didn't realize until later that on moonlit nights the sand is so brightly illuminated that being light in color, like sand, could be beneficial.

Equally fascinating that night were the orb-weaving spiders, of which two caught my fancy, because of the strength and size of their webs. One,

Argiope florida, was a species endemic to the area while the other, *Nephila clavipes,* occurred widely throughout the New World tropics. As we moved close to examine these spiders, some of the insects that had been attracted to our lamps flew into the webs and got caught. We watched as the spiders subdued them and readied them for eating. How erroneous the notion that flight in the night was safer for an insect than flight in the daytime! There were webs to be seen in virtually every direction. True, insects could avoid these by flying higher up, but would they then not risk exposure to bats? I wondered how the threat from birds in the daytime measured up to the combined threat of spiders and bats in the night. I also wondered about the gastronomic likes and dislikes of spiders. Were there defenses that insects had evolved specifically to deter spiders?

Behind the main building of the station was a light trap for the attraction of insects. Consisting simply of a white sheet, pinned to a vertical wooden framework, and a set of lights, including an ultraviolet light that shone on the sheet, it was the sort of contraption known to be irresistible to insects. We had turned on the lights at dusk, before we went on our walk, and the trap had done its job. Insects had landed on the sheet by the hundreds, providing us with an instant glimpse of the variety of the local bug population. My mouth watered. There were insects here that I had never seen in numbers, and others that I hadn't seen at all. One by one I picked them up and sniffed them, and placed the most interesting "stinkers" in vials. There might be an occasion on the next day to try them out with a predator.

As I lay in fresh sheets that night I had trouble dozing off. We had only 3 days left at the station before we would need to head back. What a marvelous place! Would the daytime prove as exciting as the night?

Breakfast was early and we got into the field while the sun was still low. Spider webs, bedecked in dew, were visible everywhere. Surely there were insects that were protected against spiders. I did a few tests with *Argiope* that convinced me that I would need to spend time with this animal. I flipped some of the insects that I had taken the night before into webs and learned, first of all, that *Argiope* subdues prey of different kinds in different ways, and second, that there are indeed insects that the spider doesn't eat. I also learned that the spider does not reject all insects that are chemically protected. To my surprise it seemed not to be deterred by the stink of stink bugs (family Pentatomidae).

We also made some observations on antlion larvae. These remarkable little animals construct funnel-shaped pits in the sand, at the bottom of which they lie in wait, ready to feed on any ambulatory insect that slides into the pit. I found that this predator, too, had the ability to feed on protected insects. I dropped some of the "stinkers" I had taken at the trap into the pits and the larvae ate them all. And, of course, being antlions they ate all the ants that I offered them, including the formicine ants that I knew sprayed formic acid. How is it that they seem unbothered by the acid?

The next morning we dug up a colony of *Pogonomyrmex* ants, which I wanted to take back to Cornell so I would have ants available for predation tests, and managed to get our car stuck in one of the sandy lanes on the station grounds. "Don't drive on the sandy lanes," we had been warned, by the very same station personnel who were kind enough to tow us out.

The little time we had at the station went by altogether too quickly. The stay had been marvelously productive and I had uncovered leads to dozens of projects. How should I sort these out? As we drove back to Ithaca, I decided that I would not try. I would file away all the information in my memory bank for later use. The mind can be overwhelmed by the first experiences in a new habitat, but the memories are durable, and over time, cumulative. When you return to a site to explore some more, the old memories are always there to provide guidance. New facts connect to the old, and pretty soon the stories take form. There was therefore no other option. I would have to come back to Archbold's. There were discoveries there to be made, and I wanted to be the one to make them.

Since that first visit in 1958 I have returned to the Archbold Station almost every year, and in some years more than once. Eleven of my graduate students did a major part of their research there, or on animals from there, as did some of my undergraduate honors students. And I have taught field courses there, under the title "Exploration, Discovery, and Follow-up." For the past 25 years Maria has been my companion on every trip to the station, and we both remember our times there as some of the most joyous in our lives. We recall in detail what we discovered there, but more than anything else we remember how glorious it felt to make the discoveries.

The remainder of this chapter is taken up by Archbold stories—just simple stories about bugs in the scrub, none of profound implications.

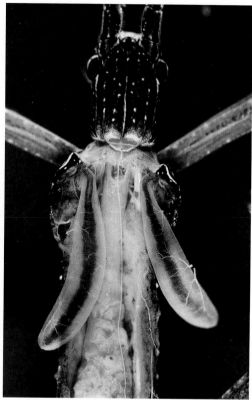

They are not presented in chronological order, or in any other order that matters. They can be read, or not read, in any sequence.

Anisomorpha buprestoides.
Left: A male astride a female.
Right: Dead female, dissected to show the two defensive glands.

■ **THEY CALL IT** the "devil's rider," which is an odd name for a walking-stick. But it is not an inappropriate name. It alludes to the fact that the insect is usually found in pairs, with the small male astride the much larger female. They are not necessarily mating when thus found, although the pairing is sexual, and the two do eventually mate and produce eggs. Exactly what the implications are of the prolonged bonding between "horse and rider" in this species appears not to have been worked out. I encountered my first *Anisomorpha buprestoides,* as the devil's rider is called, as a single pair in Myakka State Park, but at the Archbold Station I eventually found the walkingstick in abundance.

I was out one night at the station when I caught sight of a pair in the light of my lantern. They were on a bush, *Lyonia lucida* (fetterbush), and

The neck region of a female *Anisomorpha buprestoides*, showing right gland opening (arrow).

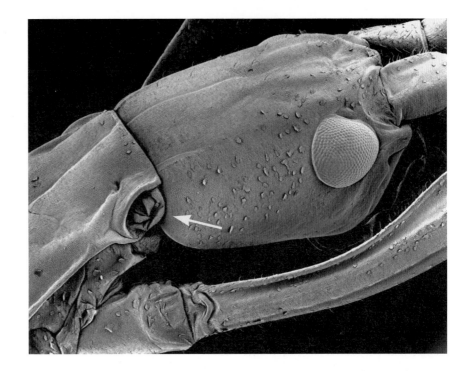

the female was feeding. I touched her and she responded instantly by spraying. The fine mist was visible, and the stench penetrating. No, it was piercing. I was standing close to the pair and got the brunt of it. My eyes hurt, as did my lungs when I got a whiff. This was evil stuff. I started coughing. I touched the female again and she discharged once more. The stuff shot out from just behind her head. It was a white liquid. There were telltale droplets on my finger where I had touched her. The male also sprayed when poked, but his output was puny by comparison.

It wasn't long before I realized that I had hit upon a population of the riders. I spotted a second pair and then a third, and eventually some more on adjacent bushes. In fact, there was this peculiar raining sound, for which initially I had no explanation, until I realized that it was from the animals' fecal droppings bouncing off vegetation.

Gingerly, so as not to cause them to spray, I coaxed some *Anisomorpha* into a cage and took them back to the station. The next day I asked Richard Archbold whether I could present the insects to the scrub jays in the late afternoon when he fed the birds, and he consented. I expected the birds to be repelled by the walkingsticks' discharges, but found instead that the birds did not approach the devil's riders but merely eyed them from a distance. It was as if they knew that proximity to that particular prey meant trouble. I wondered how other birds might behave that were inexperienced with *Anisomorpha,* and I soon found out.

At Cornell I had three caged blue jays *(Cyanocitta cristata)* that had been hand-raised and were quite tame. They were used to me and had become accustomed to having their diet of commercial bird food supplemented with occasional insects. In fact, they loved insects and usually got quite excited, in expectation of an insectan handout, when they saw me. They were also used to getting an occasional distasteful insect and appeared not to hold that against me.

I was concerned that the walkingsticks would be killed by an avian attack. They were soft-bodied and would probably be torn open by the very first pecking and as a result bleed to death. As to the jays, I had no doubt that they would be repelled.

Things turned out differently. The jays were indeed repelled, but the walkingsticks were never pecked. The birds showed immediate interest in the insects and descended from their perches, but as soon as they got close to the *Anisomorpha* they were hit by a full blast of spray. I had fed a number of spraying insects to the jays before but had never witnessed such preemptive spraying on the part of any species. Remarkably, I didn't recall it happening, not at least as a matter of routine, when I myself reached out to grasp an *Anisomorpha* in the field. I had noted that in crowded laboratory cages individual *Anisomorpha* sometimes sprayed when their containers were jolted or opened, but as a rule the animals never discharged unless touched.

The walkingsticks survived unscathed in all cases where they sprayed preemptively. I took movies of the encounters and could determine that the bird had to be within at least 20 centimeters of the insect to elicit a discharge, a distance well within the range (30 to 40 centimeters) of the spray.

The jays were visibly affected by the discharges. Typically they jumped back, shook their head vigorously while at the same time usually losing their balance, then rubbed the head in the plumage. When they got sprayed one often could see their eyeballs being wiped by the cleansing action of the nictitating membrane, that interesting extra eyelid that birds possess and use like a windshield wiper. All three jays were quick to learn to discriminate against *Anisomorpha.* They eventually refused even to descend from their perches when presented with a walkingstick. Had the

Four consecutive frames of a motion picture film, showing a blue jay (temporarily in a cage used for photographic purposes) being sprayed by an *Anisomorpha buprestoides*. Note that the bird is hit before it has actually come in contact with the insect.

scrub jays in Florida similarly learned from experience to keep their distance from *Anisomorpha*? Or had they evolved an innate aversion to the insects as a consequence of long-term coexistence with them?

I never figured out what it is about a bird that makes it recognizable as such to the walkingstick. It is clear that no crude combination of vibrational and visual cues is involved. I tried to elicit discharges by waving objects in the vicinity of the walkingsticks, or by tapping the substrate around them, but without success. *Anisomorpha* evidently is programmed not to waste its secretion, and to fire only on the "real thing." I wondered, since *Anisomorpha* is nocturnal, whether birds are in fact a real menace in the life of these insects and was able to confirm that they most probably are. By staying up all night with *Anisomorpha* at a site where they were abundant, I noted that they continue feeding until well after dawn, with the result that they are then clearly silhouetted against their food plant at a time when bird predation is at a peak. It isn't until later in the day that they

WONDERS FROM WONDERLAND

Top: A blue jay (in a photographic cage) an instant after being sprayed by *Anisomorpha buprestoides.* The nictitating membrane is drawn across the eyeball. The bird uses this membrane to wipe the eye surface clean. Bottom: The bird losing its footing as it recoils after being hit by *Anisomorpha* spray.

seek shelter from the scorching sun by descending to the base of the plants.

The *Anisomorpha* laid eggs in the laboratory—single hard-shelled pellets that they squeezed out like fecal droppings—and I was able to determine that the young are born with loaded glands. Newly emerged individuals are able to repel ants. The secretion, in fact, is potently repellent to other insects as well, as became clear from various predation tests I staged. I dissected some *Anisomorpha* and found the glands to consist of

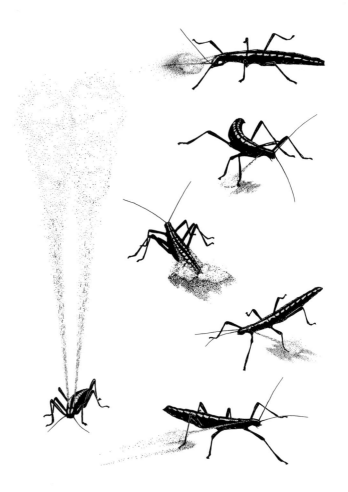

Spray aiming in *Anisomorpha buprestoides*. The insect is able to discharge from one gland or both, and it can aim its ejections in virtually all directions.

two large sacs, surrounded by powerful compressor muscles, opening to the sides of the thorax just behind the head. The glands were present and functional in males and females of all ages. I learned also that *Anisomorpha* aims its discharges. It can spray forward, upward, and backward, and can do so from one gland at a time or from both. Some of these experiments were unpleasant to perform. I got sprayed in the face more than once while working with *Anisomorpha*. It was only fair, I thought, given what the poor blue jays had gone through.

I was eager to know the composition of the secretion but was not able to "milk" enough fluid for chemical analysis from the relatively small number of *Anisomorpha* I had brought back from Florida the first time. So I took another trip, again with a group of students, this time to the vicinity of Gainesville, where we managed to milk several hundred *Anisomorpha* directly in the field. All we did was press the open end of glass vials di-

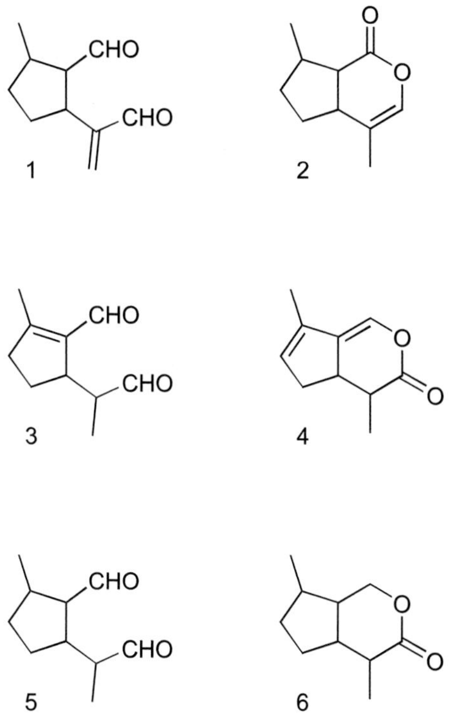

Cyclopentanoid monoterpenes:
1, anisomorphal;
2, nepetalactone (= catnip);
3, chrysomelidial;
4, plagiolactone;
5, iridodial;
6, iridomyrmecin.

rectly upon the thorax of the animals and they would comply by firing into the vials.

Jerry and his associates tackled the chemistry, and the results were most interesting. First—and this was in itself unusual—the secretion contained only one primary component. And second—and this was the rewarding part—the component was new. Not surprisingly, it belonged to a category of compounds, the terpenes (or isoprenoids), widely distributed in nature, and including many of the substances that give organisms, particularly plants, their peculiar scents. More specifically, it was a cyclopentanoid monoterpene, meaning that it contained 10 carbon atoms, of which 5 were formed into a ring. We had to give the compound a name, and Jerry called it anisomorphal. Working with my student George Happ, Jerry showed that *Anisomorpha* can itself synthesize anisomorphal from simple molecular building blocks, which means that it doesn't obtain the compound ready-made from the diet. This result was of some significance because there had been some question as to whether insects can synthesize terpenes.

Anisomorphal bore a chemical resemblance to, of all things, catnip. Formally known as nepetalactone, and produced by a plant of the mint

The larva of *Plagiodera versicolora,* a chrysomelid beetle, responding to an "attack" (the pinching of a leg with forceps). In the top picture, the larva is showing an initial localized response, in which it is discharging from glands close to the site stimulated. In the bottom picture, the larva is showing a generalized response from almost all glands at once. The larva has the ability to draw the droplets back into the glands after the attack has subsided. The secretion is potently effective against ants, and contains the cyclopentanoid monoterpenes chrysomelidial and plagiolactone.

family *(Nepeta cataria),* catnip derived its reputation from its peculiar ability to excite cats. That property, surely, had nothing to do with whatever the compound did for the plant that produced it. It occurred to me that I was now in a position to propose a natural function for nepetalactone. Could the compound not be defensive like anisomorphal, and serve to protect the plant itself? I got some pure nepetalactone from Jerry—by coincidence it was he who had determined the structure of the compound—

and did some simple tests, in which I showed the chemical to be a potent insect repellent. I found that insects would quickly fly off or walk away if I pointed at them a capillary tube filled with nepetalactone, and that ants would shy away from insect baits that I had laced with the compound. It was clear that plant and insect had hit upon a common defensive strategy here, in the sense that they had both evolved the capacity to produce similar substances for a similar purpose. Such parallel production of defensive chemicals by insects and plants is now known to be quite widespread. Insects and plants share much the same enemies and, given that on the whole they have comparable biosynthetic capacities, it makes sense that they should have evolved similar chemical weaponry.

Interestingly, anisomorphal is now known also to be produced by a plant. Not surprisingly, that plant, cat thyme *(Teucrium marum),* is also a member of the mint family (Labiatae). And nepetalactone itself has been shown to be produced by a species of walkingstick.

Cyclopentanoid monoterpenes are secreted by other insects as well. There is, for example, the larva of a leaf beetle, *Plagiodera versicolora* (family Chrysomelidae), that emits droplets of defensive secretion from segmental glands when attacked. The active principles in the fluid, which is potently repellent to ants, are chrysomelidial and plagiolactone. Other cyclopentanoid monoterpenes, for example iridodial and iridomermecin, are produced by ants. Like benzoquinones, cyclopentanoid monoterpenes are widely put to use in nature for defensive purposes.

There is still an odd fact about *Anisomorpha* for which we lack an explanation. The eggs of *Anisomorpha* contain a carotenoid—a new compound, in fact—that gives the egg contents a red color. The color does not show through the shell, so it does not seem intended as a warning color, to discourage predators. We have a hypothesis that the carotenoid could serve as an antioxidant for protection of the newly hatched young against toxins they might ingest with their food plants, but that is conjecture. At any rate, carotenoids are well known for their antioxidant capacities, and we do know that the diet of immature *Anisomorpha* includes highly toxic plants of the genus *Hypericum.*

■ **ACTUALLY** it is not quite true that I have never seen preemptive spraying in insects other than *Anisomorpha*. In southeastern Arizona, where I had been in the summer of 1959, there is a large squash bug, *Thasus acutangulus,* that lives in aggregations on mesquite trees. Squash bugs (family Coreidae), like their close relatives the stink bugs (family Pentatomidae), have dischargeable glands. Little was known at the time about the chemistry of the secretion of these insects, so I thought I'd collect some of them and "milk" them. The problem was that most were on branches too high to reach, so I decided to bag them instead. My plan was to hit the branches with my insect net and cause the bugs to fall into the net. I noticed right away that they were visually alert. As I reached up with the net they immediately repositioned themselves on the branch in such way as to present one flank to the approaching net. I knew their defensive glands to open on the flanks, so they were evidently taking aim. All I had to do to get them to reorient themselves is reposition the net and shake it from a different direction. It is while I was controlling their motions in this fashion that I noticed they were actually spraying as well. I caught a whiff of the discharges even as I stood beneath them and found that the net reeked of secretion when I held it close to my nose. I hadn't even touched them. They sprayed again later when I handled them after capture, which indicates that they had more in store than what they were ready to relinquish preemptively.

Preemptive shooting is, of course, a relatively expensive way of warning a predator, and that may be the reason why it occurs so rarely. Much more common are visual or acoustic warnings. Many protected insects are gaudily colored and by such appearance put the predator on notice. "Attack me and you are in for a mouthful of trouble" appears to be the message. Other insects emit repeated chirps when disturbed, so-called distress signals, and may thereby discourage attacks. Acoustic warnings have the advantage that they can function in the dark. There was much evidence that visual warnings work, but it took the recent studies of Mitchell Masters to demonstrate that acoustic warnings are effective as well. Mitch was a brilliant student and one of the most gifted experimenters to set foot in my laboratory. The way he recorded and analyzed the body vibrations that engender the distress sounds of an insect, and built artificial insects that could be made to emit such sounds and be tested with predators, was highly ingenious. He remained in bioacoustics and later was the first to explain precisely how orb-weaving spiders make use of web vibrations to locate prey trapped in their webs. He now is on the faculty of Ohio State University and works on bats.

My own interest in the whole matter of warnings dates back to my boyhood in Uruguay. There is one insect there that is better known to the populace than any other, and it is the moth *Automeris coresus* (it is depicted on an Uruguayan stamp). Its caterpillar, the *bicho peludo* or hairy beast, has poisonous spines that one is already taught as a child to avoid at all costs, since contact with them causes instantaneous pain and often severe systemic reactions. I used to raise *bichos peludos* and noticed that they lacked spines on the belly surface, so I concluded that I risked nothing if I could somehow get them to crawl onto my hand without touching the spines. I mastered the technique and would often induce panic among my buddies by approaching them menacingly with arms raised and a *bicho peludo* on each hand. I managed to scare even the neighborhood bully in this fashion. Evidently, if I made myself obnoxious enough, I too could emit warnings that were heeded by others. I remember experimenting with the spines. I cut them with scissors and noted that they were hollow and filled with fluid. I placed droplets of the liquid on my skin and quite foolishly also on my tongue, only to find to my surprise that this caused no pain. Needless to say, I conducted these experiments in utter secrecy.

But my interest in warnings was piqued more by the adult *Automeris* than by the larva. The adult *Automeris* is a gorgeous moth with a spectacular pair of eye images on the hind wings (the closely related American *Automeris io* is similarly adorned). Ordinarily, when the animal is at rest, it keeps these "eyes" concealed. But when disturbed it spreads the front wings and brings the "eyes" suddenly into view. When I first poked an *Automeris* and was "confronted" in this fashion, I was totally freaked out. I was sure predators would be similarly affected and I now know this is true. In a beautiful study with caged birds, in which he flashed images of various kinds at the animals as they pecked from a feeder, the British investigator A. D. Blest showed that the eye image was the one most startling to the birds. Birds are evidently programmed to shun the kind of stare that in their daily lives can signify imminent menace, as from an approaching predator. It has also been argued that fake eyes can serve to deflect an attack from a vital to a less vital part of the body. *Automeris,* for example, could easily withstand a pecking to a hind wing should a bird press the assault after being startled. A pecking to the body, by contrast, could prove fatal.

It should be noted that there are insects that "flash" their hind wings when attacked but whose hind wings are unadorned by eyes. The grass-

The io moth *(Automeris io),* a close relative of the Uruguayan *Automeris coresus.* On the left the moth is at rest; on the right it has been poked and is responding by exposing the "eyes" on its hind wings.

hopper *Romalea guttata,* a species I studied in the environs of the Archbold Station, has red hind wings that it ordinarily keeps folded under the front wings. I found that the grasshopper abruptly displays the hind wings when a bird pecks at it or merely approaches it, something that I never found it to do when I approached it. The animal is flightless and seemingly vulnerable but is protected by a pair of thoracic glands from which it emits a froth when attacked. Its wing display is therefore no fake warning. Although it seems to reserve the display for real enemies, I found that I could be mistaken for one if I adopted the pecking strategy of a bird by inflicting a series of quick pinches to its body with my fingers. As with *Anisomorpha,* I seemed never to fool *Romalea* by my mere presence. It was only when I resorted to simulated pecking that I was judged to be a threat.

I began noticing that eye markings are quite widespread among insects and decided they must be effective. They are common in butterflies, but occur also in beetles and other insects. Moreover, they do not occur only in chemically protected species. The adult *Automeris,* for instance, unlike the larva, seems to be a completely innocuous insect. Obviously, if eye displays serve primarily to startle, there is no reason why they should be restricted to noxious insects. But it certainly makes sense for them at times to be correlates of unpalatability. I developed a special fondness for one insect that is eye-adorned and chemically protected, the caterpillar of *Papilio troilus,* the spicebush swallowtail butterfly.

Caterpillars of swallowtail butterflies have long been known to possess a defensive organ, the osmeterium, consisting of two finger-like proj-

The Florida lubber grasshopper, *Romalea guttata*. Left: A mating pair. Right: A female responding to an approaching blue jay by assuming a raised stance and exposing its red hind wings. The behavior is intended as a warning. The grasshopper is chemically protected.

ections they ordinarily keep tucked away just behind the head. When disturbed, they evert the two "fingers," causing these to project outward from the front end like a pair of horns. The osmeterium is a glandular structure and the two fingers, on eversion, are coated with secretion. Jerry's group was the first to characterize the components of an osmeterial secretion. It was from a European species of swallowtail and the secretion turned out to contain 2-methylbutyric acid and isobutyric acid, two highly odorous and repellent compounds. I knew from personal experience that the osmeteria of swallowtails did not all have the same odor so I proposed to Jerry that we look into the osmeterial chemistry of a number of species.

The Archbold Station seemed ready-made for the project. Six species of swallowtail occurred there and we worked on the osmeteria of all of them. The chemistry did not turn out to be as exciting as we had hoped, although one species differed from the others in that it produced noxious sesquiterpenes instead of butyric acid derivatives. The compounds were all defensive, so it came as no surprise that some of the caterpillars should be conspicuously colored. The eye-adorned *P. troilus* was the most stunning of the lot.

The fake eyes of *P. troilus* were well known to butterfly enthusiasts, but what had apparently escaped notice is the directionality of their stare. Or perhaps I should say the nondirectionality, for the eyes appeared to look in

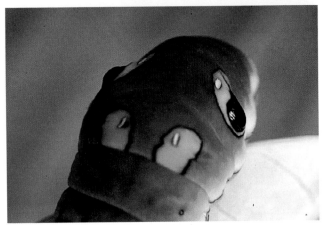

The caterpillar of the spicebush swallowtail butterfly, *Papilio troilus.* At the top left, the caterpillar is responding to being pinched with forceps by raising its front and everting its glandular osmeteria. The osmeteria are defensive devices coated with repellent secretion. The other photos show the front end of the larva with its fake eyes. Note that these "eyes" appear to stare back at you no matter from where you view the larva. The stare is presumed to forestall attacks.

all directions at once. Their stare was uncanny. If you looked at the caterpillar from directly in front, it stared right back. If you looked at it from the sides, or from behind, or from above, it likewise appeared to return the look. There was no direction from which a predator could approach the caterpillar without finding itself visually "confronted." My guess is that the confrontation works and that predators may be reluctant to press their assault on an intended delicacy that holds its ground and dares to stare back defiantly. The situation would not be exactly comparable to what it is in *Automeris.* In the latter the eyes are ordinarily concealed and are flashed only in response to the assault. In *P. troilus* the stare is on permanent display and is intended to prevent the attack.

The directionality of *P. troilus*'s stare is achieved through a very simple

WONDERS FROM WONDERLAND

A   B

Compounds in the osmeterial secretion of *Papilio troilus:* methylbutyric acid (left) and isobutyric acid (right).

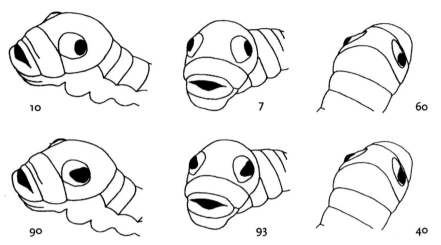

10   7   60

90   93   40

Pairs of images (left, middle, and right) presented to human viewers who were asked to choose the member of each pair that more directly appeared to "stare back." The numbers give the percentage breakdown of their responses. The top three pictures show the eyes without the pupillary triangular add-on; the lower three have the add-on.

design feature: the shape of the dark pupillary marking in the center of the eye. That marking, instead of being circular as it commonly is in actual eyes, is tear-shaped in the caterpillar's imitative version. It consists essentially of two portions: the basic circular marking, and a triangular anterior add-on that merges seamlessly with it. Leave out the add-on, and you have an eye with a circular pupil that is not nearly as able to convey the impression of looking forward or to the side. It is not needed to enable the eye to "stare" obliquely backward or directly upward, but it is essential if the stare is to cover the front and flank. Contributing to the visual fakery is the fact that the front end of the caterpillar is enlarged, as if it were a head, and that the "eyes" upon it bulge out as if they were real.

I have done a simple experiment recently that shows that the visual "defiance" of *P. troilus* is indeed a function of pupillary design. I drew images of the caterpillar in which the eyes were depicted either with or without the pupillary triangular add-on, and asked 30 students in my department, in individual interviews, to point to the images that seemed most directly to stare back at them. Where frontal and lateral stares were at is-

Left: A print by Joan Miró (in the possession of the author). Right: The same print with the circular eye image eliminated, a change that totally alters the character of the image.

sue, the actual eye image depicted on the caterpillar was favored by a large margin.

There can be little question that the eyes of *P. troilus* draw the attention of an approaching predator. They certainly draw our own, and are in fact what gives the caterpillar away when you search for it on its food plant. There is ample experimental evidence that the circular disk, highlighted by a dark pupillary center—the eye image—attracts human attention. If you trace the eye motions by which a human scans a facial image, the glances are seen to be cast back and forth from one eye of the image to the other, and to be directed only occasionally to other facial features such as the mouth and nose. The attention-gathering qualities of the eye image, or even of the circular disk itself, are certainly known to the artist. Just think of the abstract creations of a Franz Marc or Joan Miró, and of how skillfully these artists make use of circular design features to bring highlights or altered visual balance to their canvases.

In art there is also a way to impart to eyes the ability to gaze in different

directions. There are paintings that "follow" you as you walk past them, portraits with a seeming ability to maintain a visual hold on you as you pass them by. Such portraits do essentially what the *P. troilus* caterpillar does, but the effect is achieved differently. One trick, in portraiture, is to impart upon the two eyes a slightly divergent direction of view, so that one eye appears to stare at you while you are to the left of the portrait and the other while you are to the right. You are thus never out of eye contact with the painting no matter from where you view it. There are many portraits in which the eyes are thus depicted, which makes one wonder whether artists deliberately employ the technique.

When I was a child I used to love the books about Babar, king of the elephants. I was particularly impressed by the incident in which the elephants defeated their archenemies, the rhinos, without resorting to violence. The rhinos had grouped with the intent of attacking, but the elephants, which had painted huge eyes on their rumps, managed to thwart the assault simply by lining up and presenting their rears in mock defiance. Confronted by the fake stares, the rhinos fled in disarray.

In 1972 I was in Australia and had occasion to test whether one could use fake eyes to forestall a very different kind of assault. At issue were the attacks of Australian magpies (species of *Gymnorhina*), birds whose males commonly pounce on pedestrians and bicyclists during the nesting sea-

The Italian Lady, by René Auberjonois. The actual painting is in the center and shows the eyes in slight divergence, giving the viewers the impression that they are being looked at directly, whether they are to the left or in front of the painting. In the other two pictures, the eyes have been modified, by realigning either the right pupil (right) or the left pupil (left). The result is two portraits that stare in different directions.

The strategy used by Babar and his elephants to scare off their archenemies, the rhinos.

son, when they are particularly protective of their "homes." The birds almost always attack from the rear, striking the back of the head with such force that the scalp is often torn open by the bill and claws. Particularly persistent magpies have at times been shot by authorities, even though the birds are protected by law. My daughter Yvonne and I were dive-bombed on four occasions by magpies in Canberra, and I had heard of one such attack causing a fatal accident to a bicyclist in Melbourne.

It made sense for the magpies to attack from the rear and to avoid the visual field of their intended target. What apparently works for insects, I thought, might work for me, so in late spring, during the magpie season in Canberra, I decided to affix two large fake eyes to the back of my cap, with the hope of scaring off the magpies that I knew were mounting guard along my route to work. Alas, the experiment was a failure. I had thought of it too late. The initial results had been promising—two attack-free walks, followed by an assault on a third, when I had taken the eyes off—but all trials after that, whether I was outfitted with "retrovision" or not, failed to elicit attacks. The nesting season was coming to an end, I was told, and so was the bird's aggressiveness.

I still think that the experiment might have worked. The problem is not restricted to Australia. Joggers in North America and Europe are also occasionally targeted by birds that fly into the back of their heads, something that might well be prevented by an appropriately decorated cap. The tech-

The author, in Australia, with the hat he wore in attempts to ward off attacking magpies.

nique, incidentally, is used in some parts of the world to deter tigers. Villagers in certain rural areas of India wear facial masks over the back of the head for such a purpose, with what are rumored to be positive results.

When I was in Australia in 1972 I was shown a caterpillar, *Neola semiaurata,* with the remarkable ability to "wink" at you. The animal has two fake eyes built into its sides, near the rear end. These eyes are ordinarily held shut by overhanging lidlike folds of the body wall, which reduce them to narrow slits. Upon disturbance, however, the animal draws the lids upward, opens the eyes, and signals defiance.

There is a remarkable beetle in Florida that fools you into thinking its rear is its front. The beetle is a scarab (family Scarabaeidae), and goes by the name *Trichiotinus rufobrunneus*. At the Archbold Station, where I studied it, it is most commonly found inside the flowers of prickly pear cactus (species of *Opuntia*). When resting or when feeding on pollen the beetle orients itself in such fashion that its rear faces outward toward the flower

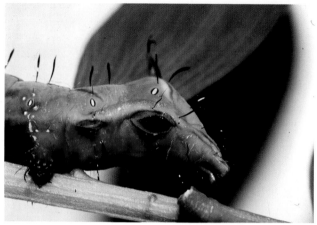

Left: The rear end of the caterpillar of *Neola semiaurata* at rest, with the fake eye closed. Right: The same but with the "eye" opened in response to disturbance.

opening. It then gives the uncanny impression of being a wasp "at the ready." The fakery fooled me totally on first glance, and could obviously fake out predators and competing pollinators as well. The notion could easily be put to the test. The beetle is relatively abundant in the spring when prickly pear is in bloom.

MILLIPEDES make up a special group within the arthropods, the Diplopoda. Although they don't have a thousand legs as their name implies, they certainly have many, but they are nonetheless, almost without exception, slow, deliberate crawlers. An ancient group, millipedes have been around since Silurian times, some 350 million years. They are also a hardy group, for they have survived the evolutionary onslaught of insects. Insects are in fact their main enemies and it should come as no surprise that millipedes have anti-insectan defenses. Primary among these are dischargeable glands, of which they have different types, producing diverse chemicals. Some millipedes, such as *Apheloria corrugata*, produce hydrogen cyanide, others, such as *Narceus gordanus* and *Floridobolus penneri*, produce benzoquinones, while still others produce phenols, alkaloids, or quinazolinones. I have over the years worked on millipedes in all these categories, and Jerry's group worked on their chemistry. In a number of cases this research resulted in the isolation and characterization of novel, highly interesting chemicals (see Chapter 5). Altogether we have worked on millipedes from New York, Florida, Arizona, Texas, the Netherlands, Panama, and Africa. In the process I have become quite fond of these animals.

The scarab beetle *Trichiotinus rufobrunneus* on a flower of a prickly pear cactus. The beetle is seen in rear-end view, in which it gives the false impression of being a wasp on the alert.

There was one group of millipedes that always intrigued me because its members apparently lacked chemical defenses. Known as the Polyxenida, and constituting no more than about 60 species, these millipedes were millimeters in length and furtive, and for that reason little known. I had for years wanted to come upon a natural source of polyxenids, but had been totally unsuccessful. If polyxenids lacked chemical defenses they had to be protected in some other way. For one thing, given their size, they had to be able to cope with ants. But how? It was thanks to a friend and frequent collaborator, Mark Deyrup, that I got my chance to find out.

Mark is one of the best naturalists I know. A native of New York City, he did his undergraduate work at Cornell and his doctoral work at the University of Washington. After a stint as professor at Purdue University, he assumed a post as naturalist at the Archbold Station, where he has been since 1982. In his new capacity he is able to combine his passions for natural history and conservation, and to exercise his extraordinary talent for discovery. Mark is the quintessential naturalist explorer. Where others see nothing, he sees novelty. Where others are baffled, he comes up with explanations. His chief strength? Bugs. He knows insects and has a very special appreciation of them. If bugs spoke English he'd be on a first-name basis with them all.

Take, for example three of Mark's recent discoveries: a new species of pygmy mole cricket that lives deep in the soil and comes up after rain to

Top: The polyxenid millipede *Polyxenus fasciculatus.* Bottom: An ant that has fallen victim to tuft contact by a polyxenid.

feed on a layer of algae that grows millimeters from the soil surface (the layer of algae was in itself a discovery); a caterpillar that lives in spider webs; and a caterpillar that feeds on the shell of dead turtles—not surprisingly, this species is a relative of the clothes moth; turtle shells have outer plaques made of keratin, the constituent of wool.

Mark had found that polyxenids hang out under the bark of the slash pine *(Pinus eliottii),* where they can be collected in numbers adequate for experimentation. The bark of the slash pine is typically flaky. The outer-most flakes can be readily peeled off without injuring the tree, and it is beneath these flakes that the polyxenids live, together with an assortment of spiders, pseudoscorpions, and above all, marauding ants. With Mark's help, Maria and I collected a quantity of polyxenids. The technique was simple and involved using a small brush to sweep the little animals into vials. Transferred into petri dishes, with pieces of bark and detritus from their habitat, they thrived. They also needed moisture, as captive arthropods typically do, so I added a small wad of water-soaked cotton to each dish.

The polyxenids turned out to be all of one species, *Polyxenus fasciculatus*. Unlike most millipedes, whose body surface is smooth, *P. fasciculatus* is densely beset with bristles. These are neatly arranged in transverse rows along the back, and in sets of flower-like clusters along the flanks. Projecting from the rear and glistening conspicuously in the light was a tuft of bristles much finer than those on the body. There appeared to be something special about that tuft, in that the animal seemed to make use of it when provoked. If I probed the millipede, say with fine forceps, it would move away, but as it did, it would rotate its rear and wipe the tuft against the probe. It was from that maneuver that the polyxenid derived its protection.

To get a real look at the tuft we had to resort to scanning electron microscopy, and that became Maria's task. I had myself learned scanning electron microscopy during a sabbatical leave in Australia in 1972, when I was a guest of the Commonwealth Scientific and Industrial Research Organization (CSIRO) in Canberra. I'd spent hours at the instrument and, being photographically inclined, realized soon that I would never again be able to do without the technique. When we returned to the States and Maria began working with us on a regular basis she was herself quick to become expert with the instrument and she is today "our" electron microscopist. If we need to examine an arthropod close up, or have a look into what we call the "inner space" of an arthropod's world, Maria is the one to take the photos.

At higher magnifications it became clear that the polyxenid's caudal tuft is really a pair of tufts, which the animal usually holds closely appressed. The tufts' bristles are slender, and consist individually of a shaft densely beset with barbs and a tip fashioned as a grappling hook.

Top: The caudal tuft of a polyxenid; note that two bristles have been partially pulled out. Bottom: The distal portion of a bristle, showing the barbs on the shaft and the terminal grappling hook.

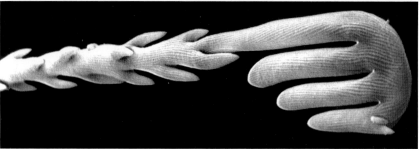

The bristles are loosely anchored. Give them the slightest tug and they detach.

To see how the tufts work I exposed the polyxenids to ants. I did not want to overdo things so I introduced only 2 to 4 ants into the millipedes' chambers. Attacks took place promptly and ranged in severity from brief contacts to deliberate probings in which the ants attempted to bite the millipedes. The millipedes responded quickly by flexing their rears toward the ants, splaying the tufts, and making their getaway.

The effect on the ants was immediate. Visibly coated with bristles, they discontinued the attack at once. Those that were sparsely coated usually succeeded in cleaning themselves in a matter of minutes. But those whose appendages had become more heavily contaminated were literally immobilized. They attempted to clean themselves, but in so doing seemed only to aggravate their plight. They wiped antennae with forelegs, drew appendages through the mouthparts, or stroked legs against one another,

WONDERS FROM WONDERLAND

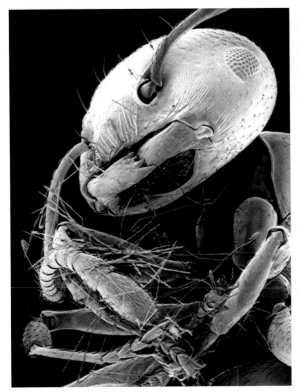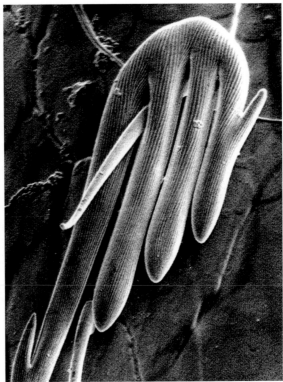

but they usually succeeded only in further entangling themselves. It was as if through preening they were spreading a glue. Many lost their footing and fell to the side, without ever recovering. Those that did disentangle themselves usually took hours to do so. The polyxenids, without exception, survived the encounters.

Pictures taken with the scanning electron microscope showed in great detail how the entangling mechanism works. The bristle tips are the functional units that ensure that the bristles become anchored to the ant. As grappling hooks they become fastened to the hairs (setae) that project from the ant's surface, with the result that the bristles are then torn from the tufts. Also of importance are the barbs that project from the bristle shafts, for these act as hooks by which the bristles become cross-linked. Fastened to one another, the detached bristles form a loose meshwork by which the ant is muzzled and its legs are strung together. The ant is literally tied up. Velcro, it would seem, had a nonhuman origin.

Polyxenids must have enemies other than ants. Since setae and other hairlike projections are a standard feature of the arthropod integument one could imagine any number of small arthropods proving vulnerable to

Left: An ant entangled by a polyxenid, showing the mesh of bristles that has immobilized its legs. Right: The grappling hook, fastened to an ant's hair.

bristle entanglement. Potential enemies of polyxenids could include centipedes, spiders, and pseudoscorpions. There is, however, one enemy that can cope with polyxenids. It is a Brazilian ant of the genus *Thaumatomyrmex,* which circumvents the bristle defense and feeds on polyxenids as a matter of routine. It has mandibles the size of pitchforks, which foragers drive through the bristles as they impale their victims for the transport home. In the nest, workers then use coarse pads on their forelegs to scrape off the bristles before settling to consume the millipedes.

I estimate that polyxenids ordinarily have enough bristles in their tufts to fend off a substantial number of ants. Amazingly, running out of bristles is not the end of the world for them. The bristles are renewable. Millipedes, unlike insects, continue to molt—shed their "skin"—as adults, and polyxenids are no exception. In a very nice study, a German investigator, G. Seifert, showed that when a polyxenid emerges from a molt it invariably has fully constituted tufts, even if it lost bristles before the molt. In fact, it is specifically the loss of bristles that hastens the advent of the next molt. Polyxenids thus have the capacity to expedite the process of rearmament precisely when they are in need of arms. That's a pretty neat trick.

■ **CHARLES DARWIN** was fascinated by carnivorous plants, so much so in fact that he wrote a whole book about them. He was intrigued by the notion that there should exist plants that eat animals and he set out to study these plants himself. Of particular interest to him was the sundew plant *Drosera rotundifolia,* to which he devoted most of the book.

Sundew plants derive their name from the glittering droplets of secretion borne by the stalked glands on their leaves. The droplets are sticky and insects that land upon a leaf become trapped in the glue, struggle, and die. Over the next hours the glandular stalks at the periphery of the leaf all bend inward, in such way that they come to deliver their droplets upon the prey. The latter, as a result, is drenched in secretory fluid and digested. Darwin was very much taken by the coordinated bending response of the glandular stalks and he did countless experiments to determine what specifically triggered them. Suspecting that *Drosera* possessed some kind of primitive nervous system, he even tried to see whether the bending mo-

Top: A sundew plant *(Drosera capillaris)*. Bottom: The leaf of the plant, with a trapped insect. The glandular hairs on the periphery of the leaf have bent inward and delivered their secretory droplets onto the dead insect, which will be digested as a result.

tions could be inhibited by neurotoxins. His account makes for fascinating reading.

At the Archbold Station there is a species of sundew plant, *Drosera capillaris,* often found in wet sandy soil at the edge of natural ponds. Back in 1965 I was at the station studying spiders when I encountered a stand of the plant. Having never seen a sundew plant in the wild, I became immediately interested. I was particularly fascinated by the stickiness of the secretion. I had some familiarity with another biological glue, that responsible for trapping insects in spider webs, so I decided I'd watch how the *Drosera* glue did its job. A former undergraduate honors student of mine, Julian Shepherd, had come along on the trip, and also became smitten with *Drosera*. With a tool I always carry in my collecting bag, a soupspoon, we scooped up a number of sundew plants and transplanted them into plastic containers at the station laboratory. We had fun with the plants, but what we discovered had little to do with what we had set out to study.

We collected some ants and some midges and released these in our sundew "gardens," and sure enough events followed the script. The insects got trapped and died, and were eventually drenched in secretory fluid and digested. It was fascinating to watch how the sticky droplets took their effect. As the insects struggled they kept stretching the glue into fine strands but these seemed never to snap loose. It was as if the victims were tied down by rubber bands. We made one observation that I thought was interesting because it related to something I had learned about spiders. We had caught some tiny moths and released these among our sundew plants, and found that they did not get stuck. They were protected by their scales, which detached readily wherever glue came in contact with their body, so that the body surface itself did not touch the adhesive. Moths, I had already learned, are similarly shielded by their scales from getting stuck in spider webs.

The bending of the glandular stalks toward the entrapped prey was a slow process that proceeded overnight, so we always checked our sundew plants in the morning. It was in the course of one of these inspections that we noted that someone was nibbling on our sundews. Whoever it was had a special appetite for the glandular hairs and an ability to consume them, secretory droplets and all. Some leaves were missing a few stalks; others were missing the entire complement. In place of the stalks our mysterious stranger left its fecal pellets, loosely scattered over the center of the leaf. Judging from the looks of the pellets we suspected a caterpillar, one that had to be active in the night. So we thought we'd become nocturnal ourselves, and catch our culprit in the act.

We figured that our stranger was probably light-shy, so we took our sundews to a place where we could observe them in red light, in an otherwise

The sundew-feeding caterpillar *Trichoptilus parvulus*. Top left: An intact sundew leaf beside a leaf that has lost most of its glandular hairs to the caterpillar. Note the caterpillar's fecal pellets in the center of the leaf on the right. Top right: A young larva feeding in the center of a leaf. Bottom left: An older larva imbibing a glandular droplet. Bottom right: An older larva eating a glandular stalk.

darkened room. Most insects are blind to red light and we assumed that our friend would be no exception. We took some goose-necked lamps that we had outfitted with red bulbs, set them up so they shined on our plants, arranged things so we could each look down on a cluster of the plants with a stereomicroscope, and watched.

We did not have long to wait. Our suspects were indeed caterpillars, and they all appeared at about the same time, shortly after dark. There were six of them, all similar in appearance and therefore probably of one species. They varied in size, and they had each emerged from beneath a plant to surface on a leaf. The littlest one fit in neatly between the short-stalked glands at the center of the leaf. The oldest ones were about as long as the longest stalked glands. They all began feeding right away.

They were consistent in how they disposed of the glands. First they imbibed the secretory droplet, then they ate the glandular knob, and finally they chewed away at the stalk. The smaller larvae concentrated on the shorter glands and sometimes ingested only part of the stalks, while the larger larvae consumed even the longest stalks in their entirety and proceeded sometimes to chew into the leaf blade itself. All deposited their waste on the leaf surface, as we anticipated.

We watched carefully and thought we had found out why they don't get stuck to the glands. The caterpillar's body is covered with long slender hairs, which we believe it uses as feelers. As a larva crawled between the glands it touched the droplets with the hairs but always managed to keep the body itself away from the glue. It was as if the caterpillars were using the hairs to gauge safe space. To serve as feelers the hairs would need to be innervated, but such is commonly the case in insects.

We watched the caterpillars night after night, as they devoured sticky lollipop after sticky lollipop. We noted that they ate even the remnants of insects that had become trapped on the plants. They consumed these carcasses skeletal parts and all. Their rate of consumption was impressive. One specimen that we watched over a period of 8 days ate several entire leaves in that time, and deglanded a number of others. Interestingly, when the larva pupates—when it becomes encapsulated in anticipation of its transformation into an adult—it picks a safe place. It ascends the long, upright floral stalk that grows from the center of the plant, attaches itself to it, and sheds its skin. The floral stalk is protectively encircled at its base by gland-bearing leaves. The pupa is a beautiful bright green. Adult emer-

Trichoptilus parvulus.
Left: A pupa. Right: An adult.

gence takes place 10 to 11 days after pupation. We waited impatiently to see who our friend might turn out to be. We recognized immediately, by its distinct narrow wings, that it was a plume moth, a member of the family Pterophoridae, but in order to find out its species name we had to consult an expert, R. W. Hodges, of the National Museum of Natural History. The moth was *Trichoptilus parvulus.* Its life history was unknown.

There are many questions left unanswered about this remarkable insect. What is it, for instance, about the larva that enables it to move about the sundew leaves without causing the glandular stalks to bend over, as they do in response to entrapped prey. Also, we learned almost nothing about the adults, which died shortly after emergence. We would have loved to see where and how the female moth lays her eggs. Does she lay them directly on the sundew plants, and is she then protected by her scales from becoming stuck to the glands?

■ **WHEN I REVISIT** a field site I usually have an agenda. Top priority is always given to projects started earlier that need to be finished. But as the time of departure approaches I usually strive to get ahead of schedule, so as to end up with one or two free days that I can devote to pure exploration. As of late, since Maria has been joining me in all field studies, we do the exploring as a pair. We simply stroll about at leisure and look around.

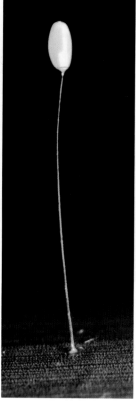
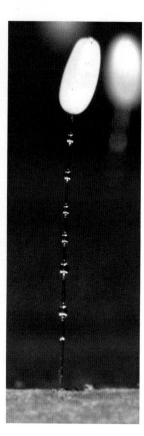

Left and middle: An adult chrysopid, *Ceraeochrysa cubana*, and its egg. Right: The egg of *Ceraeochrysa smithi*.

We start early in the morning when life awakens, and go out at dusk again, with headlamps, to catch the night life. At the Archbold Station we usually try to get Mark Deyrup to come along. Inevitably we come upon something previously unnoticed on such walks.

Ever since Frank Carpenter, at Harvard, tried to persuade me to work on the taxonomy of green lacewings, the chrysopids (family Chrysopidae), I have had a fondness for these insects. They are beautiful and there is much about them that is interesting. Their eggs, for instance, are laid on stalks. One often finds them in the field, singly or in clusters, each atop a little filament, 2 to 3 millimeters in length.

In 1965, on one of those last-day walks in Florida, at a location near the Archbold Station, I came upon a chrysopid egg cluster that was different. The eggs had been laid on a pine needle and they were typically stalked, but the stalks bore tiny droplets of fluid, neatly spaced along the stalks' length, like beads on a thread. I eventually found such eggs to be relatively common and, curious about their taxonomy, wrote the late Ellis MacLeod at the University of Illinois at Urbana, an expert on chrysopids. He had never heard of such "beaded" eggs and had no idea who might be laying them. But, driven by curiosity, he was himself eventually able to locate one such batch in Florida, and having succeeded in raising the emergent larvae, could determine the species to be *Leucochrysa floridana,* a relatively common chrysopid.

I didn't give much thought to chrysopids for a while until I found the eggs of another species that also had droplets on its stalks. This species had the additional habit of laying the eggs in a spiral, which made the clusters instantly recognizable. Whoever laid these eggs was also not uncommon. The clusters were relatively easy to come by at the Archbold Station. I contacted Ellis again and he told me he had come across this second species on his own. Moreover, he had raised it and determined it to be *Ceraeochrysa smithi.* "Would you like to work on it jointly," I asked, "I think I can get my friends in chemistry to look into the composition of the droplets." Ellis was enthusiastic about the idea.

The first thing I thought we'd do is see whether those little droplets deter ants. One of my graduate students at the time, William Conner, was at the Archbold Station, and together we designed a simple test with a local ant, *Monomorium destructor,* which we had used in experiments before. Bill was an excellent experimentalist, busy with a project on a moth at the time (see Chapter 10), but he could easily be enticed into taking on a second problem.

Monomorium, too, was a good partner. The ant had a nest near the laboratory building and it could easily be induced to lay foraging trails right into our quarters. All we had to do was put out a sugar bait, say on a laboratory bench, and within a few hours we would have a column of ants eagerly marching to and from the bait. It was a simple matter to place test items in or near the trail to see whether the ants took them or rejected them, and I had used these ants for such purposes before. To score the results we simply sat by the bench and took notes. We could even have a beer on the side. Not everyone at the station was happy with our tests, since the ants had a nasty tendency to branch out to other benches as well, driven no doubt by the very sound logic that what they found on one bench they might also find on the others. At any rate, we tried to keep the other benches clear of food to convince the ants that their logic was faulty.

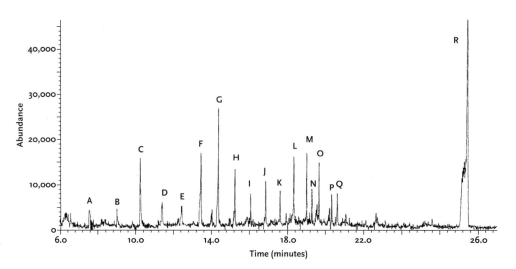

Peak label	Aldehyde
A	Butanal
B	Pentanal
C	Hexanal
D	Heptanal
E	Octanal
F	Nonanal
G	Decanal
H	Undecanal
I	Dodecanal
J	Tridecanal
K	Tetradecanal
L	Pentadecanal
M	Unidentified component
N	Unidentified component
O	Heptadecanal
P	Octadecanal
Q	Unidentified component
R	Tetracosanal

Modern gas chromatography provides an extraordinary opportunity for resolving the components of complex chemical mixtures. Here, in a chromatogram, the aldehydes in the egg stalk oil of *Ceraeochrysa smithi* are clearly depicted by a sequence of peaks. All the peaks except three have been characterized. The nonvolatile fatty acids (myristic, palmitic, linoleic, oleic, and stearic acid), which make up the bulk of the oil, did not register on this particular chromatogram.

We knew that female chrysopids tend to lay their eggs when they are confined overnight in vials. If we lined the vials with wax paper they laid the eggs on the paper, and the eggs could then be easily collected for experimental purposes. So we caught a number of female *Ceraeochrysa smithi* and got them to lay eggs in vials. We also collected females of another chrysopid species, *Ceraeochrysa cubana,* which we knew laid "normal" eggs devoid of stalk droplets, and got these to lay eggs as well. We staked out eggs of both types in the ant trail, and kept track of their fate. We separated the eggs from the clusters by cutting away a small square of paper around the base of their stalk. Each egg was thereby provided with a small basal platform by which it could be stood upright. For a test, we offered the ants five eggs of each kind, and kept track of the results for 12 hours. We repeated the test five times.

The results were convincing. Not a single *C. smithi* egg was taken, while a good fraction of the *C. cubana* eggs—2 to 4 per test to be exact—were carried off, each by an individual ant. As we wrote in our published account, when an ant came upon a *C. cubana* egg, it "ascended [the] stalk, straddled the egg, and cut the egg from the stalk with the mandibles. It then fell to the ground with the egg, grasped the egg in the mandibles, and scurried off along the trail." The ant, in other words, climbed the "tree" and sawed off the branch it was sitting on. There was something comical about the behavior, but it worked for the ants. Discrimination against the *C. smithi* eggs seemed to be contingent upon the ants' contacting the stalks. Indeed, the moment they touched a stalk the ants backed away and walked off. They never reacted that way when contacting the "unbeaded" stalks.

Collecting stalk droplets for chemical analysis posed no problem. All I had to do was draw the droplets by capillarity into pieces of fine glass tubing. The problem was that when I wanted to collect the fluid we could find only a single batch of eggs. This meant having only 21 stalks available for "milking." Incredibly, the minute liquid sample thus obtained sufficed to identify the more than 20 components present in the secretion.

The analysis was done in 1996 by Athula Attygalle, a master chemist from Jerry's group. Thanks in large measure to advances in instrumentation, small quantities suffice nowadays to characterize the components of a sample. Things have come a long way since the days some 30 years earlier when we had to milk several hundred *Anisomorpha* to obtain sufficient secretion for analysis. Today we could have characterized anisomorphal on the basis of what we could "milk" from a single gland of *Anisomorpha*.

The stalk droplets of *C. smithi* contained a mixture of aldehydes and fatty acids. The aldehydes included compounds such as hexanal, known from the defensive secretions of other insects (see Chapter 6), compounds that could in themselves account for the deterrency of the stalk fluid to ants. Not surprisingly, we found that hexanal ranked as a potent irritant in our scratch test with cockroaches.

The fatty acids included compounds commonly found in plants and even in foods, compounds such as palmitic, linoleic, oleic, and stearic acid, which we did not think would be deterrent, since they are innocuous to us. We thought that these fatty acids might serve as surfactants that made the droplets stickier for the ants, but found instead that the acids were deterrent in their own right. Oleic acid and linoleic acid, for instance, proved to be irritants in our cockroach scratch test.

We thought it interesting that E. D. Morgan and his collaborators, in England, who worked on the chemistry of a secretion used by certain wasps to impregnate the stalks (pedicels) of their nests, also found fatty acids, including oleic and linoleic acid, in their samples. These investigators described both these compounds as deterrent to ants, which are supposedly kept by the secretion from invading the nests of the wasps.

Fatty acids like those in the chrysopid egg stalk droplets and wasp nest pedicel secretion are found also, oddly enough, in ear wax. To my knowledge, no proof exists for any function that has been proposed for this se-

A newly emerged *Ceraeochrysa smithi* larva, pausing to ingest one oil droplet after another as it descends along the egg stalk.

cretory material. I think the possibility is worth entertaining that ear wax serves to keep insects out of the ear, or at least that it served in that capacity during our early evolutionary history. I admit, however, that the idea seems a bit far-fetched.

We wandered how the *C. smithi* larvae might make it down the stalks when they hatched, and by patiently keeping a watch on clusters so we wouldn't miss the moment, made an interesting discovery. When they come down the stalks the larvae eat the droplets. They come down slowly, head first, and pause by each drop for as long as it takes them to suck it up with their pointed hollow jaws. When they get to the bottom of the stalk they briskly walk away, in seeming readiness for what could be a much more risky existence than the protected one they led atop the stalks.

The stalk fluid of *C. smithi* is unusual in that it serves both for defense of the egg and as the first meal for the emergent young. I know of no other "mother's milk" in which the attributes of "guns and butter" are so elegantly combined. It is also surprising that the *C. smithi* larvae should be able to take by mouth a fluid that is so patently offensive to other insects.

As to the other chrysopid with egg stalk fluid, *Leucochrysa floridana,* it is out there, available for anyone to study. We know that in its case the secretion is pasty in consistency rather than oily, and that the newly emerged larvae ignore the droplets when they crawl down the stalks.

■ **ANIMALS DEFECATE.** They cannot utilize everything they take in as food, so they void as waste—as feces—what is useless. Feces can be hazardous to an animal. They can be the source of parasites or infective microbes, and most organisms are behaviorally programmed to put a distance between themselves and their waste. "You don't defecate where you eat" seems to be a rule by which animals pretty generally abide.

There is a flip side to the coin. Feces can amount to a substantial quantity of material, and since they are, so to speak, at hand and produced on a regular basis, one can imagine there being opportunists that put feces to use. Such opportunists exist, and there is one in residence at the Archbold Station.

Although it doesn't say so in my notes, I'm sure I noticed *Hemisphaerota cyanea* on my very first trip to Archbold's. I don't see how I could have missed the little blue beetle and its larva. Both occur on the two palmetto species that figure so prominently in the scrub—*Serenoa repens* and *Sabal etonia*—and both are highly conspicuous on these plants. *H. cyanea* is a leaf beetle, a member of the Chrysomelidae, one of the largest families of beetles, and like many chrysomelids is in the habit of feeding on the same plants as larva and as adult. That is unusual for an insect with complete metamorphosis,—an insect having a larval stage in its development. Larvae usually have a diet different from that of the adult and therefore do not compete with the adult for food. Chrysomelid beetles prove that the rule has its exceptions. In the case of *H. cyanea,* however, the issue of competition hardly comes into play, for both larvae and adults occur rather sparsely on their palmetto hosts.

I got interested in *H. cyanea* because both the larva and the adult have unusual defenses. Maria joined me in a study of the larva, which we found to be a slow deliberate feeder that did everything at leisure. The larva was so slow, in fact, that we could not help wondering how it ever survived attack. It simply had to be protected.

At first glance the larva is confounding. What you see when you spot it

Hemisphaerota cyanea, adult (top) and larva (bottom). The yellow soft-bodied larva is ordinarily invisibly hidden under its thatch.

is not the larva itself but a kind of straw hat that it maintains as a cover. It did not take us long to figure out that the straw hat, or thatch as we called it, was made of feces. *H. cyanea* extrudes not fecal pellets but fecal strands, and it is these that it retains to build the thatch.

One thing that we liked about the larva is that we could take it indoors and it would continue to behave normally. All it needed to survive in a petri dish was a fresh supply of palmetto fronds. It fed by carving narrow linear grooves into the fronds with its mouthparts, advancing at the barely detectable pace of a millimeter or two per hour. While feeding it kept itself anchored to the leaf, thanks to sharp claws on the tips of its legs. We set up dozens of larvae in dishes and observed them for hours with the microscope. We set up adults and eggs as well, and watched the larvae hatch and develop.

The eggs of *H. cyanea* are large, ovoid, and laid singly. They are embed-

WONDERS FROM WONDERLAND

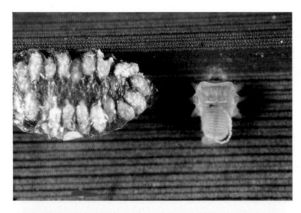
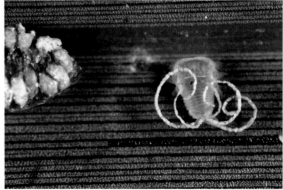

Thatch construction by a newly emerged *Hemisphaerota cyanea* larva. At the top left, the larva is beside the egg (encrusted with the mother's fecal pellets) 40 minutes after hatching. The first fecal strand is being formed. At the top right, 1.6 hours after hatching, four strands have already been produced, and the fifth is in the making. At the bottom, 12 hours after hatching, the thatch is virtually completed.

ded in a hardened matrix and encrusted with fecal pellets. The purpose of this fecal embellishment remains a mystery. The larvae remain thatch-covered throughout development and they retain the thatch when they pupate. Pupation occurs on palmetto fronds. A tiny droplet of glue anchors the pupa to the frond surface. The adult, when it emerges, stays under the thatch until its skeleton has hardened.

The larva begins feeding within minutes after it emerges from the egg. The first fecal strand makes its appearance shortly thereafter, slowly squeezed from the anal turret. Strands are then produced one after the other. Several are in place by the end of 2 hours, and after 12 hours the thatch is nearly or entirely complete. The larva attaches each strand to a forklike projection that sticks up from the rear of its abdomen, just in front of the anal turret. When it has completed a strand, the larva rotates the anal turret upward until it contacts the fork. It then constricts the anus and pinches off the strand, at the same time that it squeezes a droplet of quick-hardening glue form the anal turret. The strand is thus cemented to the fork. While producing a strand, the larva keeps the anal turret bent to one side, causing the strand to curve around the larva on that side. Con-

The process of strand production and attachment in the *Hemisphaerota cyanea* larva. For visibility purposes, all previous strands have been removed from this larva. At the top left, the larva has nearly completed the first strand (which curves around the larva's body). Note that the anal turret is deflected to the right. At the top right, the larva has just pinched off the strand and has fastened it to the base of the caudal fork with a glistening droplet of glue. At the bottom, the larva has turned the turret in the opposite direction, and has begun to produce the next strand.

secutive strands are produced with alternating curvatures, with the anal turret bent in alternating directions, so that as the thatch builds up it does so evenly on both sides.

As the larva grows it molts. Remarkably, as it sheds its skin, it does not discard the fork with its attached strands. Instead, it retains the fork, and adds it to the one that forms with the new skin. The thatch is thus kept intact, with the newly acquired fork providing a place of anchorage for the next set of strands. The process repeats itself with each molt. The larva, as it grows, is thus never deprived of cover, and the thatch keeps pace with the larval increase in size.

We wondered why the fecal strands were so resistant to breakage and found the reason. They are membrane-enclosed. Insect feces are often wrapped in a sheath, the peritrophic membrane, formed in the midsection of the intestinal tract. In the case of the *H. cyanea* strands, the membrane showed up clearly when we examined strands with the scanning electron microscope. We were reminded of strings of sausages encased in skin.

To see whether the thatch provides protection, we exposed larvae to two

Top: A *Hemisphaerota cyanea* larva in ventral view, showing the fecal strands, including the latest one, emerging from the anal turret. Bottom: A fully grown larva (minus its thatch), showing the caudal fork, to which the fecal strands are ordinarily attached. The fork is a composite of the individual forks from the first four larval instars stacked one on top of the other.

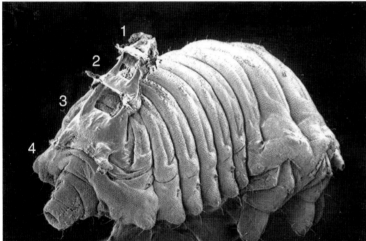

species of predators, a carnivorous stink bug *(Stiretrus anchorago)* and a ladybird beetle larva *(Cycloneda sanguinea)*. The tests involved presenting the predators with a choice between normal, thatched larvae and larvae from which the thatch had been removed. Both predators ate only the denuded larvae. They made contact with the covered larvae but spent no time probing the thatch. They seemed completely oblivious to the morsel that lay concealed beneath the cover.

We experimented with the thatch itself and found that the larva can re-

A predaceous stink bug *(Stiretrus anchorago)* sucking out a thatchless *Hemisphaerota cyanea* larva. The bug ate only such "naked" larvae, and avoided those that were thatch-covered.

pair it. We cut away the top of the thatch and noted that within 24 hours the hole was repaired. The larva seemed aware of what needed to be done and curved its new strands in such a fashion that they fit neatly into the opening. Damage to the front of the thatch or to its sides was similarly repaired by the addition of new strands.

One day Maria and I had the incredible luck of coming upon a predator in the act of eating a *H. cyanea* larva. It was a beetle, and it had its head buried in the thatch. It was so eager that it didn't even withdraw from the thatch when we brushed it, together with the larva, into a vial. We took the beetle into the laboratory and over the next few days presented it with a total of 15 larvae, all of which it ate. It had two strategies. Either it forced its way beneath the margin of the thatch, or it chewed its way to the larva through the top of the thatch. Either way, the larva had no way of escaping. The beetle was thorough. It ate every part of the larva except the anal turret.

It was not surprising that there should exist a predator able to cope with the *H. cyanea* larva. Every ploy in nature has its counterploy and after years of study of insect defenses I had gotten used to the notion that there is always at least one specialist able to circumvent a given defense. What was surprising was that we should have discovered our specialist in the act. In my experience such events are rarely witnessed, probably because by our mere presence we often keep natural events from happening.

With Mark Deyrup's help we eventually identified the beetle. It was

WONDERS FROM WONDERLAND

Repair of a hole cut into the top of the thatch of a *Hemisphaerota cyanea*. The first new strand produced, seen initially to the right of the thatch (top right photo), has been laid in place (left center). A second strand is then produced and laid in place, following which a third one is begun (right center). After 23 hours (bottom) the thatch has been almost completely repaired.

Left: The predaceous carabid beetle *Calleida viridipennis* feeding on a *Hemisphaerota cyanea* larva. Right: After the meal, all that is left is the thatch and the anal turret.

Calleida viridipennis, a member of the family Carabidae, the familiar ground beetles, a largely predacious group that also includes the bombardiers. We later caught two more *C. viridipennis* beetles and found that these too ate the *H. cyanea* larvae we offered. I have little doubt that this beetle habitually preys on this larva. The larva, incidentally, falls victim frequently to an entirely different enemy, one against which the thatch is useless. I often found *H. cyanea* larvae dead, their bodies intact but darkly discolored, in a condition suggesting that they died of microbial infection. Microbes could well be the *H. cyanea* larva's worst enemies.

H. cyanea is by no means the only chrysomelid larva making use of an overhead sewer system. Other members of its subfamily, the Cassidinae, also construct fecal shields as larvae, although in their case the shields are usually compacted from pasty rather than filamentous wastes. These species also have an abdominal fork upon which they form the shield. By revolving the abdominal tip and thereby rotating the fork, they are able to orient the shield in any direction. Such shield maneuverability is known to be used to advantage by some chrysomelid larvae in blocking the attack of ants.

Also of interest is the finding by other investigators that in some species fecal shields contain chemical deterrents derived from the plant diet. The shields in such animals could act not just to block enemies but to repel them.

■ **THANK HEAVEN** for dental wax. Dentists use it to make dental casts. I find that I too need to have it within reach. I keep it in the laboratory, and

Two fecal shield–carrying beetle larvae from the same subfamily (Cassidinae) as *Hemisphaerota cyanea*: *Gratiana pallidula* (left) and *Chelymorpha cassidea* (right).

keep a sample of it in my collecting bag, together with—aside from the soupspoon—plastic vials and containers, a hand lens, and a stopwatch. I also carry along a pair of no. 3 watchmaker's forceps—they are fine-tipped and come in handy in all kinds of situations—plus a pair of scissors, a fine brush, and a pair of "soft" forceps. These forceps, made out of spring steel, are useful for grasping insects without hurting them. I carry the forceps, brush, and scissors in a small leather holster attached to my belt so that I have them at the ready.

So why the dental wax? Because it sticks to insects. I don't know its composition, but I know that it is better for the purpose than bee's wax or paraffin wax. I use dental wax to fasten insects to a tether or to attach something to their body. The wax has to be melted to be applied but the melting temperature is low enough so the insect is not hurt. The wax also hardens quickly on application. But most important, it comes off easily after the experiment, so one can return the insects to their cage or to mother nature. Carrying dental wax means carrying matches as well, in addition to wire, thread, and whatever else one might want to use to tie an insect down. It all fits in a little box in the collecting bag.

The dental wax proved indispensable when I did my first experiments with *H. cyanea* adults. I had noted something odd about the beetles. When I tried to pick them off their palmettos they offered resistance. Being hemispherical in shape, they were rather difficult to grasp in the first place, but even when I succeeded, I had to apply considerable force to pry them loose. It wasn't that they were mechanically anchored to the leaf surface, like the larvae. If I sneaked up on them and tried to sweep them into a vial with one abrupt stroke of the brush, I usually succeeded. It was

A *Hemisphaerota cyanea* adult withstanding the pull of a 2-gram weight. The brush seen to the left was used to tickle the beetle intermittently, thus subjecting it to simulated attack. Unless stimulated in that fashion, the beetle does not clamp down.

when I proceeded gradually, by first touching them and thereby giving them a warning, that they responded by clamping down. Whatever the mechanism, the beetle could activate it on demand, and it seemed to use it for defense.

I took some beetles and tried to hang weights from them. I used standard weights that came with a balance and tried to attach these with strings and dental wax to the backs of beetles that were clinging to the underside of pieces of palmetto frond. Easier said than done. The beetles didn't like being upside down and I had trouble attaching the strings to them when they were restless. So I changed my tactics. I attached the strings to the beetles (with dental wax) while they were positioned upright on a piece of frond and, by leading the threads over a pulley, ensured that they were pulled up as the weights hung down. As long as I tickled the beetles intermittently with a brush they held their ground and resisted. I tried low weights at first. The beetles had no difficulty withstanding pulls of up to half a gram. They weighed only 13.5 milligrams on average, so 0.5 gram was the equivalent of 37 times their body mass. I tried 1 gram and even 2 grams, and found that some could resist such weights, even if only for a short time. Two grams amounts to 148 times their body mass. Think of it. In human equivalents—relative to a 155-pound human—this amounts to about 23,000 pounds, or 7.5 automobiles (Subaru Legacy station wagons, 1998 model).

Dan Aneshansley was at the station at the time, and so was Jim Carrel, one of my graduate students, and the three of us did some more tests with the beetles. We devised a way for pulling on them by using a water receptacle instead of weights. We would hang the receptacle from a beetle and fill it at certain rates with water, by delivering the water one drop at a time with a syringe to the rhythm of a metronome (there was a piano at the station and with it came a metronome). We got some results by this technique that told us that when we pulled on the beetle it made a difference whether we pulled quickly or slowly. The beetle could tell the difference between a rapidly accelerating force pulling on its body and a slowly accelerating force. Dan Aneshansley, forever the perfectionist, thought that our technique was too crude and that we should take some beetles back to Ithaca where we could do things right.

And so we did. The beetles traveled well and as long as we had fresh palmetto fronds for them they seemed happy. We actually had to arrange to have fronds sent to Cornell at intervals since they didn't keep long, and this transportation itself had to be done right. The fronds had to be packed with plenty of moisture or they would arrive dried and useless.

The apparatus that Dan rigged up was beautiful and simple. It consisted of a platform on which the beetle was positioned, and which could be subjected to a downward pull, applied either electronically with a device called a solenoid or by hanging weights from directly beneath it. The beetle was connected, by way of a hook glued to its back with dental wax, to a force transducer—an instrument that registered the downward pull on the platform—positioned directly above it. The arrangement was such that when a downward pull was applied to the platform, the force was sensed by the transducer and relayed electronically for visual display on the screen of an oscilloscope.

With the solenoid we could apply a force at a controlled rate of increase (100 milligrams per second) and determine the beetle's adhesive strength: the force in grams at which the beetle relinquishes its grip on the platform. By hanging fixed weights from the platform we could measure the

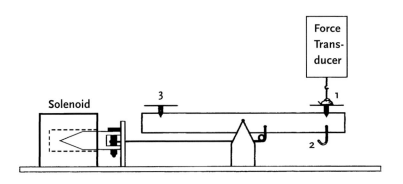

Diagram of the apparatus for application of pulling forces to *Hemisphaerota cyanea*: 1, beetle; 2, hook for suspension of weights; 3, pan for placement of balancing weights.

beetle's clinging endurance, the length of time that it withstood pulls of different magnitudes without detaching. We were also able to put different surfaces on the platform to see whether such changes affected the tenacity with which the beetle maintained its hold.

Dan and I worked together and we adopted a standard procedure for making measurements. We first fixed up the platform by covering it with the particular surface we wanted to put to the test. We used four surfaces: natural frond, glass, aluminum foil, and a waxy material marketed under the name Parafilm. We then brought the platform into horizontal equilibrium by adding counterweights to the side of the balance beam opposite the platform. Next we hung the beetle by its hook from the sensing element of the transducer and lowered it (by raising the entire balance beam) until its feet contacted the test surface. To ensure that it clamped down, we tickled the beetle's front end gently with a brush, making believe we were an attacking ant. We then applied the force, by either of the two methods.

The first thing we found is that the beetle clung with unequal strength to the different test surfaces. That fact in itself told us something about the clinging mechanism. The beetle was evidently not adhering by use of suction cups. If suction were involved the adherent strengths would have been the same on the various solid surfaces.

We also found that the beetle clung best to its natural substrate, palmetto frond. Its grip on the other three surfaces was significantly weaker. The adhesive strength on palmetto frond was on average nearly 2.5 grams, well in line with what we had found using our initial crude methods. The highest adhesive strength we recorded for any one beetle on a palmetto frond was 3.2 grams, or nearly 240 times the body mass. We celebrated that performance with a glass of wine.

As to the beetle's clinging endurance, we found, not surprisingly, that it

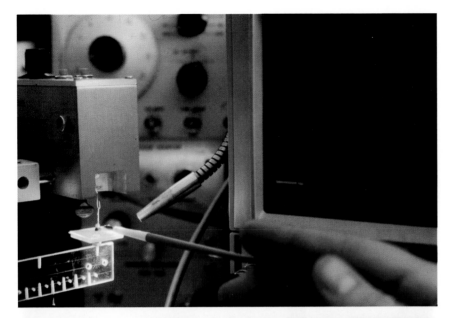

The apparatus diagrammed on the facing page at work. Top: The beetle is on the platform before lift is applied (horizontal trace on the oscilloscope). Bottom: The lift has been applied (the ascending green trace), beyond the point where the beetle has detached (return of the trace to the baseline).

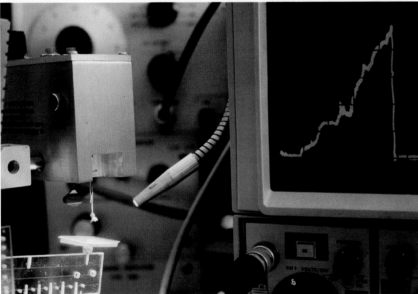

depended on the force applied. The greater the pull on the beetle, the shorter its endurance. We applied the pull for a maximum of 2 minutes. If the beetle held out that long, we discontinued the experiment. The reason we picked 2 minutes is that we thought that ants—which we suspected to be the beetle's main enemies—would not individually persist longer in efforts to pull the beetle loose. We thought we would eventually measure the

WONDERS FROM WONDERLAND

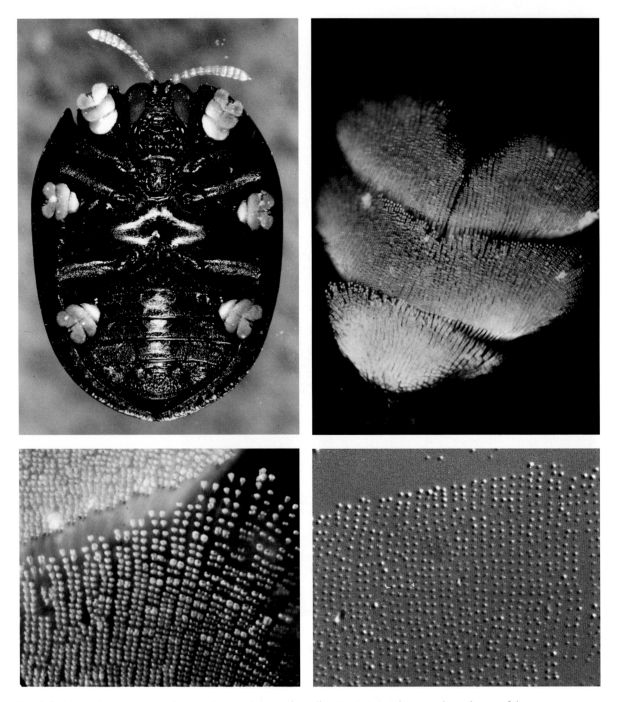

Top left: *Hemisphaerota cyanea,* in ventral view, showing the yellow tarsi. Top right: An enlarged view of the tarsus. Notice the three subdivisions of the tarsus, the tarsomeres. Bottom left: An enlarged view of a portion of the tarsus. The bristle pads are clearly resolved. Bottom right: Droplets left on glass as part of a tarsal "footprint."

WONDERS FROM WONDERLAND

persistency of ants, but until we did we had to chose an arbitrary cut-off time for out tests and 2 minutes seemed long enough. In any case, at pulls of low magnitude the beetles seemed able to cling indefinitely, so we had to chose a time at which to call it quits.

The measurements came out nicely. At pulls of up to 0.8 grams, amounting to nearly 60 times their body mass, the beetles held out for the full 2 minutes. At pulls beyond 0.8 grams the beetle's endurance time gradually shortened, to the point where above 3 grams it was reduced to a few seconds.

There was a conclusion to be derived from the fact that the beetle's endurance decreased with increasing pull. Whatever the beetle's attachment mechanism, it was evidently subject to fatigue. Clamping down, in other words, was tiring to the beetle, which means that the beetle was probably using muscles to keep its hold. That fact, we thought, was worth keeping in mind.

Next we subjected the beetle to attack by ants. We did some experiments in Florida with a Florida ant *(Camponotus floridanus)* and some experiments in Ithaca with a New York ant *(Formica exsectoides)*. Both sets of experiments involved presenting individual beetles to the ants and monitoring the events by eye. It became patently clear that the beetle derived protection from clamping down. The ants in most cases failed to pull it loose. With the New York ant we actually timed how long the individual ants persisted in their assault before giving up. The average time was 22.8 seconds, well short of the 2-minute cut off time we had used in the endurance tests.

It remained to be seen how the beetle pulled off its feat. Or rather how it kept from being pulled off its feet. We would obviously have to take a good look at those feet, which meant resorting to microscopy.

I looked at the beetle in ventral view and saw that its feet—the tarsi in entomological parlance—were unusually large. They were also yellow, in sharp contrast to the uniform blue of the rest of the beetle. Beetle tarsi usually have a pair of claws at the tip, which a beetle may use to hook into the leaf surface. But for such an anchorage to work the leaf has to be fleshy and penetrable, which the palmetto frond is not. The frond surface is smooth and tough, such as one would expect to be unyielding to tarsal claws. True, the larva of *H. cyanea* does anchor itself to the frond surface, but in its case the anchoring devices are short bladelike structures quite unlike the claws of adult beetles. At any rate, the *H. cyanea* adult could not possibly use tarsal claws for anchorage since in its case the claws are atrophied. What the beetle does have on its feet is bristles, thousands of bristles, sticking out from the sole of each foot—the ventral tarsal surface—like hairs on a brush. Bristles are a common feature of beetle tarsi generally, but I had never seen them in such quantity per foot. *H. cyanea* was a champion of sorts, and there seemed little question that the bristles were somehow involved in the clinging mechanism.

Scanning electron microscope photographs—SEMs as they are called—resolved the tarsal structure in exquisite detail. Each tarsus was subdivided into three bristle-bearing subsegments, called tarsomeres. We counted the bristles and found that there were about 10,000 per tarsus, making a total of 60,000 per beetle. Each bristle was forked at the tip, which means, if we assumed the bristle endings to be the contact points with the substrate, that the beetles had the option of relying on 120,000 such points. Could these bristle endings be points of attachment? If so, how exactly did they function in that capacity?

A possibility that occurred to us is that we might be dealing with physical adhesion. Take, for instance, a familiar example of such adhesion. Imagine pressing two pieces of glass together. They won't stick. But if you press them together with a drop of water in between, so the drop is squeezed into a thin film, the two pieces will adhere tenaciously. The water is not a glue, but pressed into a thin film it can act as such. And so can other liquids, with respect to many kinds of surfaces. The only requirements are that there be physical affinity between the liquid and the two connecting surfaces and that the liquid be reduced to a thin layer. We thought that the bristle endings of *H. cyanea* might be wetted, and that they might serve individually as adhesive pads. Collectively, by the thousands, they might provide the necessary contact surface by which the beetle secures adhesion.

The bristle endings of *H. cyanea* did indeed turn out to be padlike, and they were wetted. Examination of the portion of a glass surface upon which a beetle tarsus had trod revealed a footprint of tiny droplets that, to judge from their spacing, could only have come from the bristle pads. Similarly, if a beetle was placed on the underside of a piece of glass, so that its tarsi could be examined with a microscope as they were in contact with the glass, it became apparent that the pads bear fluid.

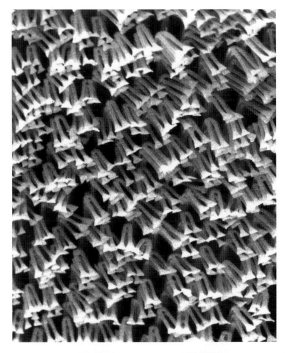

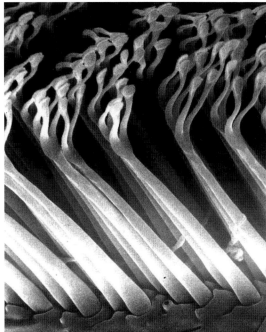

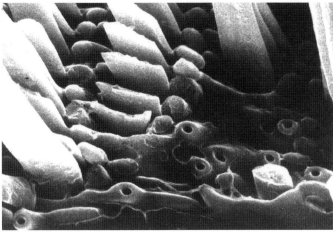

Hemisphaerota cyanea tarsi. Top left: A close-up view of the tarsal bristles. Top right: The bristles in lateral view. Note how they branch near the tip. An oil pore is seen at the base of a bristle. Bottom: A portion of a tarsus, with bristles cut away to expose the oil pores.

Thanks to analytical work done in Jerry's lab by Athula Attygalle, we now know that fluid is an oil, consisting of a mixture of long-chain hydrocarbons (specifically C_{22} to C_{29} n-alkanes and *n*-alkenes). An oil is ideally suited to secure adhesion to a leaf, since the outermost surface of leaves is waxy (palmetto fronds are no exception), and waxes make good contact with hydrocarbons.

When we looked carefully at the tarsi with the scanning electron micro-

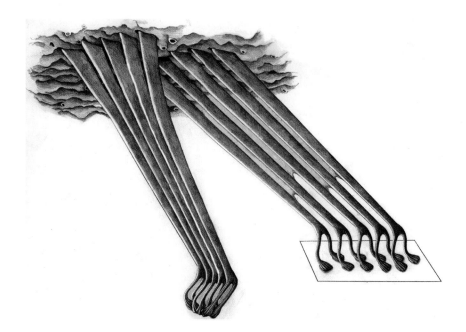

The postulated mechanism by which the tarsal bristles of *Hemisphaerota cyanea* secure their adherence (right) and become pre-wetted for adherence when the foot is lifted (left).

scope we found tiny pores on the "soles," between the bases of the bristles. We presume these pores to be the openings of microscopic glands that produce the oil. We assume that the oil, on emergence from the pores, seeps by capillarity into the narrow clefts between the flattened bases of the bristles, and onward to the bristle pads themselves.

We thought initially that walking could be an expensive proposition for the beetle. If every time the beetle pressed its tarsi down it lost 120,000 droplets of oil to the substrate, it might never be able to walk very far. It turns out that walking is not nearly as costly to the beetle as one would guess. Dan and I did a simple experiment. We rigged up a small chamber, with a glass top, in which the beetle could be confined, walking upside down with its feet on the glass. We installed this chamber on the stage of a microscope so we could observe the beetle's tarsi through the glass. We outfitted the beetle with a small piece of metal that we glued to its back, and placed an electromagnet under the chamber. This made it possible for us to "attack" the beetle by turning on the magnet and exerting an abrupt pull on it. What we found is that ordinarily, when the beetle is undisturbed, it walks with only a fraction of its bristles. It then commits to contact only the first few rows of bristles of each of the three tarsomeres. But when attacked—when we turned on the magnet—the beetle responded immediately by pressing its six tarsi down flat, so that virtually all its bristles were brought into contact. The beetle is evidently programmed to re-

WONDERS FROM WONDERLAND

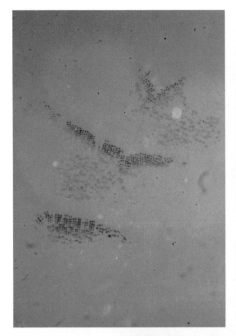
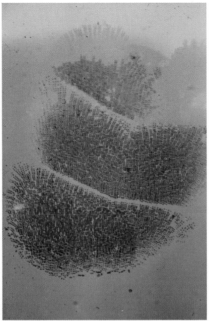

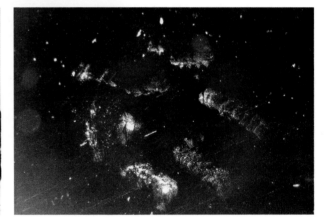

The top two pictures show a tarsus of *Hemisphaerota cyanea* photographed through a glass plate. At the left, the beetle is walking and only a few bristles at the leading edge of the tarsomeres are in contact with the glass. At the right, the beetle is under "attack" (being subjected to an electromagnetic pull), and virtually all the bristles have been brought into contact with the glass. Bottom left: An ant attack on *Hemisphaerota cyanea*, staged on a glass surface. Bottom right: The dark-field view of such a glass surface, showing the oily prints left by the beetle's tarsi. The beetle shifted the position of its legs during the attack; hence the drawn-out foot-markings.

A wheel bug *(Arilus cristatus)* feeding on a *Hemisphaerota cyanea.*

linquish little oil during locomotion and to put its full complement of bristles to use only when attacked.

Dan and I thought we'd examine sites where *H. cyanea* did battle with ants, on the assumption that such sites should bear evidence of the oil lost by the beetle when it was assaulted. We examined a number of glass surfaces on which engagements with ants had taken place and found the battle sites to be heavily smeared with droplets and streaks of oil.

We calculated from photos of footprints that the volume of the droplet of oil relinquished on contact by an individual bristle pad is 1.5 cubic micrometers. The total amount of oil lost by a beetle as a consequence of contact committal of its entire bristle complement would thus be 0.00018 cubic millimeters. Such an amount is hard to envision, but we calculate it to be equivalent to approximately 0.001 percent of the beetle's body mass.

WONDERS FROM WONDERLAND

That is not very much oil. It is as if a 70-kilogram human were to expend 0.7 gram of its mass. It should also be noted that the oil may be rather cheap for the beetle. Palmettos, like many other palms, are rich in waxes, and deriving long-chain hydrocarbons from waxes could be a relatively simple matter for the beetle, given that wax molecules themselves incorporate long carbon chains. The metabolic cost of oil production could be low for *H. cyanea*.

How does the beetle detach itself once it has clamped down? We think it does so by rolling its feet off the substrate one tarsomere at a time. The beetle presumably tenses its legs in order to press its tarsi down. Rolling the tarsi off should pose no problem once the beetle relaxes the tensing muscles.

One would predict from the preceding that the beetle would be vulnerable to a predator able to inject a muscle relaxant. Under action of such a relaxant the beetle would be unable to sustain the leg contraction needed for maintenance of tarsal appression, and it could then be vulnerable to being "peeled off." Interestingly, there is a predator that appears to use precisely that strategy against *H. cyanea*. We encountered that predator in southern Georgia, where we found it in the field, in the act of feeding on the beetle. It is a well-known predator, *Arilus cristatus,* the wheel bug, a member of the family Reduviidae. We collected a number of these bugs and fed them individual *H. cyanea*. They all attacked in the same way. They approached the beetle and straddled it, causing it to clamp down. They then proceeded to prod the beetle with their oral beak, in evident efforts to find a site for beak insertion. As soon as they pierced the beetle, the latter went limp, upon which the bug lifted the beetle up and pulled it from the substrate. With its legs gone limp the beetle seemed unable to hold on. Reduviid bugs are known to kill their prey by injection of neurotoxins, and *A. cristatus* is doubtless no exception.

Adhesive bristles occur commonly in insects, particularly beetles, and there is evidence that foot adhesion in these species is also mediated by oils. What is unusual about *H. cyanea* is that it depends on bristle adhesion for defense, and that it has bristles in vastly increased numbers for the purpose. Bristle proliferation may in fact be *H. cyanea's* primary evolutionary achievement. Beetles that do not use their tarsal bristles in defense may have hundreds of bristles per tarsus only. One species in which we made actual counts had 1,000 bristles per tarsus, one-tenth the number in *H. cyanea*.

4

Masters of Deception

James Lloyd was, first and foremost, a firefly enthusiast. Like many an entomologist with a special fondness for one particular group of insects, he could not easily be distracted with stories about other bugs. There are many anecdotes about such devotees, for example the one about the butterfly lover who as a graduate student was asked by one of the professors in his qualifying exam to comment on the vertebrate liver. "Butterflies have no vertebrate liver," was the reply, uttered in a tone that implied the question was meaningless. But Jim Lloyd was different. He had come to me as a graduate student, and he was to write a brilliant thesis.

Building on the work of the pioneer Frank A. McDermott, Jim Lloyd had cracked the communication code of fireflies (which are not flies at all, but beetles). It was known that the fireflies flying about flashing in the night are, by and large, the males. It was known, further, that males of different species have different flash patterns, different "songs in lights" as it were, to which the females respond by flashing themselves. As expected, the females respond correctly, each to males of its own kind. What was mysterious was that the females of different species all seemed to respond in the same way, with just a single flash. How could the males tell different females apart? How could they tell their "own" females from the others?

Jim Lloyd solved that mystery. He confirmed that the females do indeed answer with a single flash, but found that they differed in how they timed that flash relative to the male's signaling pattern. The specificity of the code lay in the duration of the interval between the end of a male's "song" and the reply of the female. Different females answered after pauses of different duration. Jim had measured the different durations by electronic means and the data were compelling. Intimacy in the world of fireflies seemed to be a matter of good timing.

Jim also showed something else. The females of one genus of firefly, *Photuris,* were known for their remarkable habit of preying on male fireflies of another genus, *Photinus.* How they managed to catch their prey had been another mystery. Jim found that the *Photuris* female can switch modes, as it were. To lure her own males for mating she delays her flash for the period that is attractive to her own kind, and to bring in *Photinus* she resorts to the delay that is specific to the prey. Appropriately, Jim called

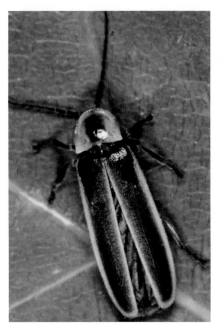

Photinus ignitus (left) and *Photuris versicolor* (right).

the female *Photuris* a femme fatale. It was a lovely story. I had no idea, when he published it in 1964, that I myself would be looking into fireflies and the ways of the femme fatale.

I had a bird as a pet in the mid-seventies, a Swainson's thrush *(Hylocichla ustulata)* that we called Phogel. Phogel was a voracious insectivore. She could eat 30 insects in a sitting, and she liked them best in the morning, but she did not necessarily like them all. Phogel was, in essence, a finicky gourmet. We soon learned to tell, just by watching her, how she felt about a given insect. If she liked it—if the bug was a "yum-yum"—she grasped it, spent a moment orienting it in the bill, and gulped it down. If it was a "so-so" bug, not particularly tasty, she would grasp it, drop it, then grasp it again and drop it, until she either ate it or let it be. The third category was that of the "no-nos." Bugs thus classed were rejected outright, usually after a first pecking, and were ignored when presented again at a later date, which indicated that the bird had remembered them as being noxious.

Phogel could obviously tell us something about the palatability of insects. We could put the bird to the test by feeding insect upon insect to her, while keeping a record of how she treated each. Items in the "no-no" category could then be examined chemically, to see if they contained interesting new compounds. The whole approach had a nice biorational ring to it.

MASTERS OF DECEPTION

So one summer I entered into a partnership with Phogel and fed her some 500 insects of over 100 species. She collaborated marvelously and led us down the path of discovery.

The protocol I adopted was straightforward. Every morning I'd go out into the field and collect upward of 20 insects, which I placed in individual vials. By the time I got back the bird was eagerly awaiting me, having quickly gotten used to the routine. I would then slip one bug after the other into her cage, and watch how she responded. There was a little dish in the cage into which I dropped the insect. Phogel seemed to be happiest if things proceeded predictably. I did these experiments at two locations, some in Ithaca, others at the Huyck Preserve near Rensselaerville, New York, a wonderfully secluded site where Maria and I sometimes fled in search of tranquility.

Phogel's offerings that summer were quite diverse and included some arachnids and crustaceans in addition to insects. Broken down by taxonomic category the menu makes for technical reading:

Thomisid spiders; phalangids; isopod Crustacea; Ephemeroptera; anisopteran and zygopteran Odonata; acridid and gryllid Orthoptera; Plecoptera; corixid, notonectid, gerrid, reduviid, phymatid, and pentatomid Hemiptera; corydalid and chrysopid Neuroptera; carabid, gyrinid, hydrophilid, silphid, scarabaeid, elaterid, lampyrid, cantharid, coccinelid, tenebrionid, meloid, and chrysomelid Coleoptera; bittacid and panorpid Mecoptera; limnephilid, hydropsychid, and other Trichoptera; tortricid, geometrid (adult and larvae), drepanid, ctenuchid, noctuid, hesperiid, pierid, and nymphalid Lepidoptera; tipulid, tabanid, asilid, syrphid, sciomyzid, muscid, tachinid, and other Diptera; tenthredinid (larvae and adult), formicid, and other Hymenoptera.

Phogel passed "no-no" judgment on eight species. With one exception, a sciomyzid fly, these were all beetles. Four were of families known to be chemically protected (blister beetles, soldier beetles, carrion beetles, ladybird beetles). The other three were fireflies (family Lampyridae). Nothing chemical was known about firefly palatability so this was exciting. I called Jerry right away.

"We'll need to get our hands on a lot of fireflies," he said, implying that one handful would not be enough. So I turned to market economics. I took out a want ad in the *Ithaca Journal* in the midst of the firefly season, offering to buy fireflies at 5 cents a piece.

We nearly went broke. We had expected a few children to catch the fireflies, but it turned out that at that rate the *parents* went out to collect the bugs, and they were better at it than we had bargained for. We quickly changed the ad, and at 1 cent a piece lured the kids into the action. We kept open house for firefly collectors on weekends and took great pleasure in watching the little ones come in with their catch to claim their reward.

As it turned out, we ended up getting most of our fireflies from another source. Fireflies are collected in large numbers by outfits interested in extracting their light organs, which contain an enzyme-substrate complex—the luciferin-luciferase complex responsible for the light production—useful in biochemical research and diagnostic medicine. Only the rear ends of the fireflies are needed for that purpose. The front ends are ordinarily discarded, but we managed to persuade one laboratory to let us have the unwanted parts. We knew that whatever was making fireflies distasteful was not restricted to the rear end.

So how do you go about isolating the compound or compounds responsible for the distastefulness of an insect? In the bombardier beetle things were made easy because the desired chemicals were bottled up in glands. But in fireflies, as in so many other insects, there are no special defensive glands, and the compounds are a part of the complex "soup" that makes up the body fluids. To fish out the active components from the soup can be difficult. At the very least it may require a bit of luck.

The standard fishing procedure is to use different solvents to extract the original insect sample, and then to check the extracts for activity. You could use one solvent for extraction of oil-soluble materials, and another for extraction of water-soluble materials, and hope that the active substances go into one solvent and not the other. To check on this you would bioassay the two extracts. In a test with a predator this would involve presenting the predator with an edible item treated with one extract or the other, and keeping track of which additive renders the item distasteful. Under the best of circumstances the active substances partition neatly into one solvent. At worst they don't, and the activity becomes difficult to follow. It can also happen that the activity is due to the combined action of substances of different solubilities, so that upon extraction activity is lost altogether. If you are lucky, and the activity goes into one solvent alone,

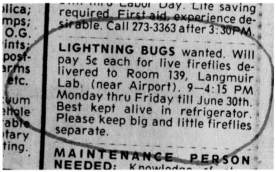

Phogel, our faithful collaborator, who alerted us to the presence of noxious chemicals in fireflies, and the first ad for fireflies in the *Ithaca Journal*.

you may still have the mighty task of separating desired molecules from all the others that went into the solvent. There are highly effective chromatographic techniques available today that make such separation possible. Even so, you may still have to bioassay the chromatographic fractions separately before you are able finally to home in cleanly on the compound or compounds at issue.

With the fireflies we were extremely lucky. The compounds turned out to partition nicely into one solvent, and because they had certain molecular absorption characteristics in the ultraviolet domain (which could be detected by spectrometric techniques) they could easily be traced in the isolation procedures. The desired compounds were steroids and they occurred in the fireflies as mixtures. They were very interesting new steroids, and all of us, including Jerry's two associates, David F. Wiemer and LeRoy H. Haynes, who had worked on the isolation, were delighted. We had to find a name for the new compounds and chose to call them lucibufagins (from *lucifer*, Latin for light bearer).

We next checked the lucibufagins for activity. Thrushes that we had in the laboratory—we had several of them—loved to eat mealworms, the larvae of certain beetles *(Tenebrio molitor)*. When we added lucibufagins to the mealworms, the birds no longer liked them. One bird that ate two

One of the lucibufagins (lucibufagin C) from *Photinus ignitus* (left) and ouabain (right).

treated mealworms responded by vomiting shortly thereafter. In previous tests we had found that these laboratory thrushes shared Phogel's aversion to fireflies. In only one instance did a thrush actually swallow a firefly, but the bird regurgitated the insect within a minute. We concluded that lucibufagins are distasteful to birds, and emetic if swallowed. We did experiments with jumping spiders and found that these were sensitive to lucibufagins as well. We offered the spiders freshly killed fruit flies, which they readily took, but found that they would reject such flies if we treated them by adding lucibufagins.

Lucibufagins are unlike any other steroids isolated from insects, although they bear some chemical similarity to certain steroids isolated from oriental toads. That similarity was the reason for our incorporating the syllable "buf" in lucibufagin (*Bufo* is a major genus of toad). But more interesting was the similarity of lucibufagins to certain plant steroids called cardenolides, including compounds such as ouabain or the widely used heart drug, digitalis. Could lucibufagins be of medicinal use? We thought we'd find out and sent a sample of lucibufagin to a pharmaceutical firm for testing. The results were both encouraging and discouraging. The lucibufagins did indeed have cardiotonic potency—they strengthened the heartbeat without changing the beat rate—but they seemed to offer no improvement over existing drugs.

We were not entirely disheartened. There were, after all, other fireflies. Our chemical work had been on *Photinus*. What about *Photuris*? Might it have "better" lucibufagins? We were very much taken by the idea of drugs from bugs and didn't want to pass up any opportunities. As things turned out, we did not find medicinal agents in *Photuris*. But we did uncover some unexpected facts about what it is that the *Photuris* female gets out of being a femme fatale.

When we looked at *Photuris* fireflies chemically we found that they did contain lucibufagins, but not consistently. The picture was baffling at first. The males seemed to contain only trace amounts of the compounds, and the females contained the substances in highly variable quantity. We also raised *Photuris* in the laboratory. The grown larvae, it turns out, are not difficult to collect in nature. They have a light organ themselves, like the adults, from which they emit a faint glow, so if you visit a site where they occur, and walk about in darkness, you can become pretty adept at spotting them. We now know where to find them in the environs of Ithaca, but some years back, before we did, a group of us, including almost everyone in my lab at the time, went all the way to Washington, D.C., on the advice of a firefly expert and friend, John B. Buck, who told us that *Photuris* larvae were abundant on the grounds of the Congressional Country Club. After getting permission, we spent one night roaming over the greens of the club and did indeed come up with a catch of several hundred larvae, together with several golf balls, a set of car keys, and a fountain pen.

We extracted a sample of the larvae and found that they were essentially lucibufagin-free. It seemed that only the adult female came to possess lucibufagins, and the question of course became *how*.

An exciting possibility was that the *Photuris* female appropriated the lucibufagins from her prey. "You are what you eat," the saying goes, and the *Photuris* female could be playing out a strategy whereby, by eating *Photinus,* she took on *Photinus*'s mantle of invulnerability. I thought I'd put the hypothesis to the test.

A most interesting young researcher had joined our group at that time. Fresh out of graduate school, David Hill was an expert on spiders. I have wondered often why there are so few spider experts, given that spiders are so incredibly interesting. The answer, I think, lies in the fact that spiders have eight legs rather than six, and therefore don't qualify for study by entomologists. Arachnologists are scarce because they must come into being on their own. There are no arachnological equivalents of entomology departments to churn them out in numbers. Just as well, I suppose. An arachnologist is thus more likely to be an original.

David's thesis had been on jumping spiders, which, as their name implies, capture their prey by pouncing upon it from a distance. David had

The femme fatale, *Photuris versicolor,* devouring a *Photinus ignitus* male.

shown that jumping spiders make use of some pretty clever trigonometry to get their fix on the prey before the jump.

David was willing to work on a six-legged creature for a change, and we got busy on *Photuris*. Jerry's group was right with us, and now included Michael Goetz, a new postdoctoral fellow, who was to do the dozens of lucibufagin analyses that made the whole project possible. For a start we showed that the only way *Photuris* females can acquire lucibufagin is by feeding on a lucibufagin source. We raised adult females from the larvae we had collected and showed, by analyzing females of different ages, that they do not on their own produce lucibufagins. Only if given the opportunity to feed on male *Photinus* or on pure lucibufagin did they end up containing lucibufagin.

It was spectacular to witness the eagerness with which the *Photuris* females devoured their prey. They usually pounced on *Photinus* males the moment these were introduced into their petri dish enclosures and in no time began chewing away at them. Aside from pieces of legs and wings, little remained of the prey after the meal. At the very least, the females were getting their nutritional fill. The females were equally eager to partake of the pure lucibufagin, which we offered, mixed with water, in glass micropipettes. All we had to do was hold the pipette to the female's mouth and she would drink the full measure of its contents. Individual females that we fed in the laboratory routinely took more than one male *Photinus*. It was easy to get them to take as many as three males. Champion femmes fatales devoured as many as six *Photinus*.

We were now in a position to test whether the acquired lucibufagin provided the female *Photuris* with protection. We used jumping spiders as predators. We had collected these in numbers, which took hours of enjoyable fieldwork, and had set them up nicely in individual petri dishes. To 29 of these spiders we offered *Photuris* females that had had no access to prey. Fifteen of these females were eaten. To another 29 spiders we offered females that had each eaten 2 *Photinus* males. Not a single one of these females was eaten. We already knew lucibufagin itself to be deterrent to spiders, so there seemed little doubt that the *Photinus*-fed *Photuris* were rejected on account of their lucibufagin content. The acquired chemical evidently did its job.

Fireflies have a special way of externalizing the lucibufagin when attacked—they bleed. Their thin outer cuticle is easily ruptured, and the blood droplets that ooze from sites of injury are laden with lucibufagin. Fireflies readily recover from such bleeding, which seems to be a normal concomitant of the defensive response. Because the bleeding is so readily induced, it is sometimes called reflex bleeding. Other insects that store their defensive chemicals in the blood also sometimes reflex-bleed. I learned early, just from handling insects in the field, that there are those that give off blood droplets when disturbed and that such droplets have a terrible taste. Now that I know that insects can be toxic, I don't sample them by mouth any more. Incidentally, when an insect bleeds, it does not suffer respiratory consequences. Insect blood, as a rule, does not carry re-

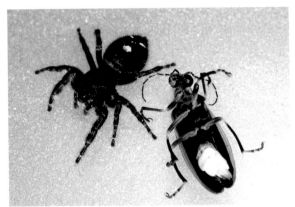

Tests with the jumping spider *Phidippus audax*. In the top picture, a dead fruit fly, suspended by a hair, is being offered to the spider. The fly was coated with lucibufagin, and was rejected by the spider. At the bottom left, the spider is eating a lucibufagin-free *Photuris versicolor* female. If such a female is fed on lucibufagin, it is rejected by the spider (bottom right).

spiratory oxygen. Insects breathe by way of special tubes, the tracheae, which convey air directly from outer openings to the tissues. Bleeding, for an insect, involves loss of nutrients and other useful chemicals, but in species that reflex-bleed, such cost appears to be bearable.

We did one additional experiment with fireflies that gave rather nice results. We knew that in nature *Photuris* females vary in their lucibufagin content, which made sense, given that we would expect them to have different individual predation histories. We also knew that we could estimate the lucibufagin content of field-collected *Photuris* females from analyses of the lucibufagin content of tiny blood droplets drawn from their wings. So we took 86 field-collected *Photuris* females, bled them, estimated their lucibufagin content, and offered them individually to spiders. Whether these females survived or not depended entirely on their lucibufagin content. Of the 19 females that we found to be lucibufagin-free, 13 were eaten. Of the 18 females that were rich in lucibufagin (1 to 10 micrograms per milligram of blood, reflecting consumption of one or more *Photinus*) none was eaten. The remaining females, bearing intermediate quantities of lucibufagin, were vulnerable to an intermediate degree. The femme fatale, it seemed, did indeed hunt *Photinus* for her own protection. The more lucibufagin she got, the better off she was.

But it turned out that the female did more than look out for herself. She also hunted for the benefit of her eggs. Andrés González, a graduate student from Uruguay, showed that the *Photuris* eggs receive a substantial endowment of lucibufagin from the mother, and that this endowment protects the eggs against such voracious egg eaters as ladybird beetles. A Canadian collaborator and gifted experimentalist, James Hare, showed that lucibufagins protect the eggs against ants as well. Andrés also showed that the *Photuris* female produces a defensive compound of her own, a quinoline derivative, that provides both her and her eggs with a certain amount of back-up protection should she not obtain lucibufagins.

We were eventually to be joined in this work by Scott Smedley, a graduate student who helped with the statistical analysis of the data, and we published the results. The study gave us something to think about. We had really enjoyed doing the work—so many of us, both chemists and biologists, had collaborated on it and the fireflies had turned out to be so immensely likable—but the findings also increased our awareness of some of the issues at stake in insect defense. The concentration of lucibufagins in the blood of *Photinus* is extraordinarily high. A single *Photinus* contains on average 60 micrograms of lucibufagin, amounting to 0.5 percent of body mass. This is particularly remarkable given that insects in general appear to be incapable of synthesizing steroids except from other steroids. *Photinus* probably produces its lucibufagins from cholesterol, a steroid that it can be expected to obtain with its food. But cholesterol is needed for many biochemical purposes in an animal, and to have to invest so much of it in lucibufagin production must impose a heavy metabolic cost. Small wonder, then, that *Photuris* has evolved a strategy for getting its lucibufagins free of charge. Defense appears to be expensive in general in the biotic world, and there are countless ways in which animals and plants make provision for reducing the cost. Acquiring defensive chemicals ready-made from the diet is a common approach, but it appears to be most prevalent in herbivores. Animals, like *Photuris*, that get their defensive chemicals from an animal source are rare. None, most certainly, are

An Australian bearded dragon *(Pogona vitticeps)*. These lizards can die from ingestion of a single *Photinus*.

known to obtain their defenses by resorting to the sort of mimetic subterfuge practiced by *Photuris*.

There is no question that lucibufagins could protect fireflies also against predators other than birds, spiders, ladybird beetles, and ants, and one could easily test for this. Lizards in the southeastern United States, for example, are said to be aversive to fireflies, and lucibufagins may well be the cause. It makes sense for American lizards, which have coexisted over evolutionary time with fireflies, to treat fireflies as "no-nos." They have had the chance, as it were, to evolve the capacity to reject these insects on the basis of taste. There are lizards, however, that have not had that evolutionary opportunity, and for them exposure to lucibufagins can be fatal. We were to become aware of this as a consequence of some peculiar events.

In the summer of 1998, I got a phone call from Dr. Michael Knight, a veterinarian at the National Animal Poison Control Center in Urbana, Illi-

MASTERS OF DECEPTION

nois. He had been called by the owner of a *Pogona* lizard who reported that his pet had died following ingestion of what he was certain were fireflies. I had never heard of anything like this, but shortly thereafter got news of a number of similar incidents. *Pogona* lizards, commonly known as bearded dragons, are easily maintained in captivity. They are highly tractable and friendly, and though native to Australia are bred in huge numbers in the United States for the pet market. What was happening to them was rather gruesome. Witness, for instance, the following case, as reported in the paper we eventually wrote with Dr. Knight:

> The owner of a healthy 8-month-old male *Pogona vitticeps* (about 100 grams body mass) captured a number of fireflies one July evening in the environs of his home (Iowa City, Iowa), and offered these to the lizard in its cage (aquarium tank). The lizard promptly ingested several of the fireflies. Within about 30 minutes the lizard began exhibiting violent head-shaking movements, followed by pronounced and increasingly frequent oral gaping. The animal seemed intent on vomiting, but no regurgitation was noted. As the gaping intensified, so did the lizard's respiratory effort, and the animal soon showed severe shortage of breath. Within the next 30 minutes, it underwent a conspicuous color transformation, its dorsal trunk and nape changing from the usual light tan to black. Within the hour after ingestion of the fireflies, and before veterinary assistance could be enlisted, the lizard died.

We examined the stomach contents of the animal, and found them to include remnants of nine *Photinus pyralis*. We knew that particular species of firefly had a lucibufagin content, on average, of 60 micrograms per individual. Having eaten nine individuals meant that the lizard had taken in a whopping dose of over half a milligram of the steroids. To judge from the potency of steroidal heart drugs similar in chemical structure to lucibufagins, 10 to 20 micrograms of lucibufagin—well under half the amount in a single *Photinus*—should have been enough to kill the lizard. The animal had clearly overdosed. Its heart may have been overtaxed.

Other reports indicated that ingestion of a single *Photinus* could suffice to kill a *Pogona*. They indicated also that other exotic lizards were at risk. Another lizard killed, for example, was a *Lacerta* from the Caucasus.

Even amphibians were endangered. We heard of a *Litoria* frog from Australia dying following ingestion of what may have been only three fireflies. The fireflies survived for a time in the frog's stomach and could be seen glowing right through the body wall. It must have been an eerie spectacle.

An undergraduate honors student in our laboratory, Richard Glor, did some experiments with *Pogona* that showed these lizards are unusually aggressive feeders. Being on guard just doesn't seem to be part of their nature. Although they appeared to be bothered by the shots they initially got in the mouth when they attacked bombardier beetles, they eventually ate even those insects. Their incautious nature suggests that in their native Australia *Pogona* don't usually encounter potentially lethal insects. Lucibufagins, I would predict, are not part of their natural gastronomical heritage.

A close colleague and herpetologist, Kraig Adler, joined us in publishing these results. Our intent was to save pet's lives. The message was a simple one. In strange territory, the evolutionarily "unprepared" could be dietarily endangered. Pet owners beware! But we felt there were broader implications. Geographic barriers are breaking down the world over, with the result that exotic species are making their presence felt virtually everywhere. In every region, resident and invader species are forced to cope with new realities. Among these must be the need to adjust to new dietary hazards. What we have seen played out in the lizard cage may therefore be symptomatic of what occurs these days on a significant scale in natural settings throughout the world.

There is one fact that has puzzled me for years. Given that lucibufagins can stimulate the heart, why haven't indigenous people discovered the medicinal or, if you wish, poisonous properties of fireflies? Is it that insects are not routinely checked for curative properties in indigenous cultures? Surely that can't be true. The toxicity of blister beetles, the source of Spanishfly (a potent poison, misused in the past as an aphrodisiac), appears to have been discovered independently by many cultures. Whatever the answer, I certainly advise against oral sampling of fireflies. Their toxicity is real, and deserves to be taken very seriously.

Incidentally, Phogel lived with us for a number of years, happily we hope. On autopsy she was found to be a male. Not knowing that *she* was a *he* does not affect the nostalgia we have for that bird.

THE KIND OF MIMICRY exemplified by the firefly femme fatale is sometimes called aggressive mimicry, the notion being that the mimic, by imitating some feature of its prey, gains access to the prey. There are variants of the paradigm, and some of these are fascinating. One of my favorite examples involves fish. Fish can't scratch their backs when they itch. I don't even know if fish ever itch, so let me put it in other terms. If a fish's slimy coating should be contaminated, say with unwanted particulate matter or even parasites, it has no way of cleaning itself. Its fins can't serve as arms. But it does have the option of turning to certain other fish for help, to fish called cleaner fish, which nibble on their surface and remove all foreign bodies. What's in it for the cleaner fish? A bit of slime, a parasite or two, and anything else nutritious that comes with the nibbles. Cleaners venture even into the mouth and gill chambers of fish to do their job. Fish are quite tolerant of cleaners, and become docile when approached by them. They recognize them visually and welcome their services. It turns out that there are certain other fish, which look like cleaner fish and are therefore also welcome, but which are anything but friendly. They are predators that, on gaining access to a fish, take out a chunk from its sides or its fins. Both the cleaner fish and its aggressive imitator are small. The mimicry enables the aggressor to attack fish much larger than itself.

In autumn of 1971, while at the Huyck Preserve, I had the luck of discovering a case of aggressive mimicry myself. The case was unusual, and I'm still not sure that it falls strictly within the definition of aggressive mimicry, but I do know that it provided me with some nifty times in the field. I'd gone to the preserve at the end of an intensive summer to do a bit of exploring and to gather my wits before the onset of the formal academic year at Cornell. The weather was crisp and the fall colors were at their peak. With me was my close friend and former student Robert Silberglied, who at the time was a doctoral candidate at Harvard. As was often our habit—we'd been in the field together in Ithaca, Florida, and Arizona—we were simply strolling about with collecting gear and cameras, having a look. I'd known Bob since his freshman days at Cornell. A gentle, extremely funny, and considerate person, Bob was a naturalist through and through, at once observant and inquisitive. He was great to be with in the field because he was always attentive. If you walked side by side with him outdoors and were observant yourself, and if you divided the territory so one of you looked mostly to the right and the other to the left, you were bound to make a discovery.

Some years earlier, Bob and I had made an interesting observation on a walk in Arizona. The desert was in bloom at the time and there were countless pollinators on the wing. Some of these we had thought were wasps, because of the color pattern on their bodies, but they turned out to be harmless beetles of the genus *Acmaeodera*. The beetles had fooled us completely. Instead of flying beetle-like with their wing covers (the elytra) spread apart and held out to the sides, they kept the elytra folded over the abdomen and flew with the membranous hind wings only. This in itself was misleading, since wasps also fly with what appears to be a single pair of membranous wings. But most deceptive were the colored stripes on the unfolded elytra, which made them look very much like the abdomen of a wasp. *Acmaeodora*, of course, as a harmless insect, profits from being the mimic of a model that stings. It was a classic case of what is called Batesian mimicry, where the model is noxious and the mimic isn't. No one had noted that case of mimicry, so we published our observations.

Bob and I are both Jewish and we often mused over this because there are so few Jewish naturalists. We wondered why that might be and concluded that it was because Jews were by and large an urban lot, at least until relatively recently in history, and therefore deprived of exposure to nature as children, an exposure that is so formative for a naturalist. It would have been much more logical for us to have become molecular biologists. The reason we didn't is that Bob used to be sent to a summer camp in the Catskill Mountains north of New York City as a boy, and I had the good fortune of having been chased by Hitler all the way to South America.

At the Huyck Preserve, as we wandered outdoors on that crisp autumn day, our attention was focused on colonies of the woolly alder aphid, *Prociphilus tesselatus*. So named because of their white flocculent covering, these aphids did indeed have the appearance of little sheep. Their whiteness made their colonies extremely easy to spot. We had earlier worked on the chemistry of the "wool" and shown it to consist of a waxy substance, a long-chain ketoester. The aphids secreted this wax in the form of extremely fine filaments, neatly grouped into tufts.

Our interest at the time was not so much in the aphids themselves but in the ants that stood guard over them. Aphids have a very special relation-

Wasp mimicry. The yellow stripes on the abdomen of the tethered wasp (*Dolichovespula arenaria*, top photo) are diagnostic for many wasps. The beetle *Acmaeodera pulchella* (bottom photos) displays these yellow stripes on its wing covers. Beetles ordinarily fly with the wing covers splayed. *Acmaeodera* is exceptional in that it flies with the membranous hind wings only, while keeping the wing covers locked together over the abdomen. The result is a beetle that looks like a wasp.

ship with ants. Except when dispersing by flight, aphids are relatively immobile and need little carbohydrate. But they are extremely good at producing baby aphids, and for that purpose require nitrogenous materials—essentially amino acids for synthesis of proteins—in disproportionately large amounts. They feed by pumping large quantities of plant juices through their bodies, thereby meeting the nitrogenous demands but also imbibing excess carbohydrate. They void this excess as part of their excreta, a sweet fluid appropriately called honeydew. As is well documented, aphids do not necessarily squander this fluid, but may present it as a gift to ants, which drink it and in exchange shepherd the ants and provide them with protection. Predators intent on feeding on aphids must have ways of coping with ants, which are ferociously aggressive when guarding their aphid "flock."

Virtually every colony of *Prociphilus* that we observed was tended by ants, which were busily feeding on the honeydew. They begged for the fluid by "tickling" the aphids with their antennae, and eagerly lapped up the anal droplets that the aphids squeezed out in return. Every ant seemed to be getting a share of the rewards.

We challenged the ants by poking them and watched how they held their ground and attempted to bite whatever instrument we used to provoke them. If we touched them they would turn and try to bite our fingers. We also saw how this defensive action worked in a real-life situation. Wasps—the familiar kind known as yellowjackets *(Vespula maculifrons)*—which themselves had been drawn to the honeydew, were actively prevented by the ants from getting to the aphids. They were individually confronted and shooed away, and forced to restrict their drinking to excess honeydew that had dribbled from the aphids to leaves lower down on the tree, where there were no ants.

I decided I wanted to watch the aphids close up, so I cut some colony-bearing branches from alder trees, and took these indoors, where we had our microscopes. We had a small cottage available to us at the preserve, with enough table space to set things up for experimentation. We lit the wood in the fireplace and went to work. At higher magnifications the aphids were interesting to watch, particularly because the females were giving birth. They were delivering their young live, as aphids often do, and I thought I'd photograph the action. I was about to take my first shot when I noticed something that took me completely by surprise. There, plain as day, making its way across the viewing field of the camera, was a running aphid. Ridiculous, I thought, aphids don't run! So I looked again and realized that what I had taken to be an aphid was not an aphid at all, but the larva of a green lacewing, so similar in appearance to the aphids that it could easily pass as an aphid itself. The resemblance was uncanny.

I had recognized the larva because I had some knowledge of the family Chrysopidae, the green lacewing family, to which it belonged. Frank Carpenter, my mentor in graduate school at Harvard, was a world authority on the Neuroptera, the insect order that includes the Chrysopidae. He had suggested initially that I work on the taxonomy of the chrysopids, and I was at first tempted to do so. Chrysopids, as adults, are stunningly beautiful, pale green with delicate translucent wings, and I was instinctively drawn to them. I ended up reading as much as I could about the group, and even kept several species in living culture, together with their aphid

A colony of the woolly alder aphid, *Prociphilus tesselatus,* on an alder tree by a pond at the Hyuck Preserve in Rensselaerville, New York. The ant in the bottom left photo is drinking honeydew, totally oblivious of the *Chrysopa slossonae* larva (arrow) beside it. The ant in the bottom right photo is performing its guard duty.

food, but I realized quickly that taxonomy was not for me. But on that day at the preserve I remembered something I had read. Some chrysopid larvae make it a habit to cloak themselves with debris. They were called trash carriers, and they make use of the most diverse material for the purpose, including vegetable matter and arthropod remains. Chrysopid larvae are carnivorous, and the trash-carrying kind often load up with the leftovers of their meals.

The close resemblance of our newfound larva to the woolly aphids—the

A larva of the green lacewing, *Chrysopa slossona*, reloading with wax, after having been plucked clean with forceps. In the top photos, it is shown picking up a load of wax with the jaws and applying it to its back. In the bottom left photo, it is adding wax to a nearly completed packet. At the bottom right, it has finished reloading.

larva bore a wooly coating indistinguishable from that of the aphids—suggested that we might be dealing here with a case of deliberate deceit. I called Bob to the microscope and he too marveled at the resemblance. We scanned the aphid colonies and found several more of the larvae. Did they feed on the aphids? Yes indeed they did, and they seemed to be quite persistent. One larva that we watched for a while appeared to have a predilection for newly born aphids, but others fed on older individuals. The larvae killed and ate their prey as chrysopid larvae typically do, by impaling them on their pointed hollow jaws and sucking them out.

We spent a single evening with the aphids on that particular occasion, but we did manage to confirm what we suspected, that the larva was of the trash-carrying kind, and that it obtained its wooly covering by plucking it from the aphids. By tearing away at a larva's coating with forceps, I had discovered that the "wool" was only loosely held by bristles on the larva's back. So I took a larva, denuded it with forceps, and put it back among the aphids. It wasted no time rebuilding its cover. It shoved it jaws into the coating of the nearest aphid, and using the jaws as a two-pronged fork, plucked away and transferred tuft upon woolly tuft to its back. To deposit the tufts the larva arched its front end sharply over the back, doing so with such precision that no place on its back remained uncovered. By the time it was finished it looked almost exactly like one of the aphids.

It took several return trips to the preserve, some with Maria, another with my research assistant Karen Hicks, an enthusiastic and very competent young ecologist who was part of our group for several years, to find out why the larva goes to so much trouble to disguise itself. First, we had to find out whether the larva was known. It turned out that the adult was known, and had been named *Chrysopa slossonae,* but the larva was not. The loading behavior, and more important to us, its function, seemed to be unknown.

It seemed to us that by coating itself the larva might be masquerading as an aphid, and that by so doing it might fool the ants. Like a wolf in sheep's clothing, it could be mistaken for a "sheep," and thereby be spared the attack of the "shepherds." The analogy turned out to be appropriate.

We did experiments with ants outdoors as well as with aphid colonies that we brought indoors. The first thing we showed is that the larvae, if plucked clean of their wool, are attacked by the ants. We collected 27 nearly mature larvae, brought them into the laboratory, and denuded them by wiping the wool off their backs with a fine brush. We then released them into natural colonies of the aphids and watched to see what happened. The colonies were carefully selected to make sure there were ants in attendance. All of the larvae were quickly discovered by the ants and were bitten by the first ants that contacted them. Only 4 got away. Fourteen of the remainder were seized in the mandibles by individual ants, then lifted over the margin of the branch and dropped, so that they fell to the ground. Another 2 that were similarly grasped fell to the ground with the ants. The remaining 7 were carried off the plant by individual

The head of a *Chrysopa slossonae* larva. The jaws (arrows) serve both to impale aphid prey and to grasp loads of wax.

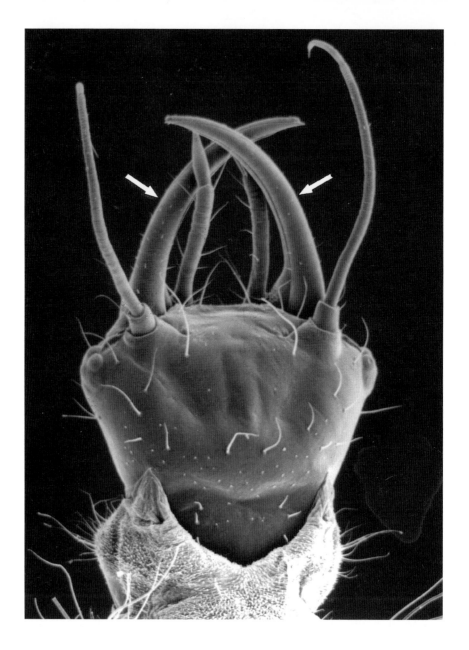

ants that held them tightly in their mandibles as they descended along the branches to the ground. Two of these larvae were actually pierced by the ant's mandibles, and they died as the ants proceeded to feed on their body fluids. Naked larvae evidently don't stand much of a chance with the ants.

We also released 23 larvae with intact woolly covers. These were also encountered by ants, but only 8 were bitten and the bites were fleeting. No

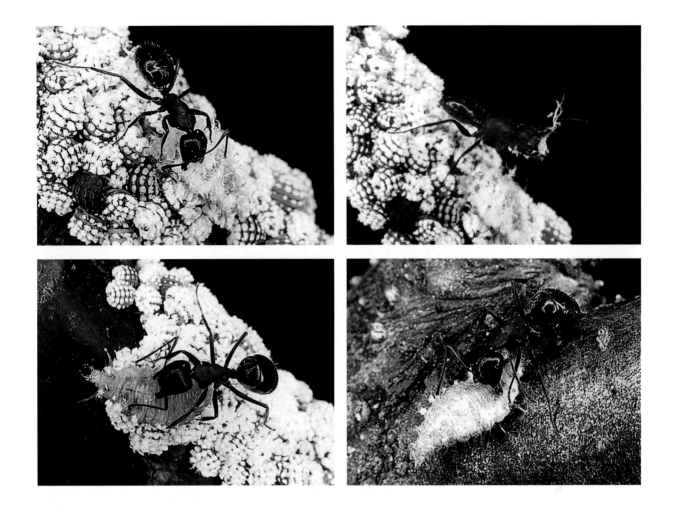

sooner had an ant clamped down on the back of a larva than it let go and, with its mouthparts visibly contaminated with wool, backed away. It then proceeded to clean itself by wiping its mouthparts with the forelegs, but this took time and kept the ant from resuming the attack. A second and even a third ant sometimes bit the larva, but with identical results. Undeterred, and usually deprived of only a small fraction of its cover, the larva walked away, settling eventually at a site where it fit inconspicuously into the colony. The remaining 15 larvae were merely inspected by the ants and then ignored. The ants palpated them with antennae and forelegs but refrained from biting. They had evidently "passed" as aphids. Cloaked in wool, they had fooled the prey's guardians.

I was curious about the behavioral priorities of the larva. What if the

Ant attacks on *Chrysopa slossonae* larvae. Attacks upon fully cloaked larvae (top photos) result in the larva's escaping, and the ant's being left with its mouthparts contaminated with wax fibers. Denuded larvae are persistently attacked and often killed (bottom photos).

larva was deprived of food for a while, and at the same time deprived of its cover? How would it respond? Would it give priority to eating or to rebuilding its cover? It was the classical guns-versus-butter issue. Under the worst of circumstances, how does an animal decide between the fundamental needs of food and defense? We took larvae of four categories—satiated and cloaked, hungry and cloaked, satiated and denuded, hungry and denuded—and released them into aphid colonies (without ants) we had taken into the laboratory. We monitored the behavior of the larvae for the first hour following their release and recorded what they did. As expected, the larvae of the first category, which had neither need, did little except rest. Those of the second and third categories, having one need only, dedicated themselves to remedying that need. Those of the fourth category, which had the dual need and therefore the conflict, divided their time between feeding and rebuilding their cover. They rated the two activities as about equal in importance.

We found the *Prociphilus* colony a microcosm of fascinating complexity. Several other predators associated with the aphids are also intriguing. There are, for example, the maggot-like larvae of certain flies called hover flies (family Syrphidae), which when abundant can literally wipe out entire colonies of the aphid. These larvae, too, are protected against ants. When attacked they regurgitate a droplet of salivary secretion, a gluey fluid that solidifies quickly on exposure to air and acts to immobilize ants. There is also the caterpillar of a butterfly, *Feniseca tarquinius,* unusual because carnivory is so rare among caterpillars. It lives beneath a loose silken sheath that it spins amidst the aphids and subsists by devouring them in huge numbers. The ants appear never to be aware of this predator. Also associated with *Prociphilus* is a fungus, *Scorias spongiosa,* an ascomycete of the family Capnodiaceae, a ragged entity resembling a worn kitchen sponge, which grows on branches of the alder wetted by the drippings of excess honeydew. The fungus at times may be drenched in honeydew and may then be visited by wasps, which imbibe the liquid from its surface. Wasps prevented by guarding ants from drinking honeydew at the aphid source appear to learn quickly that the fluid is available also on the fungus. There is no question that there are still aspects of this extraordinary scenario that remain to be uncovered. On a small scale, the various interactions that bind *Prociphilus* to its guardians and enemies are illustrative of the relationships, both antagonistic and mutualis-

A syrphid fly larva, well concealed among its aphid prey, using its sticky salivary secretion in defense. The larva is about to be pinched with forceps (top left), and when subsequently pinched, revolves its front end (top right) to deliver a droplet of defensive fluid onto the forceps (bottom).

tic, that hold together the biosphere itself. The story of *Prociphilous* and its entourage could be told marvelously in pictures. Film makers take note.

    I BECAME INTERESTED in mimicry on the very first trip that I took to the Southwestern Research Station in Arizona. The first director of the station, Mont Cazier, a fabulous naturalist, was to remain a close friend until his death in 1995. He was the sort of person you went to see right away when you thought you had made a discovery in the field. Chances are he had made it himself, and he would then always tell you freely about his own observations. His knowledge of desert biology, and of desert insects in particular, was prodigious.

On that particular summer there were two other renowned entomolo-

The harvester, *Feniseca tarquinius,* is exceptional among butterflies in that its caterpillar is predaceous. The caterpillar (bottom) feeds only on *Prociphilus tessela-tus* and is ordinarily well concealed among its prey.

gists at the station, E. Gorton Linsley, a world authority on long-horned beetles and dean of the College of Agriculture at Berkeley, and Alexander B. Klots, a specialist on butterflies and professor at the City College of New York.

On the grounds of the station, directly opposite the main entrance, there was a meadow dominated by masses of white sweet clover *(Melilotus alba).* The plants grew shoulder high and were in full bloom. Over the years Mont had noticed that certain beetles formed aggregations on these plants and he had made a point of showing the beetles to visiting scien-

The fungus *Scorias spongiosa* (arrow) grows only in association with *Prociphilus tesselatus* colonies.

tists. Gort and Dr. Klots had begun looking into the biology of these beetles and I was invited to join. What we discovered that summer, and on a later visit to the station in 1961, when my student Fotis Kafatos joined us, was intriguing. The aggregations on the clover turned out to be complex assemblages of models and mimics, including not only beetles but also moths, and featuring one set of mimics with the extraordinary habit of feeding on the models.

The chief protagonists in this scenario were beetles of the family Lycidae, which served as the models. Lycids were known to figure as models in mimetic assemblages the world over, and it had long been suspected that they are chemically protected. Australian investigators have since shown them to contain phenolic compounds that make them distasteful

A lycid *(Calopteron terminale)*. Lycids are called net-winged beetles because of the distinct arrangement of their elytral veins into ridges and cross-ridges. The veins, shown close up in the photo at bottom left, are swollen and hollow. They are filled with blood, and rupture readily when lycids are attacked. The emergent droplets of blood are deterrent to predators. In the photo at bottom right, a lycid is bleeding in response to having an elytron seized by forceps.

and pyrazines that give them an unpleasant odor. On that meadow in Arizona the lycids were of two kinds. One kind, the concolorous kind, was evenly orange-brown in appearance, and the other, the black-tipped kind, was also orange-brown but had black-tipped wing covers. Both types were of the same genus, *Lycus,* and could therefore be expected to have had a common evolutionary ancestor.

The lycid beetle *Lycus loripes* (left) and two of its mimics: the cerambycid beetle *Elytroleptus ignitus* (center) and the geometrid moth *Eubaphe unicolor* (right).

Each type of lycid had a set of mimics. These were present on the clover plants amidst the lycids and were of several kinds. There was a concolorous moth that was clearly imitative of the concolorous lycid, and a moth with black-tipped wings that was a mimic of the black-tipped lycid. Most interesting, however, were two beetles of the genus *Elytroleptus*, one concolorous, the other with black-tipped elytra, which were obviously mimetic of the two types of lycid. Indeed, each tended to occur in the aggregations of their respective models. Being of the same genus, the two *Elytroleptus* could themselves be expected to have evolved from a common ancestor. That ancestor, conceivably, could already have been associated as a mimic with the common ancestor of the two lycids. When the ancestral lycid underwent its evolutionary split, the ancestral *Elytroleptus* could have followed suit.

Things got interesting when we realized that *Elytroleptus* was sometimes found in the field seemingly hugging the model lycid. Close observation showed that in such acts the *Elytroleptus* was in fact attacking the

The lycid beetle *Lycus fernandezi* (left) and two of its mimics: the cerambycid beetle *Elytroleptus apicalis* (middle) and the pyromorphid moth *Seryda constans* (right).

lycid, with consequences that were usually fatal to the latter. We took specimens of *Elytroleptus* into the laboratory and found that they actually fed on the lycids, and did so as a matter of routine. The lycids are naturally sluggish and take virtually no evasive action when attacked. As is often the case with chemically protected insects, they act as if they have nothing to fear.

It is remarkable that *Elytroleptus* should be a carnivore. *Elytroleptus* belong to the family Cerambycidae (the long-horned beetles), the species of which are virtually all plant feeders as adults. The single exception is a European species that has been reported to hunt young spiders. *Elytroleptus* is also remarkable in that it is able to feed on *Lycus,* a model insect that is chemically protected and therefore generally shunned. What is it about the biochemistry of *Elytroleptus* that makes this beetle tolerant of the defensive chemicals of its prey? Does it inactivate the chemicals or does it tolerate them, and if it tolerates them does it store them in its body for its own defense? Mimics are usually classified as Batesian or Müllerian depending on whether they are undefended and palatable or defended and unpalatable. Does *Elytroleptus* belong alternatively to one mimetic category or the other depending on how recently it devoured a lycid? In other words, is it a Müllerian mimic after feeding on a lycid, and a Batesian mimic if it goes unfed for a period of time? How long do the acquired chemicals persist in *Elytroleptus*' body, if indeed the beetle acquires the chemicals from its prey? *Elytroleptus* is clearly engaged in a risky strategy.

MASTERS OF DECEPTION

Elytroleptus ignitus feeding on a lycid. Ordinarily, in nature, *Elytroleptus* are usually found feeding on the particular lycids they imitate. Here, in a laboratory experiment, the *Elytroleptus* is attacking a black-tipped lycid *(Lycus fernandezi)* from the other mimetic complex, which shows that it is able to feed on these *Lycus* species as well.

The numerical ratio of lycids to mimics in a *Lycus loripes* association. The area sampled was about 30 meters square. The model lycids (57 individuals) make up the upper four rows. The fifth row includes two *Elytroleptus ignitus* on the left and three *Eubaphe unicolor* on the right.

A lycid *(Lycus fernandezi)* at one of its diurnal drinking sites.

If it is to profit from its imitation of the lycid model, it can't afford to eat excessive numbers of the model, or predators won't have the opportunity to learn to avoid the model. *Elytroleptus,* one would predict, must at all times be present in small numbers relative to the lycids, and this turns out to be true. Actual counts that we made showed the lycid model to outnumber the *Elytroleptus* by a large margin.

Further work that we did with taxonomists showed that we ourselves had been blind to certain aspects of the mimicry. What we had thought to be single species of model lycids in the two associations were in fact, in each case, two species. In other words, the concolorous lycid was really a pair of look-alike species (*Lycus loripes* and *L. simulans*), and so was the black-tipped lycid (*Lycus fernandezi* and *L. arizonensis*). The look-alike species in each case can be expected to be closely related, and to have been derived evolutionarily from a common ancestor that resembled them. That original look had survival value, and it made sense that it should have been retained. The look-alike species, in each case, are in effect mimics of each other.

We made another discovery that raises a further question. Lycids are heavy drinkers. Their outer skeleton is relatively thin and they are subject to drying out. We found that at our Arizona site the lycids take periodic trips to the edge of ponds or running streams to drink. They alight on wet

Fotis Kafatos beside one of the lycid-stocked nets used as a lure.

soil beside the water and "tank up." They sometimes congregate in numbers at such places, with the result that they form conspicuous aggregations. Remarkably, in these aggregations there are sometimes visiting moths, which also come to drink and are mimics of the lycids. They are not the species that occur on clover, but ones that mix with the lycids only at the drinking sites. It is not unusual for moths to drink from moist soil. Such behavior is called puddling and is carried out by moths primarily to fulfill a need for salt (see Chapter 10). Most moths puddle at night. The ones that joined the lycids may have become diurnal in their drinking habits because as lycid mimics they could afford to do so. I know of no natural predators other than *Elytroleptus* that routinely feed on lycids.

Fotis Kafatos and I got interested in a specific question. What causes the lycids to aggregate? Do they spot one another visually and congregate by sight, or could chemical attraction be involved? By simple experimentation we showed that they are drawn together by chemical means. We took groups of 150 lycids, placed them in clusters on clover, and then put bags of white porous cloth over them, so that they were inescapably and invisibly confined. We kept males and females separate. The results were clearcut. Lycids of both sexes were attracted to the bags with males, but none

MASTERS OF DECEPTION

were attracted to the bags with females. Evidently only the males had attractive power and they attracted both males and females. We concluded that the males are probably the first to emerge from their pupae early in the season, and that it is the males that establish the aggregations. They produce an attractant—an aggregation pheromone—that initially brings in males, but that with time, as the females emerge, brings in females as well.

But what is the purpose of the aggregations? One is tempted to think of them as mating aggregations, since the assembled males and females are indeed commonly found copulating. But if mating is the purpose, why should the males attract other males in addition to females? Unless it can be proved that by pooling their pheromonal resources males improve their own individual chances of mating, it would seem counterproductive for males to lure other males. It is possible, though, that the aggregations play a defensive role. Lycids are aposematic—conspicuously colored. Many chemically protected insects share that distinction, which supposedly is to their advantage. Predators that learn from direct experience that a prey item is distasteful apparently remember it better, and are then more likely to discriminate against it, if it is gaudily colored. An aposematic insect can thus afford to flaunt itself, as it surely does if it is part of a conspicuous aggregation. Moreover—and this may be of critical importance—by being part of an assemblage an insect reduces the chance that it will itself be the first of the group to be encountered by the predator. Being the first entails risks, since the predator may well inflict injury even in the course of a thwarted attack. It is therefore good strategy for the individual male lycid to contribute to a collective effort that fosters crowding.

How the lycid males and females sort one another out within the aggregations remains unknown. We never looked into their mating rituals, but suspect that the courtship is fairly competitive. I am particularly curious as to how the sexes in the look-alike species avoid confusing one another. Unknown also is the chemical nature of the attractant pheromone. Our discovery of its existence, in 1961, had been a first of sorts. On no previous occasion had aggregation in an insect been shown to be mediated by a male-produced pheromone attractive to both males and females. We don't even know that all four lycids at our site produce aggregation pheromone. Our experiments were done with only one of the concolorous species. And does *Elytroleptus* also home in on the attractant pheromone?

I know of no mimetic associations that offer more opportunity for study than those of lycids. There are countless questions that could be raised pertaining to the Arizona populations alone. I would love to return to the site and have another look, but it is unlikely that I will have the chance. Doctoral students take note.

5

Ambulatory
Spray Guns

In the year 1910, on a field in the Paris suburb of Issy-les-Moulineaux, there took place a little-remembered incident that was to have an impact on aeronautics and fluid dynamics. It was also to have a bearing, albeit a barely perceptible one, on entomology. To the protagonist of the incident, the young Romanian engineer Henri Coanda, the event was doubtless unforgettable.

Coanda was a student at the Ecole Supérieure Aéronautique in Paris, where he was a disciple of Alexandre Gustave Eiffel, designer of the famous tower and an expert in aerodynamics. Young Henri had built an aircraft with his own hands and of his own design, and he had taken it to that suburban field to try it out. One can only imagine his excitement. The machine was, amazingly enough for its time, jet-propelled. It engendered thrust from two engines, which spouted a flaming mixture of compressed air and fuel. Henri had built his plane of plywood and he had worried that the jets might set it on fire. So he had installed a pair of curved metal plates on the fuselage that were to provide protection by deflecting the exhaust flames.

Alas, as soon as he started taxiing for takeoff he realized he was in trouble. The metal plates, instead of diverting the jets, were actually causing them to bend inward toward the fuselage. Distracted by this frightening development, he did not realize until the last moment that he was about to ram into a wall at the end of the field. He yanked up on the controls and managed to clear the wall, but then immediately stalled and crashed. He survived the crash, but it was a high price to pay, especially considering that his might have been the first successfully completed jet-powered flight, a possible Guinness record.

Coanda could not get over the peculiar behavior of those jets. Why had they been wrongly deflected? He discussed the event with the leading expert on aerodynamics at the time, Theodor von Kármán, who, realizing that a new phenomenon had been uncovered, named it the Coanda effect, in due recognition of its discoverer.

So what is the Coanda effect? Simply put, it is the propensity of a liquid or gas, when flowing along a curved surface, to cling to the surface and follow the curvature. In everyday life the effect accounts for the annoying tendency of liquids to curve around the spouts and lips of vessels when be-

The Coanda effect, exemplified by the annoying propensity of fluids to curve around snouts when being poured.

ing poured and to trickle down the outside of the containers (and onto the tablecloth, of course). Coanda had evidently shaped his deflectors the wrong way and that had nearly cost him his life. He did, incidentally, remain in aeronautics and would eventually rise to the post of chief engineer at the Bristol Aeroplane Company in Britain. As to the Coanda effect, it is still being invoked in the design of hovering vehicles, propulsion systems, and nozzles. As I was to find out, the effect also explains how certain beetles aim their defensive glandular discharges.

The beetles in question are the Ozaenini, members of a subfamily (Paussinae) of the ground beetle family (the Carabidae). I had never seen a live ozaenine. I knew they were primarily tropical in distribution, but had no idea that some species ranged northward into the United States. I did suspect that they discharged benzoquinones. Older publications made mention of the fact that one's fingers become darkly stained if one handled ozaenines, and I knew this to be a sure sign of benzoquinone production. I was therefore quite eager to get my hands on an ozaenine.

In the summer of 1967, on a field trip to Portal, Arizona, I got my wish. I had noticed a small, unfamiliar brown beetle at the light trap I had set up outside my cottage, so I picked it up to see if it gave off a stink. It did better than that. It sprayed, and it did so with an audible pop. It also left brown markings on my fingers and it smelled unmistakably of quinones.

Ozaenines have two little flanges that stick out from the edge of the elytra, which are diagnostic for the group. No one knew what the flanges were for, but they were unique. I looked at my little beetle with a hand lens, and there they were—two little protuberances sticking out from the wing covers just in front of the abdominal tip. Initially I gave little thought

AMBULATORY SPRAY GUNS

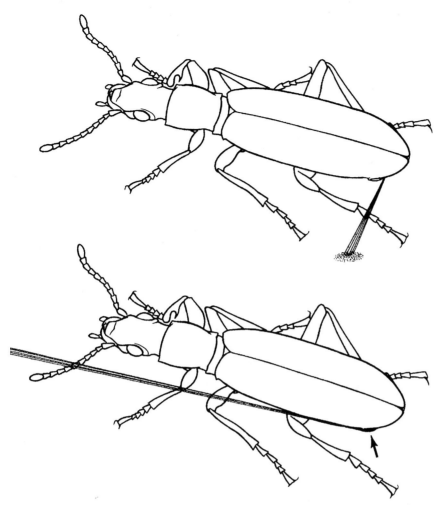

An ozaenine beetle, discharging from the left gland, toward the side (top), and forward (bottom). To discharge forward the beetle directs the jet of fluid over the elytral flange (arrow).

to the flanges. I was more interested in the bombardier-like qualities of the beetles. Were ozaenines, essentially, bombardiers? Were the pops an indication that ozaenines ejected their spray explosively, from two-chambered glands, at high temperature, as bombardiers do? There was no way that anyone would be able to get all the answers from a single ozaenine, but that little beetle did survive the return trip to Ithaca, which made it possible for Jerry's group to confirm that it sprayed quinones, and for me to dissect it and find that its glands were two-chambered. While still alive it also showed us that it was a good marksman. Tethered and placed on indicator paper, it proved able to aim its discharges accurately in the direction of individual legs that were pinched. It fired forward as well as to the sides.

AMBULATORY SPRAY GUNS

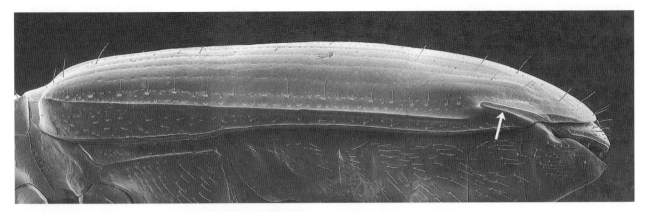

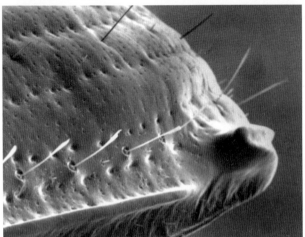
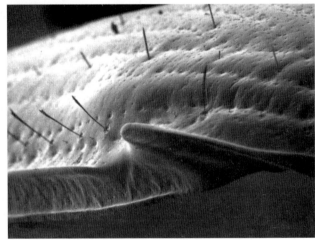

Top: Lateral view of the abdomen of an ozaenine beetle, showing the elytral flange (arrow). Bottom left and right: Two views of the elytral flange, showing its curvature.

Within the year of that first encounter I took a trip to Panama, where to my great delight I found ozaenines to be relatively plentiful. I returned to Ithaca with over a dozen live ones, which survived nicely on mealworms and water, and provided us with answers to some of the remaining questions. Dan and I did thermal measurements and found the ozaenine spray to be ejected at 60° to 80°C, not quite as hot as the spray of "genuine" bombardiers, but hot nonetheless. Everything seemed to suggest that the defensive apparatus of ozaenines was basically similar to that of classic bombardiers. But there were differences. We recorded the pops that accompany the ozaenine's ejections and found the sound to be nonpulsed, which indicated the spray ejections were themselves nonpulsed. And there were anatomical differences. Whereas classic bombardiers have their gland openings on the abdominal tip, ozaenines have their openings on the sides of the abdomen just anterior to the tip. This meant that ozaenines probably had their own way of aim-

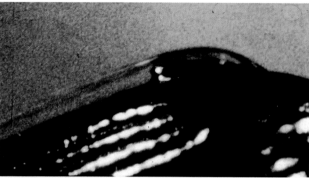

Consecutive frames from a motion picture, showing the elytral flange just before discharge (left) and at the moment of discharge (right). Note that the stream of fluid follows the contour of the flange.

ing their spray, and when we looked into that question, things got interesting.

It will be recalled that classic bombardiers—which henceforth I will call brachinines (they belong to the subfamily Brachininae within the Carabidae)—direct their spray by revolving the abdominal tip. The tip projects beyond the posterior margin of the elytra and is therefore free to rotate in all directions. The gland openings are automatically aimed as a consequence of these motions.

In ozaenines, the abdominal tip can flex only up and down, and within a limited range. It cannot rotate or be bent sideways. How then, given this limited capacity, are ozaenines able to discharge both forward and backward? The answer lies in the very special service provided by the flanges.

The flanges, it turns out, act as launching guides for anteriorly aimed ejections. They are curved and grooved, and positioned directly in front of the gland openings. Jets of fluid, on emergence from the openings, are inevitably routed onto the grooves and forced to follow the curvature of the flanges. They are thus bent sharply in their trajectory and directed forward, parallel to the body. Motion pictures that we took of ejections elicited by pinching forelegs showed how precisely the jets are directed by the flanges. We destroyed the flanges in a number of ozaenines, by shaving the structures away with a microscalpel, and were able to show from motion pictures that this caused the ejections to be misdirected at angles away from the body.

Our most convincing evidence came from a high-speed motion picture film that we took of a flange in close-up during an ejection. At 400 frames per second the discharge took up only a few frames of film, but these illustrated in vivid detail how the emergent jet of fluid is bent into its final trajectory by adherence to the outer curvature of the flange. The angular deflection of the jet was over 50 degrees! Here, quite clearly, was proof that

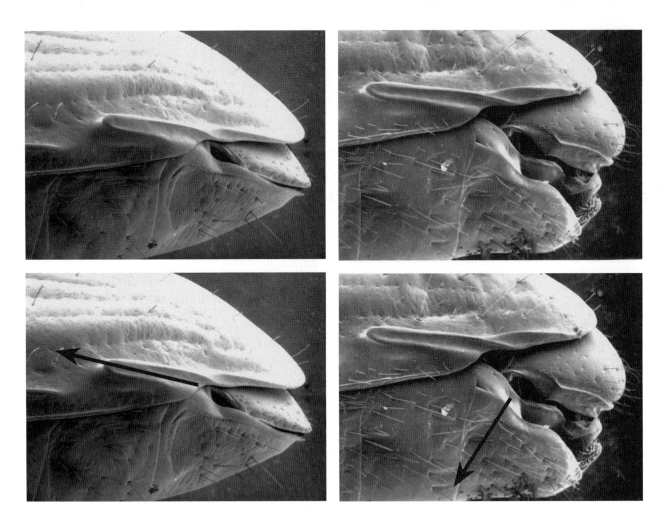

Left: The relationship of the abdomen to an elytral flange during a forward discharge; the gland opening is aligned with the flange, which causes the discharge to be directed along the flange (arrow, bottom left). Right: The relationship of the abdomen to the elytron during a lateral discharge; the gland opening is not aligned with the flange, with the result that the discharge is directed to the side (arrow, bottom right).

the Coanda effect really works. It occurred to us that Coanda himself might have profited if he had based his jet deflectors on the exquisite model provided by ozaenines.

Incidentally, ozaenines can also discharge without use of the flanges. They do so, for instance, when they spray toward the hindlegs. To this end they simply deflect the abdominal tip ventrally a bit, thereby causing the gland openings to become disengaged from the flanges, with the result that if they then discharge, the jets are directed obliquely downward.

The existence of two types of bombardiers, the brachinine type and the ozaenine type, raises questions about their relationship. Are the two groups closely related, in the sense that they have evolved from a common ancestor that was also a bombardier, which would imply that the bombarding mechanism evolved only once in carabid beetles? Or have the two groups acquired their bombarding apparatus independently? Scholars are divided on the issue and the beetles have so far proved unwilling to provide the clues that might settle the question.

■ **ON NOVEMBER 4, 1970,** I found myself in Denton, Texas. I had been invited to be a Sigma Xi National Lecturer, which meant giving a series of talks at a number of colleges in some region of the country. I was given a choice of regions, and I had picked the Southwestern Tour, which included Oklahoma, Texas, and Louisiana, primarily because I had little field experience in those states. It was an exciting and tightly scheduled trip—10 lectures at 10 locations over a span of 12 days. Jim Carrel, one of my graduate students, had come along, and we made the trip by car. Jim was an enthusiastic naturalist, patient and of great physical endurance, and we got along famously. A Harvard graduate, he had been inspired by Ed Wilson, and had fallen in love with bugs.

My lectures were scheduled 1 or 2 days apart at colleges separated by 100 to 200 miles. We therefore adopted a fixed routine. My lectures were in the evenings, following a more or less formal dinner. After the lecture I would return to the motel, where we had checked in earlier, and where Jim had set things up for experimentation. The stereomicroscopes would be in place, together with dissecting equipment and photographic gear, and we'd get to work. On the day's drive to that particular location we would have made as many stops as time permitted at field sites that we had judged promising. We would usually have found insects of interest at such stops, which we collected live and would then study after the lecture in our makeshift laboratory. Depending on what we discovered, we sometimes worked through the night. As a result, by morning we were often in no mood to travel, but by the time we resumed exploration at wayside stops we were usually in full swing again.

At one stop in Texas we found a caterpillar, *Heterocampa manteo,* that sprayed an acid secretion when disturbed. It had a large gland that opened ventrally just behind the head. It aimed its spray by revolving the front end and did so accurately in all directions. We dissected a number of glands to collect secretion for later analysis by Jerry's group, and found out eventually that the fluid contained in the order of 20 to 40 percent formic acid. In addition it contained a little over 1 percent of a mixture of two ketones

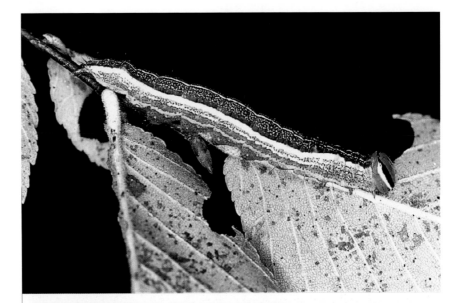

The caterpillar *Heterocampa manteo* (top) on its food plant and (bottom) beside its formic acid–containing defensive gland, dissected from the neck region.

(2-undecanone and 2-tridecanone). The ketones, which have repellent potential of their own, may function much as caprylic acid does in the acid secretion of whipscorpions (see Chapter 2). They may act as surfactants that promote both the spread and the penetration of the secretion on its target. We have since found other caterpillars that spray acid mixtures as well, also containing low concentrations of surfactants.

But most memorable was our discovery, in the outskirts of Denton, of

Compounds from the defensive secretion of *Heterocampa manteo*. From the top down: formic acid, 2-undecanone, 2-tridecanone.

the daddy-long-legs *Vonones sayi*. That arachnid was, of course, known to naturalists, although little had been learned about its behavior, let alone its defenses. I had lectured the day before at Texas Women's University and we had a whole uncommitted day available for field work. We spent most of our time at one shrubby site by a lake, which had much to offer, particularly when we looked under rocks and in rotting wood. I encountered my first phengodid beetle larvae on that occasion, bizarre wormlike creatures that make their living feeding on millipedes. I knew about phengodids and had always wondered how they got around the millipedes' defenses, and here, right before my eyes, was an individual in the process of consuming a hydrogen cyanide–producing polydesmoid millipede. I retained that image for years, as a reminder that I would some day have to look into phengodid behavior, which I eventually did (see Chapter 7).

But most numerous in all those hiding places were the *Vonones*. We could have taken them by the dozen, and did collect a number in vials, with the intent of studying them that night. They reminded me a bit of the daddy-long-legs I knew from Uruguay, of those very animals that had introduced me to the subject of benzoquinone production, so there was something nostalgic about our find. I noticed that the animals discharged some fluid when I handled them, but without the benefit of a microscope, I could not make out where the liquid was coming from. But I did think I recognized the odor of benzoquinones, and there were those tell-tale markings on the fingers.

We worked on *Vonones* that evening and on the next several nights. Their defense mechanism was unlike any I had seen. It was clear that

The daddy-long-legs *Vonones sayi.*

they produced benzoquinones. In close quarters the odor was unmistakable. But it was in how they went about readying their defensive fluid for use and in how they administered it that they were unusual. *Vonones* stores its benzoquiones as crystals in two small saclike glands that open at the anterolateral margins of the carapace. The first thing the animal does when attacked is regurgitate a drop of gut fluid. This drop emerges from the mouth, which is located midventrally, but it does not long remain there. There exist two deep grooves on the ventral side of the body that lead from the mouth to the sites at the edge of the carapace where the glands open. The regurgitant, almost as soon as it is produced, seeps into these grooves, and is conveyed to the gland openings, where it forms two droplets. The animal then injects some benzoquinone crystals into the droplets, causing these to swirl as the crystals go into solution. Having thus concocted the defensive fluid—which may take the animal no longer than about a second—*Vonones* is now ready to use it. To this end, it dips the tips of its forelegs into the two droplets, and using the tips as brushes, wipes them in a series of rapid strokes against whatever agent is being used to inflict the assault. In our motel experiments we ourselves were the offending agents. If we held *Vonones* in forceps, the animals would wipe their forelegs precisely against whatever site of the body was being grasped. It was a very efficient way of administering defensive fluid.

I have vivid recollections of our last motel session with *Vonones*. The

AMBULATORY SPRAY GUNS

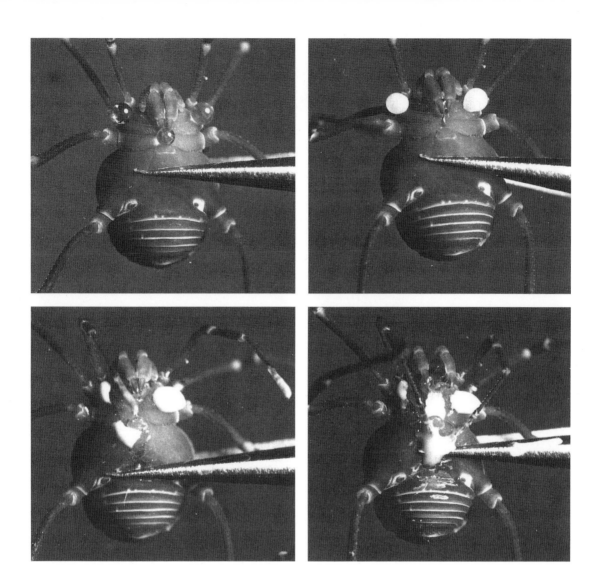

Vonones sayi discharging in response to being held in forceps. Top left: The clear regurgitated liquid initially discharged stems from the mouth (where excess fluid has accumulated as a median droplet), from where it is conveyed to the base of the second pair of legs. Top right: Quinonoid secretion is then ejected by the animal into the droplets, which tarnish as a result. Bottom left and right: The fluid is then administered to the offending forceps by brushings of the forelegs. Note (bottom left) that the tips of the forelegs are covered with defensive fluid.

date was November 9, and I had just lectured at Sam Houston State University in Huntsville, Texas, where we were locally billeted. It was also the day that France lost Charles DeGaulle, so we had kept tuned to the television news coverage all night while we studied and photographed our little arachnid. It was a most unusual scenario for learning French history, but

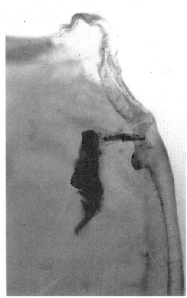

Left: The carapace of *Vonones sayi*, showing the two defensive glands at the anterolateral corners. Right: An enlarged view of the right gland.

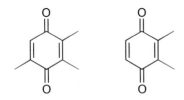

Quinones in the defensive secretion of *Vonones sayi*: 2,3,5 trimethyl-1,4-benzoquinone (left) and 2,3 dimethyl-1,4-benzoquinone (right).

by dawn we were much enlightened about everything from the instability of the French government to the stability of French wine.

The *Vonones* survived the return trip to Ithaca and lived with us for months. We were therefore able to do much work with them. One of Jerry's associates, Arthur Kluge, found the glands to produce two benzoquinones (2,3-dimethyl-1,4-benzoquinone and 2,3,5-trimethyl-1,4-benzoquione), stored in the glands in the combined amount of 0.5 microgram. The quantity of benzoquinone that the animal actually mixes with a load of regurgitated fluid is very small, so small in fact that the 0.5 microgram suffices to provision more than 30 such loads. After an attack, *Vonones* is therefore unlikely to be left with its quinonoid supply depleted. To rearm after an assault it may need to do little more than drink its fill of water.

We exposed *Vonones* to ants and found that the leg-brushing technique lent itself admirably to defense against such predators. Ants that ventured to attack were literally brushed off. It was also clear that the *Vonones* seemed to buy time with each individual assault, in the sense that they were never attacked shortly after having repelled an ant. They seemed to be protected from renewed assault by residual defensive fluid remaining on their bodies.

I have since worked on a number of tropical relatives of *Vonones,* and so have other investigators. Some of these arachnids produce benzoquinones, others produce phenols. Several apparently administer their secre-

tion by leg-dabbing, as does *Vonones*. I cannot help but wonder whether the *Gonyleptes* I remember from Uruguay are in that category.

A swarm of whirligig beetles *(Dineutes hornii)* at the edge of a pond.

■ **OF THE 300,000** or so beetles that have been described, the majority are terrestrial. But there are some that are aquatic. Among these are the Gyrinidae, whose common name alone I had always found to be irresistibly charming. Gyrinids are known as whirligigs, in reference to their frenzied swimming habits.

Everyone with access to a pond as a child has seen whirligigs. They are usually found in swarms near the shore, rapidly gyrating on the water surface like little torpedo boats. They actually use the water surface for support. If detergent is added to the water so as to reduce the surface tension, they sink. They are not permanently restricted to the surface. They can dive on their own, and when they do they carry along a bubble of air which serves them as a scuba tank. Their eyes are specialized for life on the water surface. Each is divided into an upper and a lower half, the former for

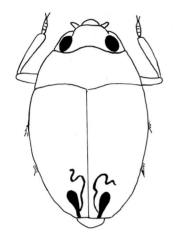

Diagram of a whirligig beetle, showing the defensive glands in the rear of the abdomen. Each gland consists of a sac-like reservoir, in which secretion is stored, and an attached strand of glandular tissue.

vision in air, the latter for vision in water. When whirling about they are thus able to keep track of what goes on both above and below the water line. Adult whirligigs are scavengers. The larvae are also aquatic, but they are predaceous.

Imagine for a moment being a whirligig. You are on the surface of the water, silhouetted in dark brown against the sky, and down below in the depths, bigger than you and hungry, and equipped with marvelous eyesight, are fish. Imagine the predicament.

I've observed gyrinids for hours, both the little ones of the genus *Gyrinus* and the larger ones of the genus *Dineutes,* and have yet to see a fish break the surface to snap one of them up. There had to be a reason for this, and the logical one was that the beetles were protected. They were either hard to catch or chemically defended or both, and the fish acted as though they knew that. Anyone who has tried to catch gyrinids by hand knows that it is virtually impossible. They are too quick and too evasive. One needs a good-sized net to scoop them up. And there was ample evidence that they were chemically defended. Even the field guides made reference to that effect, and I had noted on my own that both *Gyrinus* and *Dineutes* emit peculiar odors when handled.

The defensive glands of gyrinids had been described and were easy to find. Located in the rear of the abdomen, like the glands of bombardiers, they consist of two sacs, each with an attached strand of secretory tissue. The sacs were good-sized and usually full of fluid. I decided we would work on *Dineutes.* It would be easier to get enough secretion from the larger of the two types of beetles.

A group of us, including Jim Carrel, packed up collecting gear and drove to the Huyck Preserve for a weekend, where we knew we would have access to a pond and a boat. The *Dineutes* were out in droves and we collected several hundred of them by scooping them up in insect nets. It was a matter of sneaking up to the swarms by boat. One of us would frantically row, while the others would madly scoop away. It was a winning strategy but barely so. The percentage taken was minuscule.

We took the beetles to the cottage alive and "milked" them one at a time by holding them in forceps and using small pieces of filter paper to wipe away the white, yogurt-like ooze that welled from their rears the moment we picked them up. We liked the technique because it enabled us to return the beetles to the pond after we were through with them. But it was inef-

Gyrinidal.

ficient. We were losing too much secretion in the milking process. So we adopted an alternative procedure. We had noted that one could collect the secretion efficiently if one froze the beetles immediately after capture. When one then thawed them out, the secretion would ooze from the glands spontaneously and exhaustively, so that one could collect it simply by rinsing the beetles with solvent. Unfortunately, the beetles did not survive this procedure. But the effort paid off. We milked a large number of *Dineutes,* presented Jerry with a good sample of secretion, and in due course his group came up with the answer. There was a single major component in the secretion and the compound was new. It was a terpene—technically a nor-sesquiterpenoid aldehyde—and we called it gyrinidal. The question now was whether this substance repelled fish.

To find out we set up some aquaria, borrowed some fish nets, and made arrangements to collect a number of large-mouth bass *(Micropterus salmoides)*. We used small specimens, about 20 to 25 centimeters in length, and gave them each their own "personal" aquarium. We kept them in one room, but set up cardboard dividers so they would not be able to see one another. We weren't sure this would be necessary, but for good measure thought we'd prevent them from sharing visually in one another's experiences.

Our first tests were intended to see how the bass would react to the beetles themselves. We had trained the fish to come and get food items that we presented on the water surface in one corner of the aquarium. They liked mealworms, and we had for several days routinely given these to the individual bass to get them used to us. When, in due course, we thought they'd be ready for the real thing, we offered them live *Dineutes.* Of 96 beetles that we offered to six bass, 76 were ignored. Another 17 were rejected after being taken in the mouth, and only 3 were eaten (all 3 by one fish). They evidently didn't like *Dineutes.*

Most interesting was the behavior exhibited by the fish when they rejected a *Dineutes.* The beetle was always offered to them at the water surface and they were quick to snap it up, but they usually did not spit it out right away. Instead they held the beetle in the mouth for a period, while at the same time opening and closing the gill covers (the opercula), in a rhythmic action that we took initially to be a sign of distress. But if it was

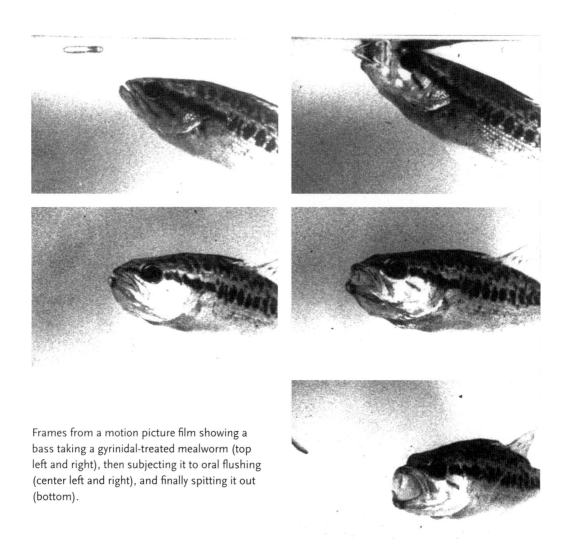

Frames from a motion picture film showing a bass taking a gyrinidal-treated mealworm (top left and right), then subjecting it to oral flushing (center left and right), and finally spitting it out (bottom).

distress, why did they not reject the beetle outright? We eventually learned that the opercular flapping had a very special function, but it took additional experiments to find that out.

Jerry's group eventually synthesized gyrinidal and we were able to test for the activity of the chemical itself. It turned out that gyrinidal stuck nicely to the surface of mealworms. So we pipetted gyrinidal at various dosages onto mealworms and offered these to our fish. They utterly disliked the treated items, which they rejected by spitting them out. Even as little as 0.5 microgram of gyrinidal—half of 1 percent of the 100 micrograms present on average in a *Dineutes*—could render a mealworm unacceptable. The sensitivity of the fish to gyrinidal varied in accord to the

AMBULATORY SPRAY GUNS

A bass subjecting a mealworm coated with charcoal and gyrinidal to oral flushing; the direction of water flow, alternatively out through the opercula (top) or out through the mouth (bottom) is rendered visible by the charcoal marker. (From a motion picture.)

fish's state of satiation. The hungrier the fish, the higher the dosage of gyrinidal tolerated. Just as in humans—driven by hunger we too will eat what we ordinarily reject.

We learned from these tests that the opercular flapping is really a cleansing behavior intended to flush the oral cavity. The gills' covers, by opening and closing, pump water back and forth through that cavity, in such a manner that anything within it is rinsed. When the gill flaps open, water is sucked into the mouth and forced out through the gills, and when

AMBULATORY SPRAY GUNS

they close, it is taken in through the gills and pumped out through the mouth. We fed the bass mealworms coated with an edible paste containing both gyrinidal and powdered charcoal, and could see from the darkened jets that squirted out from the mouth and the gills how this pumping action operates. The fish flushed their mouths consistently when they were given treated mealworms. The behavior, it seemed, was intended specifically to rid noxious food of its noxiousness. It represented an incredibly elegant way of taking advantage of the fact that the very fluid in one's surrounds can be used to cleanse food.

We found that the bass engaged in oral flushing for lengths of time that depended on the dose of gyrinidal applied to the mealworm. The higher the dose, the longer the flushing. The fish sometimes rejected a mealworm despite prolonged flushing. This happened particularly often with mealworms bearing the highest dosages of gyrinidal. We also found that the fish flushed more persistently when they were hungry. It all made sense. Noxiousness drove them to try to clean the ingested item, and the length of time that they invested in cleaning varied as a function of the degree of noxiousness of the item and of their appetite.

In light of these results there was also something we could predict about the gyrinids themselves. The beetles should have a way of doling out their defensive secretion over time, so as to make themselves taste bad for longer than the fish are willing to try to flush them clean. Indeed, the beetles have that ability. Their whitish, pasty secretion is not squirted out in a single burst, but is emitted as a trickle, over a period that we found, on average, to be in the order of 1.5 minutes. The fish, on average, flushed only 1.3 minutes with actual beetles. Hence the frequent rejection of the *Dienutes* in our initial predation tests with the bass.

It is interesting how *Dienutes* achieves the slow deliverance of its secretion. When the secretion first becomes visible at the rear of the beetles upon discharge, it takes on the appearance of a white band around the margin of the pygidium (the beetle's last abdominal segment), as if somehow it had been trapped there. With time the viscous fluid then drifts gradually into the water. The beetle has two sets of interdigitating hairs that serve specifically to retain the secretion when it is discharged. The hairs form a kind of sieve, into which the glands void their contents, and through which the secretion gradually percolates.

For the present, it seems, *Dineutes* has it made, at least vis-à-vis such

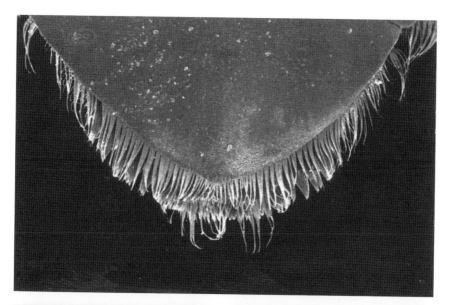

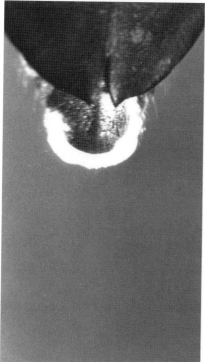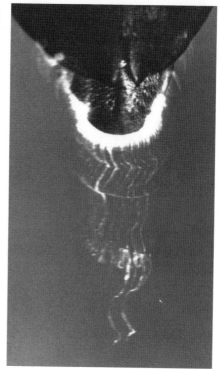

Top: The rear end of *Dineutes hornii,* showing the marginal hairs that form the "sieve" in which the secretion collects when discharged. Bottom: The tip of the abdomen of a *Dineutes* in the process of ejecting secretion. The white fluid is retained initially by the sieve (left), then drifts backward into the water (right).

fish as large-mouth bass, with limited flushing endurance. Given evolutionary time, however, fish might well come up with a counterstrategy of their own for dealing with these beetles. In fact, I wouldn't be surprised if there is already a fish somewhere on this planet able to circumvent the defenses of gyrinids.

■ **I HAD DELIBERATELY** chosen not to work on the defenses of social insects. I seemed to be allergic to wasp stings, didn't want to compete with the many specialists working on ants, and had no particular fondness for termites. So I thought I'd leave social insects to others. But then, in 1968, I had an experience that changed my mind about termites.

I was on a visit to Panama at the time, a guest of the Smithsonian Tropical Research Station, on Barro Colorado Island. I hadn't been back to Latin America since 1947 and it felt good to be able to speak Spanish again. Barro Colorado Island is an artificial island created in the Panama Canal when the canal was built. Visiting scientists are provided with living quarters and laboratory space, but most important, are given the chance to explore, undisturbed, at any time of day or night, the utterly magnificent rainforest that covers the island. During the 3 weeks I was in Panama I hardly set foot on the mainland. I spent most of my time roaming through the island wilderness, feasting on the unfamiliar. I collected the ozaenine beetles there that enabled me to work out their spray mechanism, and made countless observations that I hoped eventually to pursue. There was the little dung beetle, for instance, that rolled its dung balls out of monkey dung, and produced a defensive secretion that differed drastically in odor in the male and the female, which suggests that the fluid may function also as a pheromone. And there were the heliconiid butterflies, the most graceful insects on the wing in the rainforest. What was it that they achieved by their slow bobbing flight? They could afford to fly slowly because they were distasteful, but was there an additional reason? I noted that they took evasive action whenever they came close to a spider web. They seemed able to spot the web somehow, and then fly up and over it without touching it, thereby routinely avoiding capture by what must be one of their major enemies. These butterflies have especially long antennae. Do they use these to "feel" the webs? At any rate, it was clear that if they flew faster they wouldn't be able to avoid entanglement.

And then there was *Rhynchotermes perarmatus*. I'd come upon a nest of

AMBULATORY SPRAY GUNS

Rhynchotermes perarmatus on the march. The soldiers with their pointed snouts are in the foreground, guarding the flank of the marching column.

this termite and noted, with fascination, the foraging trails that radiated outward from the colony. Most interesting was the fact that these trailways were closed over, so that each was essentially a tunnel. Wonderful protection against ants, I thought. I wondered what would happen if I breached one of the tunnels. Surely that must happen in real life. I knew that termites had soldier castes that helped in emergencies, so I thought I'd create an emergency.

I took a pair of forceps and carefully laid open a section of trail by pealing away a portion of the cover. I expected the termites to pour from the opening and to spread helter-skelter in all directions. Instead they marched right on without deviating from the track, as if nothing had happened. But mostly it was the worker termites that took part in the march. The soldiers, easily recognized by their pointed snouts, tended to align themselves neatly along the flanks of the column, one beside the other, with their snouts pointed outward in defensive readiness.

Rhynchotermes is a member of the subfamily Nasutitermitinae, the soldiers of which are little more than ambulatory spray guns. The "gun" of the soldier is its frontal gland, which takes up much of the cranial cavity and opens at the tip of the pointed snout, or nasus, that characteristically projects from the head. The soldiers are programmed for instant action. They converge on sites of trouble, and singly or in groups discharge their cephalic secretion on target. The fluid is sticky and is ejected as a fine strand. A Swiss investigator, E. Ernst, had reported that the secretion had the makings of an alarm substance in the sense that, once discharged by a

Christina Eisner beside a *Coptotermes* termite nest in Australia. The mounds of *Nasutitermes exitiosus* are dome-shaped and not as tall.

soldier, it induces other soldiers to converge on the site and join in the defense.

Using forceps, I poked the column of *Rhynchotermes* and found that the soldiers did indeed rally into action. They sprayed the forceps and converged in numbers upon the instrument. The worker termites, in contrast, remained for the most part doggedly committed to the march. I could disrupt the workers as well, by poking them directly, but they seemed able to reassemble for the march as soon as I left them alone. There was an orchestrated flavor to the termites' response. The soldiers held their ground or rallied in groups to where they were needed, while

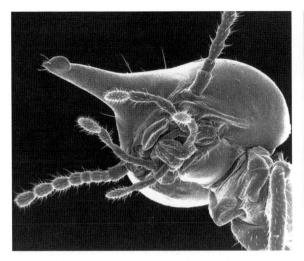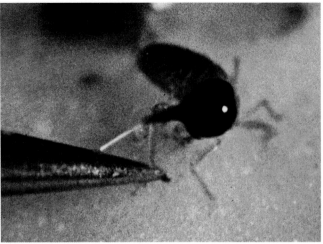

the workers stuck to the march. When I returned to the site hours later, the trail had been repaired and enclosed again.

I took a liking to termites that day but wasn't able to follow up on my observations until 4 years later, when I had the immense good fortune of receiving a Guggenheim Fellowship for a year's study in Australia. I was hosted by the Division of Entomology of the CSIRO, in Canberra. Australia is prime termite country and it wasn't long before I made the acquaintance of *Nasutitermes exitiosus,* a dominant species in the Canberra region. A typical nasute termite, *N. exitiosus* had the characteristic soldier and worker castes, the former with snouts and atrophied mouthparts, the latter with functional mouthparts and no snouts. Its mounds, dome-shaped, durably constructed, and interspersed closely with ant nests, dotted the landscape. There were areas, in fact, where ant and termite were in intense competition for space. I had heard that Australia was the kingdom of the ant. What I saw convinced me that it was more in the nature of a domain that ants shared with termites.

Their coexistence notwithstanding, ants and termites were probably constantly at odds. I did an experiment in the field that gave some idea of what happens when ant and termite are brought into confrontation. I had located a *Nasutitermes* mound that projected from the very midst of a nest of the Argentine ant, *Iridomyrmex humilis.* I brought ants to the surface by tapping the entrance to their nest, and induced termites to emerge by laying open their mound near its base. To make sure I would get as precise a record as possible, I had placed a tripod-mounted camera at the scene so I could photograph the ensuing events. Things happened fast as

Nasutitermes exitiosus. Left: A scanning electron micrograph of the head of a soldier, showing the nasus, or snout (note the droplet of secretion at the tip). Right: A frame from a motion picture showing a soldier spraying in response to having a leg pinched by forceps.

Two scenes from the "battle" staged between *Nasutitermes exitiosus* and the ant *Iridomyrmex humilis*. Top: A loop of secretion is being ejected by the termite soldier, whose reddish-brown head is visible just beneath the head of the ant. Note that the loop ends in a stretch of thread connecting the two mandibles of the ant. Bottom: An ant, covered with secretion and debris, surrounded by three of the termites that have incapacitated it. The two termites on the right have a droplet of residual secretion at the tip of the snout, evidence that they have discharged.

AMBULATORY SPRAY GUNS

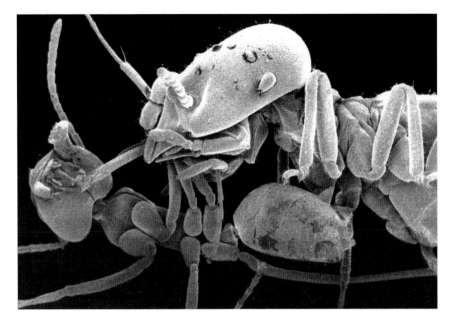

A worker of *Nasutitermes exitiosus* biting an offending ant.

ants and termites, in large numbers, came into contact. Many termites were bitten and killed, and dozens of ants were sprayed and left visibly contaminated with secretion and debris. I tried to watch things as best I could and just kept pressing the shutter at random intervals. The photography paid off. In one picture the emergent filament of spray is actually seen as it shoots forth from the soldier's snout and loops over the targeted ant. In another photo spray is being seen just as it emerges from the snout of a soldier. Things soon calmed down, however, when ants no longer ventured over the ground held by the termites. They milled beside the site and swarmed over my hand if I so much as touched the ground, but they stayed away form termite "territory." At close range that territory had the distinct terpenoid odor of the soldier's spray, and it seemed that it was this residual odor that was keeping the ants at bay.

I have mixed memories about getting on my knees on that occasion to smell the battlefield. *I. humilis,* quite frankly, is one hell of an ant. There is no such thing as being casually inspected by it. Step on its nest and it overwhelms you, crawling up your legs, into your clothes, and before you know it, over your entire body. Stripping down to your underwear helps on such occasions, inasmuch as the ants are then more readily swept away, but it can leave you groping for words if you should be unexpectedly surprised by hikers and asked to explain your activities.

I took a *Nasutitermes* colony into the laboratory and staged some en-

Results of a laboratory test in which a centipede was pitted against *Nasutitermes exitiosus*. The centipede has become incapacitated by spray, and is surrounded by a contingent of guarding soldiers.

counters with ants. I placed groups of termites into individual plastic containers, released an ant into each, and watched what happened. The ants didn't stand a chance. They were sprayed as soon as they came in contact with a soldier termite and the spray was quick to incapacitate them. The ants engaged in cleaning activities when hit, but as they wiped antennae with forelegs, and rubbed legs against one another, they succeeded only in spreading the secretion and gumming themselves up. The worker termites occasionally took part in the defense by biting the ants. They crushed small ants outright, and helped disable larger ones by clamping themselves onto them and rendering them more vulnerable to being sprayed. I also noted something that I thought I had already seen in the field: that soldiers converge upon sites of trouble where other soldiers had discharged. As soon as an ant was sprayed, other soldiers rallied to the site, ready for action. The ants were literally encircled by soldiers.

None of the ants used in these tests survived. Having been sprayed even once had lethal consequences for an ant. Other predators such as spiders and centipedes that I also put to the test fared no better.

My experimental set-up so far had been rather crude and I looked for ways to improve it. I wanted to study the defensive strategy of *Nasutitermes* soldiers under more controlled conditions and I thought the way to do that was with an artificial ant, a sort of robotic enemy that I could pit against the termites and activate by remote control.

I had noticed that if I took a group of termites and released them into a petri dish they adapted quickly and established a trail along its periphery. On and on they would march, circling around for hours on end, providing an ideal scenario for putting their state of alertness to the test. The robot I wanted was to be the size of an ant and have the capacity to come in from the center of the dish and attack the flank of marchers. What we eventually came up with was a small metallic bar, cut from an ordinary paper staple, that could be activated with a magnetic stirrer. I say *we,* because I was no longer alone on the project. I had been joined by two collaborators, our daughter Vivi and Irmgard Kriston, a behavioral biologist from Germany. Vivi was barely fifteen, but she was smart, a born researcher, and incredibly funny. Because of school she could only join us at odd times. Irmgard had just obtained her Ph.D. with my friend Martin Lindauer, the noted bee expert, and she turned out to be the best possible partner, intelligent, good-natured, and infinitely patient.

Our artificial ant was our pride and joy. The magnetic stirrer by which we brought it to life was of the conventional type, such as is used routinely to mix reagents in a beaker. With the help of Colin Beaton of the CSIRO we had modified it so that it could be made to rotate slowly at controlled speeds. The "attack" protocol we adopted consisted of placing the metal bar in the center of the dish, and the dish itself on the platform of the stirrer, making sure that the bar was lined up vertically with the magnet beneath the platform. We then set the magnet into motion, causing the bar to twirl, upon which the dish was slid slowly sideways by hand until the bar came upon the column of termites. "Assaults" thus inflicted had consistent effects, which we filmed and photographed. We also videotaped the encounters for later frame-by-frame analysis.

At first, while the rotating bar—or twirler as we called it—was still at some distance from the column, the termites ignored it. But once it reached the edge of the column it began to attract notice, first usually from soldiers that stood along the flanks of the trail as guards. Singly at first, and then sometimes in numbers, the guards positioned themselves around the twirler, their snouts at the ready. A further incursion of the twirler brought about the anticipated discharges, but only from soldiers that were directly struck by the bar. The discharges, though, brought about a massive change. Additional soldiers that arrived on the trail rerouted themselves toward the twirler, with the result that the latter was quickly encircled. Workers were also distracted and if they came upon the twirler

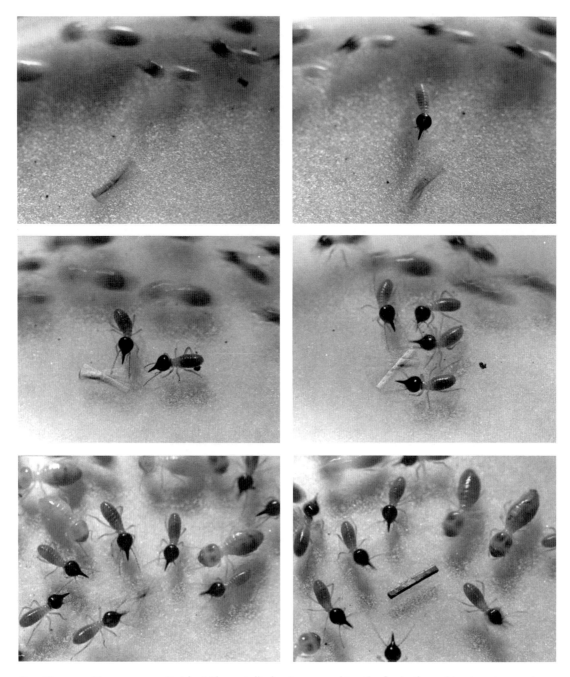

Nasutitermes exitiosus versus a "twirler." The metallic bar is approaching the flank of marching termites, and is encountered first by one termite soldier, then by another. Spray from one or both of these attracts additional soldiers, which add their own spray to the twirler, with the result that the bar is eventually glued down by the secretion.

Limonene   Trinervitane   Kempane

they attempted to bite it, which if they succeeded resulted in them being whirled around for some turns and then flung off. The trail, totally disrupted as a consequence of all the commotion, became gradually reestablished, but along a path that bypassed the site of action. The encircling soldiers held their ground for as long as the twirler remained in rotation. Individual soldiers sometimes departed to join the trail but they were quickly replaced by new arrivals. If the twirler was kept in place, the encircling soldiers would refrain from discharging upon it, but if it was "revived" by sliding the dish, so as to cause it to "attack" the trail again, it was sure to get sprayed by any soldier that was struck by it. The discharges sometimes "killed" the twirler by causing it to become immovably glued down by the secretion. When this happened the encircling soldiers did not depart, but seemed only to relax their surveillance. Neatly oriented with their snouts pointed at the twirler for as long as the twirler twirled, they now assumed less directed stances and did not remain as long in attendance. The soldiers are evidently able to detect the motions of an encircled enemy without directly touching it, and they are programmed not to let down their guard until the enemy has ceased moving. We were able to show in other experiments that the guarding soldiers gauge target "liveliness" by the air motions engendered by the target while it remains active.

We showed further that it is the secretion itself that acts as the signal that alerts the N. exitiosus soldier to trouble nearby. If we took a twirler, had it presprayed by a soldier, and then placed it beside a column of termites in a petri dish, the soldiers on the trail responded as though they had been given marching orders. They left the trail, converged upon the twirler, and positioned themselves neatly in a circle around the "enemy."

Isoprenoids from the secretion of the cephalic defensive glands of *Nasutitermes exitiosus* soldiers. Two volatile additional terpenes present in the secretion are α- and β-pinene.

Twirlers not so sprayed, presented as controls, were ignored by the soldiers. The pheromonal function of the spray seemed established.

When we did this work, the composition of the soldier's spray had not yet been completely elucidated. It was known to contain some simple terpenes, including familiar odorous compounds such as α-pinene, β-pinene, and limonene. We showed that these terpenes act as irritants to insects, a property that doubtless accounts in part for the defensive effectiveness of the secretion, but there was more to the story. In a very nice series of publications, Glen D. Prestwich was able to show that the spray of nasute soldiers also contains a number of complex diterpenes (tricyclic trinervitenes and tetracyclic kempanes) that account for the stickiness and toxicity of the secretion. All and all, nasute soldiers seem admirably fit, both behaviorally and chemically, for fulfillment of their defensive role. Survival without nasutes, for species such as *N. exitiosus,* is unimaginable. The Nasutitermitinae make up one of the largest groups of termites. Possession of nasutes must in no small way have contributed to the evolutionary success of these termites.

There is one small mystery that remains unsolved about *N. exitiosus.* This termite has a second soldier caste, consisting of nasutes with larger bodies and heads than the other soldier caste, present in fewer numbers than the primary soldiers. We found that these larger soldiers were rather aloof, in that they did not take part in the action with the other soldiers. They tended to ignore the twirler and failed to be attracted by spray. We don't know what the larger soldiers ordinarily do, but suspect that they may carry out some defensive function in the deeper confines of the nest. But that is pure conjecture.

The films I took of *N. exitiosus* being attacked by twirlers as they are marching around in dishes are treasured mementos that I often show in the classroom, where they invariably elicit a partisan response from students. Without exception viewers root for the termites. I have never had anyone root for the twirlers, not even now that cyberorganisms are moving into center stage and replacing the real thing.

■ URUGUAY IS ARMADILLO COUNTRY but it doesn't treat armadillos kindly. I don't know if it's still the custom but when I lived there the armadillos were eaten. It is a price armadillos are not supposed to pay. Nature

Top: A sowbug (probably an *Armadillidium* species) under attack by *Pogonomyrmex* ants in Arizona. Bottom: A cockroach from Thailand *(Perisphaerus semilunatus)* coiling in response to an attack by *Formica exsectoides* ants.

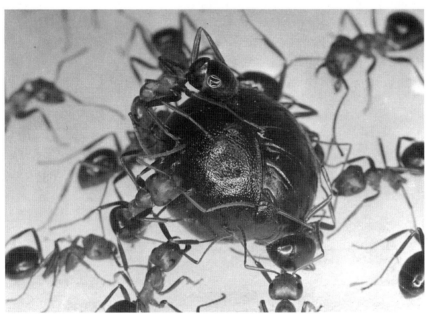

has endowed them with a hard shell and the capacity to coil into a sphere, and therefore with the ability to withstand ordinary assaults. Humans, of course, are no ordinary enemies.

Coiling is a strategy practiced by other animals as well. Among mammals there is the African pangolin, which can tighten up much like an armadillo. But invertebrates are also known to pull the trick. Everyone knows the little gray multilegged creatures called sowbugs or pillbugs, found under rocks or in leaf litter or most commonly in your basement or garage. They are actually Crustacea, related to shrimps and crabs, rather than to the millipedes they resemble. Touch them, and chances are they too will coil. There are even cockroaches that coil, such as a species I once had from Thailand that looked remarkably like a pillbug. I showed in experiments that by coiling both the roach and the pillbug can survive ant attacks. Ants find it difficult to deal with a spherical shape from which nothing sticks out that can be grasped.

The most interesting coiling arthropod I ever worked with is a small millipede I encountered in Holland. I had spent a sabbatical leave there, in 1964–1965, at the Laboratorium vor Entomolgie of the agricultural university in Wageningen. It was a good year in that it gave the family a chance to travel, but it was frustrating in that it made me come to realize how little of Europe remains in a state that can be called truly wild. I sought out open spaces as best I could, and it was on a walk with Maria through tracks of preserved woodland on the outskirts of Wageningen that I found my first specimens of *Glomeris marginata,* the coiling millipede.

Glomeris was not particularly spectacular looking. It was similar in size and appearance to an ordinary pillbug, and I was at first fooled by the similarity. But one major difference soon became apparent. *Glomeris,* in addition to coiling into a ball when disturbed, gave off a secretion. The fluid emerged from a row of eight middorsal pores, in the form of beautifully clear droplets that were both viscous and sticky. There seemed no question that the discharge was defensive. Ants and spiders were plentiful in the ground litter inhabited by *Glomeris* and they had to be a major hazard.

Secretion can be shipped across the ocean, so I decided to send Jerry a sample of *Glomeris* secretion for a first chemical look. The millipedes were easy to milk, and we could obtain the fluid by taking it up in capillary tubing without injuring the animals. So Maria and I collected a handful of *Glomeris,* milked them, sent the sample to Cornell, and waited for news from Jerry. It was quick to come and it was encouraging. There were definitely interesting compounds in the fluid. "Get more secretion," was the message that came through, and we did.

In the meantime I carried out the usual tests with ants, and found that

the coiling response itself could suffice to thwart the attacks. But when the *Glomeris* did emit secretion and the fluid contacted an ant, the effect was devastating. The sheer stickiness of the fluid, and the fact that the stickiness increased with time, had an immobilizing effect on the ant, causing it to become caked in debris. The millipede did not usually "bleed" from all pores at once but seemed capable of restricting its response to those pores closest to the region of the body subjected to attack. This became clear also when I touched *Glomeris* with a warm needle. The millipede tended to emit fluid regionally, from a few pores at a time. I dissected *Glomeris* and found that each pore is the shared opening of a pair of identical glands. The millipede had eight such pairs.

We eventually returned to Ithaca to find that beautiful work had been done on the chemistry of the secretion. The fluid contained two major compounds. They were similar in structure and belonged to a category of substances called quinazolinones, of which none were known from ar-

The millipede *Glomeris marginata*. Top left and right: An ambulatory millipede before and after coiling. Bottom left: A millipede discharging secretion from its eight mid-dorsal gland openings, in response to being seized by forceps. Bottom right: A droplet of secretion being pulled into a thread with a needle.

Defensive chemicals in the secretion of *Glomeris marginata:* glomerin (left) and homoglomerin (right).

thropods. Hermann Schildknecht, who had independently characterized the two chemicals, named them glomerin and homoglomerin. We liked the terms and adopted them.

The problem became interesting when Jim Carrel, to whom I had suggested that *Glomeris* might make a good experimental "friend," offered the millipede to wolf spiders. "Come see what happened," he said one day as he called me into his lab. "Take a look at these spiders." The wolf spiders were from the Archbold Station and I had made it a point to keep a number of them in our laboratory for predation tests. They were easy to maintain in small individual containers and usually attacked promptly when given prey. Jim had found that the spiders rejected the *Glomeris,* but that following the attack they sometimes became paralyzed. "See? They can't right themselves," he said, as he flipped them on their back with a glass rod. It was a stunning result. The spiders were motionless, flaccid, and totally unresponsive. We had never known a defensive secretion to affect a predator that way.

Jim synthesized glomerin and homoglomerin, so it became possible to determine that the paralysis—or sedation, as we had come to call it—was caused by these compounds rather than by other factors in the secretion. It took very little material to induce the sedation. A medium to fully grown *Glomeris* contains 6–90 micrograms of quinazolinone (that is, of both glomerin and homoglomerin). A small fraction of that amount—0.7 microgram, taken by mouth—is all that is needed, on average, to knock out a spider. That means that there is more than enough quinazolinone in one drop of *Glomeris* secretion to induce the effect.

Interestingly, the sedation is not of immediate onset. For the first hour after an attack upon a *Glomeris* the spiders usually remain symptom-free. They begin to show the first signs of sedation within 4 hours and by the twelfth hour are usually fully immobilized. Also extraordinary is the duration of the sedation. Although some spiders recover within 24 hours, others take up to 5 to 6 days to do so.

The fact that the spiders do recover after such long periods of induced immobilization is interesting in itself. But in nature sedated spiders

AMBULATORY SPRAY GUNS

Top left: A wolf spider with a *Glomeris marginata* in its fangs. Top right: A wolf spider that has been sedated as a result of having attacked a *Glomeris marginata*; the spider has been placed on its back, and is unable to right itself. Bottom left: A sedated spider being attacked by ants; the spider is flaccid and unable to offer resistance. Bottom right: An assessment of the deterrency of glomerin; a solution of the compound is being administered with a microcapillary tube to the mouthparts of a spider that is feeding on a mealworm. If the dosage of glomerin is high enough, the spider will relinquish the mealworm.

would probably be doomed. They could easily fall victim to ground predators such as ants or, unable to seek shelter as they typically do in the daytime, die from desiccation.

Not all spiders that attack a *Glomeris* become sedated. Some, including those that release the millipedes outright or after holding them only briefly in the fangs, tend to remain free of symptoms. By responding quickly, it would seem, they avoid exposure to hazardous levels of the toxins. By and large it is the spiders that hold the millipedes the longest, often killing them in the course of the attack, and in some cases ingesting them in part, that suffer the ill effects.

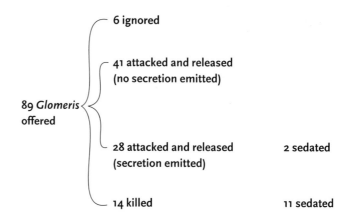

The results of a feeding test of *Glomeris marginata* with wolf spiders.

	Fate of *Glomeris*	Fate of spiders
	6 ignored	
	41 attacked and released (no secretion emitted)	
89 *Glomeris* offered	28 attacked and released (secretion emitted)	2 sedated
	14 killed	11 sedated

The results of one experiment will give some idea of the kind of data that we dealt with. Of 89 spiders that were given one *Glomeris* each, 83 attacked. Sixty-nine of the millipedes survived the encounters. Of these, 41 were released without being caused to emit visible quantities of secretion, and 28 did emit secretion before being freed. Of the 28 spiders thus exposed to secretion 2 became sedated. The 14 millipedes that were killed by the spiders had in all except one case given off secretion. Eleven of the spiders involved in these fatal encounters were sedated.

One might argue that the secretion of *Glomeris* is at best imperfect since the millipede is sometimes killed in the course of an attack. But is the millipede's death under such circumstances really in vain? If a millipede, as a consequence of being killed, kills a spider, other *Glomeris* in the spider's territory would clearly benefit, at least for as long as the defunct spider is not replaced by a successor. Millipedes are flightless and hence more likely to coexist regionally with relatives than if they were able easily to disperse. The death of a millipede, which by dying kills a spider, could thus be viewed not as a waste but as an altruistic act, a sacrifice on behalf of surviving kin. Being lethal to spiders could thus be part of the normal function of the secretion, even if the bearer of the secretion itself loses its life in the process of putting the secretion to use. The argument, of course, remains speculative, particularly since we used wolf spiders from Florida rather than spiders from *Glomeris*'s own habitat. I would dearly like to know if the "local" wolf spiders in Europe that coevolved with *Glomeris* are less sensitive to quinazolinones or, as is also conceivable, avoid *Glomeris* altogether.

Three large oniscomorph millipedes from South Africa, shown beside the tiny *Glomeris marginata*. All have coiled in response to handling.

Two quinazolinones are known that are related structurally to glomerin and homoglomerin. One is arborine, from an Indian medicinal plant, and the other is methaqualone (or Quaalude), a synthetic drug with psychoactive properties. Both substances are sedative to vertebrates, but, surprisingly, as Jim Carrel was able to show, not to spiders. The mode of action of glomerin and homoglomerin on spiders therefore remains something of a mystery. I am surprised that no one has so far bothered to look into the mystery. Compounds capable of "knocking out" an animal for days on end, without inducing permanent after-effects, should, at the very least, awaken the curiosity of pharmacologists and physicians.

Glomeris has some relatives in South Africa the size of grapes or even larger, which I managed to obtain in numbers at one time to see if they too had a defensive secretion. They were gorgeous animals, but turned out to lack defensive glands. Instead, they had an extraordinarily tough shell so that when coiled up they were as hard as marbles. I took these millipedes to New York, to the Zoological Gardens in the Bronx, to see whether I could test them with some predators native to South Africa. None of the predators in my own laboratory would have anything to do with them. Jo Davis, my graduate assistant during my first year at Cornell, was now curator at the zoo and he had promised to be of help.

One predator, *Mungos mungo,* a mongoose, turned out to have a rather

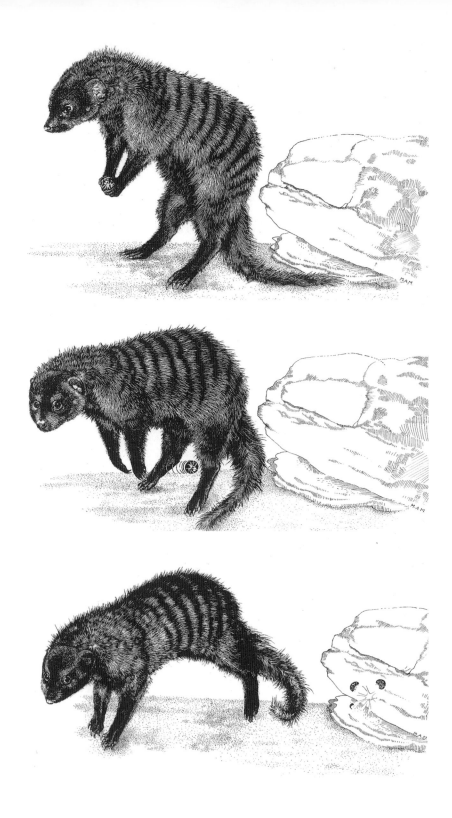

A mongoose smashing a coiled oniscomorph millipede by hurling it backward against a rock (based on a motion picture sequence).

AMBULATORY SPRAY GUNS

outlandish strategy for dealing with the "marbles." It picked each up the moment it was offered, grasped it in the front paws, and then, while standing upright with its back against a wall, hurled it through its legs against the wall with all its might. By doing so, it always succeeded in smashing the millipede to bits. The bits were then avidly eaten by the mongoose.

Jo and I published a short note in *Science,* to mixed reviews. One letter was particularly vexing, because it accused us of perpetrating the myth that anything useful could be learned from animals imprisoned in zoos. We had the last word, though. A colleague in South Africa called our attention to the following excerpt from a book by D. Wagner called *Umhlanga,* which I submitted to *Science* later, where it was published as a letter:

Mongooses . . . in captivity . . . eat almost anything, but in their wild state they live mainly on insects. A friend of mine recently told me a strange tale about one of these creatures. He's an old man, and he's more or less grown up in these wild stretches of Natal. He said that one morning when he was sitting quietly under a tree in the bush hoping to see some birds, he spotted a colony of mongooses nearby. Suddenly one of them climbed a short distance up a tree and knocked down a pill millipede. The mongoose jumped down after it, grabbed it between his front feet, and hurled it through his backlegs against the tree. The impact smashed the otherwise-impregnable ball, and before any of his friends could cheat him of his prey, he ate it.

6

Tales from the Website

The occasion was an exploratory walk at the Archbold Station. I was in the company of George Ettershank, a graduate student from Australia, and Rosalind Alsop, my new technician, and we were strolling about on the sandy fire lanes of the station looking for "action." The year was 1963. It was a particularly balmy August night and the insects were out in droves. The littlest among them, including the most diverse assortment of midges, beetles, and winged ants, were aggregating in swarms about us, getting into our eyes and ears, and of course, our mouths. They were attracted to our headlamps, and wherever we went, they followed. We learned to breathe and speak with our lips barely parted.

Spiders were out as well that night, particularly web-spinning spiders (also known as orb weavers) and we watched with fascination as they constructed their webs. We placed insects into their orbs and learned how they subdued prey. It was while watching spiders close up that I noticed that you can use a spider's web to clear away the insects around your head. All you have to do is tilt your head toward the web and let the sticky strands do the job. Insects by the dozen would be sifted out by the orb, to the benefit of the spider, which would make a meal of the offerings. We made it a point that evening to return to a web periodically for a cephalic cleansing whenever we felt that the swarms around our headlamps had reached intolerable proportions.

But I also noticed something else that night. When I dipped my head toward an orb I could see that not every insect that struck the web adhered to the strands. Moths in particular seemed likely not to get stuck. They touched the webs but flew right on, as if they were Teflon-coated. Teflon, of course, had nothing to do with it, but there was something about moths that made them nonadherent. That something, as we were quick to realize, was scales. Moths didn't stick to the spider webs because they were covered with detachable "powder."

There is no question that orb-weaving spiders are among the real enemies of moths. Moths have the advantage of avoiding birds by flying by night. But the night is full of its own dangers, including, in addition to spiders, those familiar aerial acrobats and insatiable insectivores known as bats. For a moth on the wing there is no such thing as safe travel.

Top: An *Argiope florida* web backlit by early morning sun. Bottom: A moth scar (arrow) in an *Argiope florida* web.

TALES FROM THE WEBSITE

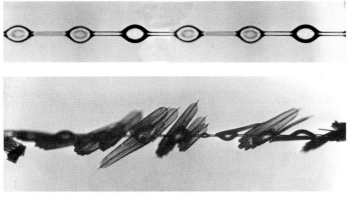

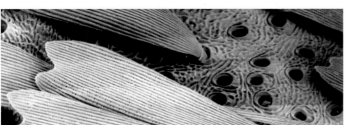

A spider's web is nothing but an insect trap. Typically it consists of a spiral of sticky silk, the viscid spiral, superimposed on a framework of nonsticky supporting strands that radiate outward from the center of the orb like spokes on a bicycle wheel. The viscid strands derive their stickiness from tiny droplets of glue, spaced precisely, like beads on a necklace, along the length of the threads. The droplets are very effective. Insects flying into the web become instantly trapped. Their struggles alert the spider, which pounces upon them, kills them with its venomous bite, and eats them. Not all orb weavers subdue prey in the same way—different spiders have different strategies—and the web architecture may differ from species to species. But ultimately all orbs are fashioned for the same purpose: to intercept insects in flight.

We collected an assortment of insects that night and dropped them into webs, only to find that most did not have a chance. It was the moths that seemed most consistently able to escape. They fluttered vigorously the moment we put them into an orb, but as a rule they were detained only momentarily. Some bounced off the web without sticking at all. Others, which did not change direction upon impact, slid momentarily over the web's surface, only to flutter free when they reached the edge. They all left impact marks on the webs where scales became detached to the viscid

Top left: A segment of a viscid strand of a spider's web showing glue droplets. Middle left: The same strand almost covered with moth scales. Bottom left: The wing scales of a moth beside empty sockets, where scales became detached. Right: Detail of a butterfly wing.

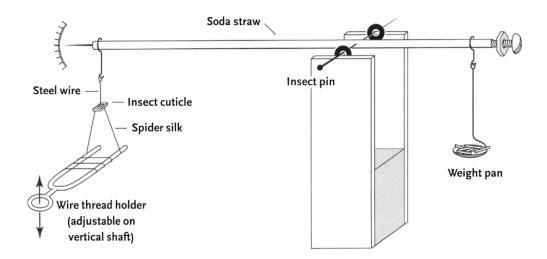

The soda-straw balance. The nut on the screw at the right serves to level the beam.

strands. Moth scars we came to call such telltale sites, and soon learned that they were common.

In the course of the next few days, we offered dozens of insects to orb-weaving spiders and learned that moths do not invariably escape. Spiders, I was beginning to realize, have a taxonomic sense. They seemed to recognize prey before they even touched it. They exercised more caution, for instance, with venomous prey such as wasps than with harmless prey such as flies. They handled wasps gingerly as if aware of their stingers, while they pounced upon flies unhesitatingly the moment these hit the web. They also treated moths as if they knew that such prey can neither sting nor bite. They responded to them instantly, darting toward them the moment they hit the web, with the result that moths did sometimes get caught. A likely victim was any moth that took time to glide over the orb before making its escape.

To obtain some measure of the adhesiveness of spider silk to insect surfaces of various kinds, we rigged up an apparatus that, although simple, served the purpose admirably. From Phillip Morison, the eminent physicist-scholar whom I had befriended in the late fifties when he was at Cornell, I had learned about a device called the soda-straw balance, used sometimes as a teaching aid in schools. I don't know who actually invented it, but knowing about this balance certainly came in handy in our spider work.

In the soda-straw balance the straw forms the beam, and the support is provided by a traversing pin, resting on two upright microscope slides. One arm of the balance bore a detachable wire hanger, the free end of which was bent into a small horizontal loop, to which were glued the

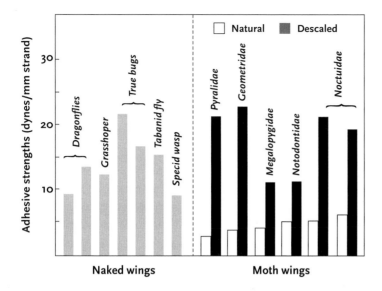

The adhesive strength of a spider web strand to insect wing surfaces. Naked wings are scale-free. Moth wings were tested in their natural condition and after removal of the scales.

pieces of insect cuticle (insect exoskeleton) to be tested. The pieces were cut from insect wings to a specific size (2 × 2 millimeters) and, to measure their adhesiveness, were brought into contact with a single strand of viscid spider thread, stretched across two prongs of a metal holder. Once a piece was fastened to a strand, weights were added to a pan hanging from the opposite arm of the balance, until the cuticle piece abruptly detached from the thread. The weights were small pieces of wire of standardized length, which we added to the pan one at a time. During the procedure we saw to it that the beam of the balance remained horizontal, which required our moving the strand holder vertically downward to compensate for the elastic stretching forced upon the strand by the rising load on the pan. The load on the pan at the point of detachment, appropriately corrected for the difference in arm length (and after calibration of the wire weights), provided the basis for calculating the adhesive strength between strand and cuticle, in dynes per millimeter of strand (essentially a force measurement).

Naked cuticles that we assessed—cuticles (wing pieces) intrinsically devoid of scales—all proved to adhere relatively tightly to the strands. Values for an assortment of cuticles from dragonflies, true bugs (hemipterans), a grasshopper, a fly, and a wasp were in the range of 8 to 22 dynes per millimeter. The adherence of moth wings, by contrast, was in the range of 2 to 6 dynes per millimeter. The same pieces of moth wings, tested after removal of the scales, checked out at values matching those of naked cuticles.

Interestingly one other group of insects, the caddisflies, which are not flies but members of the order Trichoptera, proved also to have nonadherent cuticle. Although they lack scales, caddisflies possess the near equivalent of scales in the form of hairs, which also detach with ease and can therefore protect against orb weavers. Also defended are the tiny insects known as whiteflies (members of the order Hemiptera). Whiteflies derive their name from their dense body covering of waxy powder, an investiture that also detaches on contact with spider webs. Any form of detachable surface structure, it seemed, could guard against entrapment in orbs. Protection against web-spinning spiders could in fact be one of the primary functions of such structures, of which the scales of moths would be the prime example. Butterflies have scales as well that they can use against spiders, but their need for the structures is not as acute as that of moths. The reason is that the threat from spider webs is reduced in the daytime, when butterflies are active. Most spiders keep their webs up for the night only and take them down at dawn.

Rosalind Alsop proved indispensable as a partner in these spider studies. She was to remain with us as a technical assistant for almost 10 years. An indefatigable worker, she was a person of immense charm and generosity. Extremely bright and a naturalist to the core, she had graduated near the top of her class at Cornell. She could easily have pursued a research career of her own had she not been felled by multiple sclerosis, a disease that would eventually claim her life. Roz genuinely loved insects and their kin. She had a magic touch. Animals in her care always thrived, and experiments seemed always to go better when she helped.

■ **THERE HAD TO BE** a good reason why spiders took down their webs in the daytime, and in the summer of 1982 I did some experiments at the Archbold Station that put the question to the test. My partner in this work was Stephen Nowicki, not an entomologist but an ornithologist, and a scientist of the first rank, gifted, affable, and patient. A graduate student with me at the time, he is now a physiological behaviorist on he faculty of Duke University.

For some time I had been interested in the fact that there were some spiders that did *not* take down their webs at dawn, but kept them up for days on end. Many of these spiders, members of the families Araneidae and Uloboridae, had the peculiar habit of adorning their webs with a sta-

bilimentum, a loosely spun band or patch of white silk, typically laid across the center of the orb. Stabilimenta were entirely lacking in nocturnal webs, and as additions to diurnal webs made the orbs extremely conspicuous. Stabilimenta come in different forms, and as silken artifacts can be very beautiful. Some are laid out in the pattern of an X, others in

Stabilimenta in the webs of *Argiope florida* (left), *Argiope aurantia* (middle), and an immature *A. aurantia* (right).

a vertical strand, and still others in a quasi-circular patch. If you search for spider webs in the daytime, the way to spot them is to look for stabilimenta.

What was a visual marker to humans could be a marker to animals as well, and it occurred to me that stabilimenta might serve as visual advertisements to birds, to keep these from tearing into the webs in flight. It would obviously pay for spiders to protect their webs from needless destruction, just as it would pay for birds to heed a message that kept them from flying into an aerial barrier likely to leave them contaminated with sticky threads. The idea was not new, but there was no supporting evidence.

I had noticed birds taking short-range evasive action when approaching spider webs. I am a terrible ornithologist, so I never did identify the birds, but on two separate occasions, once in the Florida Keys and once in Panama, I saw a small bird that was apparently foraging change flight direction abruptly to avoid a web just before it was about to collide with it. The spiders were species of *Argiope,* which typically produce stabilimenta.

Steve and I designed an experiment to test for the web-preserving capacity of visual web markers. What we would do is take some webs that were ordinarily stabilimentum-free—webs from spiders that ordinarily took their orbs down in the morning—and see what would happen to these webs if they were not taken down. Would they then be torn away by birds, and could we prevent them from undergoing destruction if we adorned them with artificial stabilimenta? All we would have to do to put such webs to the test is remove the spiders from them before dawn so that there would be no chance of the webs being taken down (actually spiders don't strictly take down their webs; they consume them, which is a nice way for them to salvage the protein they invest in silk production).

We proceeded by finding 30 webs that we could leave unadorned. It took time, but we eventually located that many. We took the spiders out of these webs at around 2 A.M., well after they had completed orb construction, and marked the web locations on a map so we would have no trouble finding them again.

Then we searched for another 30 webs, and marked these with adornments immediately after we removed the spiders. We adorned 25 of these webs directly, with four triangular strips of paper that we laid out in the

Unadorned (left) and artificially adorned (right) webs such as were used in the web persistence experiment.

pattern of an X on the viscid threads of the orb. The remaining 5 orbs we adorned indirectly so they would not physically bear the markers. With these 5 webs, the marker was a rigid white X that we placed in close parallel to the center of the webs, suspended by four black threads tied to adjacent shrubbery. The spiders we used in this experiment were of 6 species, including the familiar large southern orb weaver, *Eriophora ravillla.* There was a purpose to using markers in the form of an X. One of the dominant orb weavers at our study site, *Argiope florida,* produced X-shaped stabilimenta.

We then adopted a regular schedule for checking on the webs. We first paid each web a visit at 6 A.M., when it was just beginning to dawn, which gave the webs 4 hours of exposure in darkness before bird predation time. After that we checked them all again at 2-hour intervals until noon. We were able to keep track in this way of which webs survived and which underwent destruction.

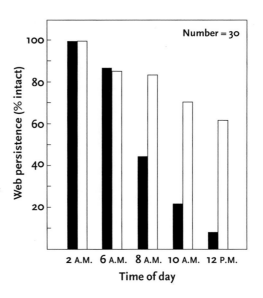

The persistence over time of unadorned webs (solid bars) and of webs adorned with artificial stabilimenta (open bars).

The results were clear-cut. During the dark period before 6 A.M. the webs sustained little damage. More than 80 percent of both marked and unmarked webs survived that period. But after dawn only the marked webs tended to endure. By 8 A.M. the unmarked webs were already reduced to less than half their original number, and by noon all but 8 percent of that category had undergone destruction. In contrast, fully 60 percent of marked webs were still in place by noon.

The results with the two types of marked webs did not differ. Those that were adorned directly endured as successfully as those marked by an X placed beside them. Survival of the web was therefore contingent upon the visual presence of the marker, not the physical presence of the marker, in the web. There being no difference in the results with the two types of marked webs, we treated all marked webs as one category.

We assumed that flying birds were responsible for the destruction of the unmarked webs, and wondered how we might obtain direct proof to that effect. From the incidence of damage to our experimental webs we calculated that if we took up the chore of web watching we should witness bird-web encounters at the rate of 0.24 per hour of web observation. So a group of us, including two volunteers from the station who gave generously of their time, took up hidden observation posts where we could keep webs in sight, and for the period between 6:30 and 8:00 A.M. one morning sat still and watched. It was Steve who saw the encounter. A rufous-sided towhee *(Pipilo erythrophthalmus)* swooped down out of nowhere and tore through a web, ripping it away. The bird faltered momentarily when it

hit the web and dropped slightly from its flight path, but it proceeded without alighting and vanished from sight. It was a single data point, but it was convincing.

Argiope florida bobbing in its web in response to being touched. The behavior is defensive and makes the spider difficult to grasp (photographic exposure = 1/30 second).

Other hypotheses had been offered to account for the function of the stabilimentum, and some of these are plausible. It has been claimed, for instance, that spiders can use the stabilimentum as a shield, for physical protection. Indeed, if you poke an *Argiope,* it typically darts across the web to the opposite side, assuming a post such that the stabilimentum is then interposed between itself and the "attacker." The spider then often also engages in a bobbing action, whereby through a quick flexion of its legs it sets the web into vibration, making itself a blurred target that is hard to grasp.

The possibility, implied by its name, that the stabilimentum somehow stabilizes the orb, is easily disproved. Steve and I singed away the stabilimenta of a number of webs by using a needle that we heated in an alcohol flame, and found that so long as we did not sever the major supporting strands of the orb, the web did not collapse. Nor did it lose its capacity to function. Insects that we dropped into webs from which we had removed the stabilimenta were located by the spiders in the usual amount of time, and were promptly subdued in the usual fashion.

Another hypothesis that has been advanced, which I never found per-

The stabilimentum of *Gasteracantha cancriformis*. The white markings are primarily on the threads that anchor the web to the ground. The orb itself, scantily marked, is at the top of the V formed by the anchoring threads.

suasive, is that the stabilimentum is somehow visually attractive to insects, and serves to lure insects into the web. Stabilimenta reflect ultraviolet light and are supposedly thereby rendered more alluring. Although I am prepared to accept the notion that some insects may occasionally be drawn to the stabilimentum, I doubt that such attraction plays a major role in prey capture by orb weavers. Why, for instance, if the stabilimentum is a lure, should the spiders apply the marker to the web's center, where there are no viscid threads? Also, how does one explain the habit of certain spiders, such as *Gasteracantha cancriformis,* of adding markers to the attachment strands of the web—to the strands that bind the web to surrounding vegetation? For what purpose, other than to protect the web from inadvertent destruction, could the spider possibly adorn such strands?

Nature itself provides evidence that the stabilimentum serves as a visual marker. It seems that on the island of Guam the spiders are becoming less fastidious about incorporating stabilimenta into their webs. The reason? They may have no cause to do so. Guam has been invaded by the Australian brown snake, which has spread over the island in recent decades, and has all but eliminated the endemic forest birds. Without birds to warn, it seems, the spiders are ceasing to put up their "placards."

■ **THERE WERE TWO** orb-weaving spiders that had become my favorites in Florida. One, *Argiope florida,* was abundant at the Archbold Station, where it was a typical inhabitant of the scrub. Its durable webs, stabilimentum-adorned, were a common sight among the palmettos. The other, *Nephila clavipes,* typically found in more wooded areas, was less common at the station, but it occurred abundantly at Highlands Hammock State Park, a preserve 25 miles north of the station where I had often experimented with it. *Nephila* is an eminently successful orb weaver with a broad New World distribution. *Nephila* was also interesting because it spun durable webs without stabilimenta. Its orbs are nonetheless relatively conspicuous in daytime because its strands are yellow. *Nephila* spins an extraordinarily tough, tight-meshed web. It is also meticulous about repairing its web. Repaired patches, debris-free and resplendent in fresh silk, are a common feature of its webs.

Argiope and *Nephila* differ somewhat in how they subdue their prey. *Nephila* is usually direct in its approach. It quickly converges upon an entrapped insect and proceeds directly to inflict its bites. It then typically spins a few loops of silk around the prey and either eats it at once or suspends it from the web for later ingestion. *Argiope* is more cautious. It also converges quickly upon entrapped prey, but as a rule does not inflict its bites until after it has restrained the prey by enveloping it in silk. To do so it engages in some fancy footwork. While hanging from the web with one pair of legs and pulling sheets of silk from the spinnerets with another, it uses a third pair to rotate the prey, so that the prey becomes wrapped. It all happens quickly and the prey is left literally encased in silk. Because of their different strategies, *Argiope* and *Nephila* are not equally effective. *Argiope,* I found, is better able to cope with chemically defended prey.

The difference became apparent in experiments with bombardier bee-

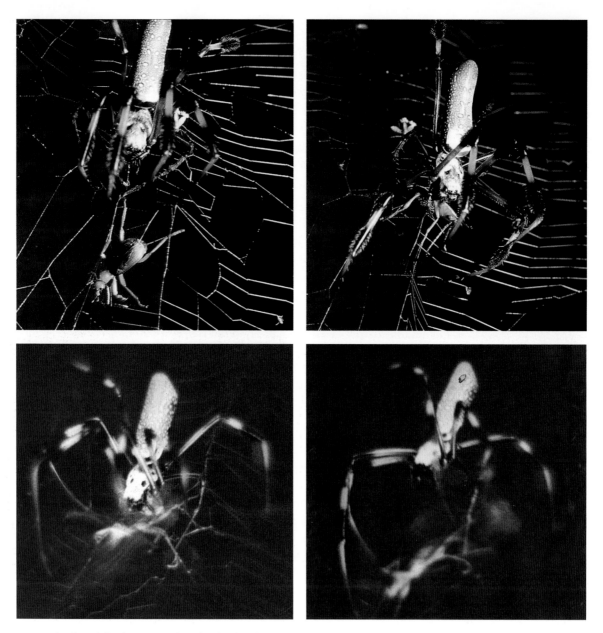

An attack of *Nephila clavipes* on a bombardier beetle. The spider is seen approaching the beetle (top left), and then proceeding immediately to grasp it in the fangs (top right). The beetle discharges in response (bottom left) and promptly repels the spider. The bottom photos are consecutive frames from a motion picture sequence. At the bottom right, the discharge is seen as a diffuse puff at the rear of the beetle. The beetle eventually escaped.

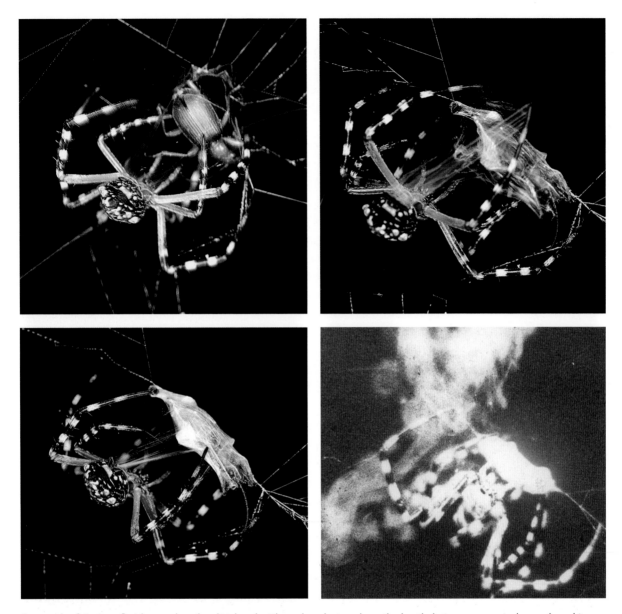

An attack of *Argiope florida* on a bombardier beetle. The color photos show the beetle being progressively enveloped in silk. Only after the beetle is wrapped does the spider bear down to bite. Encased in silk, the beetle is unable to take aim, and the discharge is misdirected away from the spider (the photo at the bottom right is from a motion picture sequence).

tles that I did in collaboration with Jeffrey Dean. We offered bombardiers to *Nephila clavipes* as well as to several species of *Argiope* and found that *Nephila* was quite unable to deal with these beetles. It failed because it did not adequately secure them. It tried to grasp and bite them the moment it came upon them in the webs, but the beetles always "fired" and by so doing brought the attack to a halt. You could hear the discharges and see the little puffs of "smoke" coming from the beetles' rears. The spider was instantly repelled. It backed away and immediately began to clean itself, drawing appendages through the mouth while vomiting, and taking considerable time in the performance of these activities. The beetles in the meantime, by persistent leg action and the occasional severance of entangling strands with the mandibles, worked themselves out of the web. The spider sometimes recovered in time to attack again, but the beetles repelled these assaults as well and eventually gained their freedom. Eleven bombardiers that we fed to three *Nephila* all got away uninjured. And there could be no question that it was because of their spray that the beetles were rejected. A single bombardier that we offered to a *Nephila* after we caused the beetle to exhaust its secretory supply was eaten by the spider.

In contrast, 18 bombardiers that we offered to the various *Argiope* were all eaten. The spiders responded at once to the beetles' presence in the web, but they spun them in without first attempting to bite them. In fact, they carried out the entire spinning procedure so gently that the beetles were never induced to discharge. The beetles did discharge, however, when the spiders subsequently inflicted their bites. But by then the beetles were encased in silk and no longer able to revolve the abdominal tip, with the result that the discharges were then partly muffled and misdirected away from the spider. The spiders at times did react to the discharges, having been "hit" in the legs, but this caused them to be distracted only momentarily and to respond by wrapping the beetles in additional silk.

I consider prey wrapping such as is effected by *Argiope* to be an enabling strategy forced evolutionarily upon spiders by the widespread presence in insects of chemical defenses. The habit of gingerly encasing a captured insect in silk, in such a way that it is not induced to spray while being wrapped, and is left unable to derive any advantage from its spray after being encased, is illustrative of evolutionary one-upmanship at its best. Over the years I have fed many insects to both *Nephila* and *Argiope,* and have noted the advantage that the latter have when it comes to dealing with noxious prey. Quite remarkable have been results I and others have obtained with stink bugs and their relatives, insects known to most of us not only for their stink, but because so many of us have inadvertently eaten an occasional one with berries we picked in the wild. How some stink bugs are subdued by spiders, and how others that repel spiders get out of the webs is a story in itself. And so is what happens once a spider kills a stink bug and finds itself crowded by unwanted visitors intent on sharing the meal.

■ STINK BUGS belong to the family Pentatomidae of the order Hemiptera, the members of which are known as the true bugs. One of the diagnostic features of hemipterans is their mouthparts, which feature a rostrum or beak, used as a piercing and sucking device. The beak contains two channels, one for the uptake of food, the other for the output of saliva. Hemipterans have diverse feeding habits. Some are insect killers, a few are bloodsuckers, and the majority, like most Pentatomidae, are plant feeders. All imbibe liquids only. Those that kill prey do so by injection of poisonous saliva.

Stink bugs derive their name from the defensive secretion they eject when disturbed. In the immatures (the nymphs) the glands open on the dorsal surface of the abdomen, an area which in the adult is covered by the wings. Not surprisingly, therefore, adults have a different glandular arrangement. They possess a single large gland, situated midventrally in the thorax, with two separate openings on the sides of the body just above the legs. Pentatomids take on an intense odor when they discharge, and the odor is persistent. One reason for this is that some secretion invariably becomes trapped upon ejection, in a specialized region of cuticle immediately adjacent to the gland opening that is minutely sculpted so as to serve as a physical sponge. If one examines that region immediately after a bug has discharged one invariably finds it to be soaked in fluid. Pentatomids can discharge separately from the right and left gland openings and they tend to restrict their discharge to one opening if the disturbance is unilateral. They can also aim the discharge to some extent by adjusting their body posture, but their control over the direction of emissions falls far short of that of bombardiers.

There is a second family of Hemiptera, as prominent as the penta-

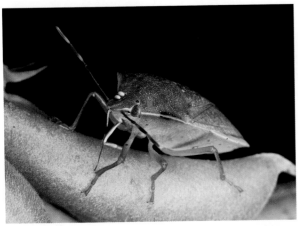

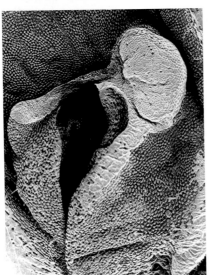
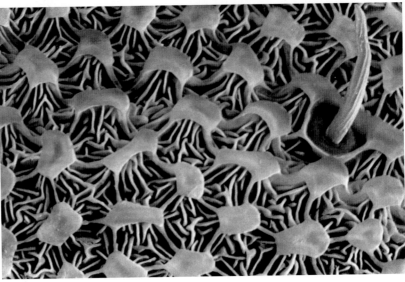

The stink bug *Nezara viridula,* family Pentatomidae (top left), beside the squash bug *Chelinidia vittiger,* family Coreidae (top right). The arrow points to the left gland opening of *Chelinidia.* In the Pentatomidae the gland openings are similarly located. Bottom left: A close-up view of the left gland opening of *Chelinidia.* Bottom right: An enlarged view of the sculpted cuticle from immediately around the gland opening in *Chelinidia.* Such sculpting, which is present in all Pentatomidae and Coreidae, serves to retain secretion after discharges.

tomids, that should really share the name of stink bugs. That family is the Coreidae, comprising the squash bugs, which have very much the same glandular system as the pentatomids and produce very much the same defensive compounds. The experiments I did with stink bugs I did with squash bugs as well, and the results with both types of bug were basically the same. The stink of Pentatomidae and Coreidae is quite characteristic, and unforgettable to the experienced. The compounds they produce are for the most part simple straight chain aldehydes and ketones—carbonyl compounds—of which *trans*-2-hexenal is perhaps one of the most common. Such compounds are potent repellents. I found *trans*-2-hexenal, for instance, to be highly effective in the scratch test with cockroaches.

Given the results with bombardier beetles, I expected stink bugs and squash bugs to have no chance of survival in tests with *Argiope*. The strategy of enveloping prey before killing it, one would think, had to be applicable to chemically protected Hemiptera as well. So it came as no surprise, when I offered bugs to *Argiope*, to find that the spider spun these in as if they were bombardiers. To wrap its prey, *Argiope* must be able to twirl it, and this requires usually that the spider cut some of the threads immediately around the prey. It leaves intact such contacting threads as are aligned lengthwise with the prey's body, and it is around this axis of threads that the prey is then rotated when it is wrapped. When handling bugs, the spiders executed these procedures without usually causing the bugs to discharge. I made it a point to sniff the bugs at close range, both while the spiders were cutting threads and during the wrapping operation, but the stink was almost never to be detected. It was usually only upon completion of the wrapping, when the spiders bore down to bite, that the bugs finally sprayed. The stink would then permeate the air, and the bug, within its silken envelope, would suddenly take on a moist appearance. As with bombardiers, the spiders were only temporarily deterred by the ejections. As a rule, they merely twirled the prey again and wrapped it some more. The bugs eventually died from the bites, and it was often only after some delay that the spiders turned to eat them. The outcome, however, was always the same. Not one of the several dozen stink bugs and squash bugs that I fed to *Argiope* over the years escaped from the orbs.

The results with *Nephila* were different. Here again I came to feel that this spider had a taxonomic sense. *Nephila* seemed to know that stink bugs and squash bugs require special handling and the spider appeared to be able to recognize these bugs on contact. It responded quickly to bugs that I introduced into the web, but upon touching them proceeded with caution. Using its legs and severing a few threads with its fangs if necessary, it would attempt to orient the bug in such a way that it would come to face the dorsal rather than the ventral surface of the bug. It did not always succeed in doing so, but the effort seemed intended to make it possible for the spider to bite without being fully sprayed in return. The bug sometimes discharged while being oriented in this fashion, but it usually held its "fire" until the spider bore down with the fangs. The discharges had a mixed effect. In most cases they repelled the spider only briefly, causing it to back away and clean itself, but they did not usually prevent it from resuming the assault. In due course, the spider did inject its venom, after which it enveloped the bug in a few loops of silk and transported it to the hub for consumption. Of 29 bugs that I offered *Nephila,* 24 were thus treated. The lot had included both pentatomids and coreids, and the spider had made no distinction between the two types of bug.

What was exciting was the fate of the 5 bugs that were not so treated. These too discharged when attacked, but in their case the discharges took full effect. The spiders reacted violently when hit, literally taking flight from the site of the action. Two spiders fled to the hub, another 2 darted to the edge of the orb, and the fifth dropped from the web so that it came to hang in free space, suspended from the hub by its drag line (the spiders remain at all times connected to the hub by such a line). To judge from the intensity with which they then engaged in cleaning, the spiders must have been hit full blast by the spray. They regurgitated and drew appendages through the mouth, taking minutes to perform such actions. This gave the bugs time, and what the bugs did with that time was remarkable. Lacking mandibles, they could not cut themselves out of the web. Instead, they freed themselves by salivating.

The escape behavior took time but was remarkably effective. Getting entangled, for a bug, meant having strands fastened to its legs and to various parts of the body. The bugs made a systematic effort to apply saliva to these contact points. The essential maneuver was the application of fluid to the legs. They did this by flexing the beak so as to bring the tip in contact with a leg or group of legs, while at the same time delivering a droplet of saliva. Saliva would ooze almost constantly from the beak during these maneuvers so that it was possible for the bugs to administer droplet after droplet on desired sites. They had a way of spreading secretion to body parts that were not accessible to the beak. Such regions could often be contacted by the legs and the bugs used the legs as brushes to wet hard-of-reach places. The saliva had two effects on the imprisoning strands. It diluted the glue responsible for their stickiness, thereby reducing their adhesiveness, and it eliminated the strand's elasticity. The viscid strands of a

web are typically highly stretchable. That is advantageous to the spider because imprisoned insects are thereby placed in the predicament of trying to free themselves from threads that resist detachment when being pulled. Imagine being imprisoned by thin rubber bands fastened to your skin with Scotch tape. As long as in your struggles you do not stretch the bands beyond their elastic limit, chances are you will not detach the tapes.

An attack of spiders on pentatomid stink bugs. *Argiope florida*, on the left, spins the prey in before inflicting its bite, while *Nephila clavipes* (right) attempts to bite immediately upon encountering the prey.

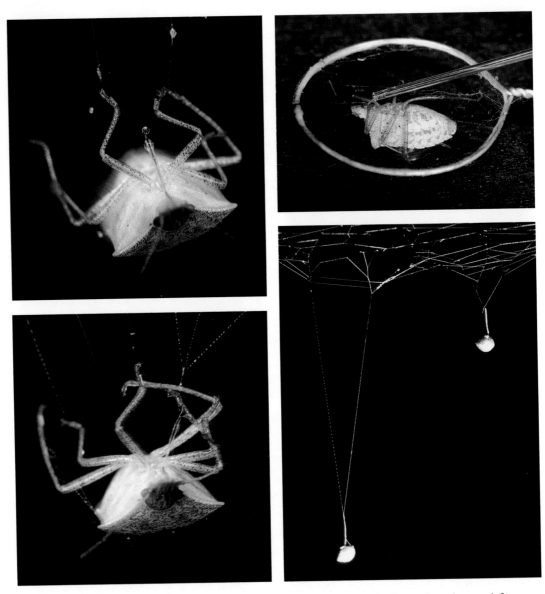

A pentatomid bug emitting saliva from its beak (top left) and applying the fluid to its legs (bottom left), in efforts to work its way out of a spider web. Top right: A pentatomid bug salivating into a capillary tube in response to having its legs contacted by spider webbing. Bottom right: A spider strand being stretched to its normal limit, beside a comparable length of strand that has lost its elasticity following treatment with pentatomid saliva.

TALES FROM THE WEBSITE

Rendered nonelastic, the spider strands become breakable, and they did undergo breakage as the individual bugs struggled to free themselves. The rubbings of legs, one against the other, that seemed so effective in mixing saliva with glue and in loosening the threads' hold, seemed also to cause threads to tear. The breakage in turn caused a bug to fall, usually by its own weight, to a point a bit lower on the web, where it would become trapped again. It then commenced the whole cycle of activity again, salivating onto its legs and rubbing these together until threads broke and it fell lower, and thus it proceeded, cycle after cycle, until it worked itself to the edge of the web and freedom. In the final stage the bug unfolded its wings as it hung briefly from the web's margin and then flew off when it fell free.

To watch this escape behavior more closely, I removed some spiders from their webs and took their orbs to the laboratory. I have a very simple way of doing this. I take a wire clothes hanger, bend it into a loop, coat it with rubber cement, and let the cement dry to a semisolid consistency. This makes the loop sufficiently sticky to adhere to the web. I then press the loop into a web, cut the strands around the loop by burning them with a hot needle, and lift the web away. I released bugs into such isolated webs and photographed them close up as they enacted their escape behavior. I also filmed the behavior.

The bugs use a large amount of saliva to work themselves out of a web, more than one would estimate could be stored in the salivary glands at any one time. The rate at which the glands refill during salivation must therefore be quite high.

I also found a simple way of coaxing bugs into delivering saliva. All I had to do was fasten them belly-side-up on a glass slide with a droplet of dental wax and press a section of spider web against their legs so the legs would become entangled. They then dutifully emitted saliva from the beak, making it easy for me to collect fluid for experimentation. I picked up droplets with microcapillary tubes and found that when I applied these to the viscid strands of an orb they would make these strands nonstretchable and breakable, just as I would have predicted from the behavioral observations.

I never published these findings, largely because there were so many questions about the escape behavior that I would still like to answer. What I did publish, however, are some entirely different observations, also on "bugs in the web," but dealing specifically with what happens when spiders do overpower a pentatomid or coreid and settle down to make a meal of it.

What I found is that when spiders are feeding on such bugs they draw visitors to the "table." Unbeknownst to me, others had already discovered the general phenomenon: that orb weavers, while feeding, attract an assortment of flies to their prey that come to share the meal. We ourselves, in our homes, also draw flies to the kitchen, so the phenomenon seemed to be just another illustration of the exploitive opportunism of flies. Kleptoparasites is the term applied to animals that make a living from stealing food, and the particular flies that had been noted to sneak up on spiders were of diverse families– Cecidomiidae, Phoridae, Chloropidae, and Milichiidae. The ones I had myself observed on hemipteran prey were milichiids. Mark Deyrup identified them for me, and with him as a partner, Maria and I decided we would study these flies. What we wanted to know is exactly how milichiids locate a spider's hemipteran catch.

There is a good reason why kleptoparasites are drawn to spiders. Spiders are slow eaters. Unable to swallow solids, they grind their prey to a pulp, leisurely drenching it in disgorged digestive fluid, then sucking up whatever is liquefied. The process may take hours, even with prey much smaller than the spider, so there is plenty of time for usurpers to share in the meal. The milichiids at our webs were landing on the prey in substantial numbers.

Although it was obvious that both pentatomids and coreids were drawing milichiids, we did most of our observations on one pentatomid, *Nezara viridula,* which was particularly abundant at our Florida study sites at the time. We fed nine *Nezara* to individual *Nephila* and by monitoring the webs at intervals kept track of subsequent milichiid visitations. The flies arrived within minutes of the spider's attack, built up in numbers over the next 2 hours, and then became fewer and fewer toward the end of the spider's meal. Fly density at peak times averaged 5 per prey item, but we recorded instances of up to 20 per item. Flies came and went during the meal but we did not keep track of the turnover. The total number of visitors per prey item must therefore have been higher than indicated by our counts.

Observation of the approaching flies showed that they always came from downwind. Detected as tiny specks from a distance, the flies zigzagged their way toward the spider's prey, making corrections in their ae-

The attraction of milichiid flies to *Nephila clavipes*. Top left: The spider has captured a pentatomid bug and the flies are starting to arrive. One fly is seen approaching from the left. Top right: Within an hour, a number of flies have settled on the pentatomid prey. Bottom left: A close-up of a milichiid feeding. The long proboscis is ordinarily kept folded beneath the head. Bottom right: *Nephila* feeding on a moth *(Utetheisa ornatrix)* that has been baited with *trans*-2-hexenal, and is attracting milichiids. The arrow points to a fly.

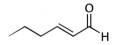

Trans-2-hexenal.

rial trajectory when gusts of breeze deflected them from their path. They flew like male moths fly when they are homing in on females along a pheromonal plume.

As the flies approached the web they ran little risk of entrapment because by the time of their arrival the spider had already taken the prey to the web center and positioned itself at the hub. The hub is constructed of nonviscid strands and poses no threat to the flies.

We found that we could observe the flies at close range with magnifying glasses without scaring them off. We could then see with the utmost clarity how they began feeding upon landing and gorged themselves to repletion. They imbibed food with a long proboscis, which they ordinarily kept folded beneath the head.

The spiders showed almost no reaction to the flies. They responded only when the flies gathered closely around the mouth, and even then did no more than attempt to brush them away with the legs. The screened-in dining room was evidently to be invented by another species.

Since the flies always approached from downwind, it seemed reasonable to assume that they were attracted chemically, perhaps by the prey itself or, more specifically, by the defensive secretion of the prey. We thought the attraction could be mediated by the very carbonyl compounds ejected by the bugs, and decided we would test for this. What we thought we would do is take a type of prey that we knew did not attract milichiids, feed it to *Nephila,* and see whether we could make such prey attractive to the flies if we laced it with, say, *trans*-2-hexenal, one of the major components of hemipteran "stink."

We knew exactly what prey to use. We were experimenting with a moth at the time, *Utetheisa ornatrix,* which we knew to be palatable to *Nephila* (when raised on a special diet; see Chapter 10), and knew also to remain milichiid-free when being eaten by a spider. So we fed two sets of *Utetheisa* to *Nephila,* a set of eight that we baited with *trans*-2-hexenal, and another of six (controls) that we kept unbaited. We added the bait to the moths after these had been killed by the spiders and taken to the hub of the web. We waited until the spiders began feeding and then simply trickled 4 microliters of *trans*-2-hexenal onto the prey, taking care not to contaminate the mouthparts of the spider.

The flies were quick to arrive, at a rate not much different from that we had recorded with *Nezara.* Unbaited *Utetheisa* remained unvisited. *Trans*-2-hexenal was evidently effective. The question was whether it was in itself

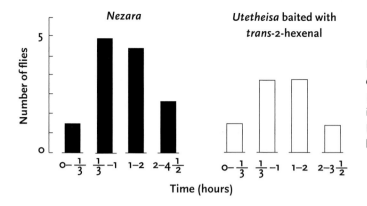

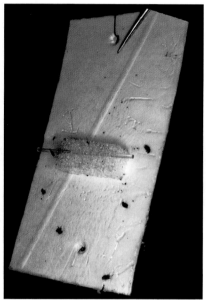

Milichiid fly visitation to two prey items of *Nephila clavipes:* a pentatomid bug *(Nezara viridula)* and a moth *(Utetheisa ornatrix)* baited with *trans*-2-hexenal. The time is given in hours after the onset of feeding by the spider. Fly numbers give the average for nine *Nezara* and eight baited moths.

Left: The site at Highlands Hammock State Park, Highlands County, Florida, where experiments on milichiid attraction were carried out. Note the two small, rectangular sticky traps, baited with *trans*-2-hexenal, suspended from the vegetation. Right: A close-up view of one such sticky trap. The microcapillary tube bearing the *trans*-2-hexenal is taped to a small piece of Styrofoam in the center of the trap. Note that several flies have already been lured.

effective or whether to be attractive it had to be mixed with some odorous factor from the spider, so we did one more test. We took pieces of glass capillary tubing, filled these with *trans*-2-hexenal, and fastened them to small pieces of cardboard that we had coated with an adhesive. The idea was to see whether such trapping devices would lure milichiids. We hung a number of the traps in the general area where we had done the *Nephila* tests and found that they did. Moreover, I discovered that with a laden capillary tube in my hand I could cause milichiids that were converging on a captured hemipteran to deviate, at least momentarily, from their flight path. This seemed to prove that it was indeed the bug's secretion that drew the flies to the spiders' catch. The defensive chemical "juice" that failed to save the hemipteran bug cued the kleptoparasites to the location of their meal.

We know from the findings of other investigators that certain prey beside hemipterans also attract milichiid flies. It is possible that these prey are also located on the basis of glandular emissions, but that remains to be proven. Milichiids could be attracted also to the catch of predators other than spiders, and there are reports of such attraction. I have myself photographed an assemblage of milichiids that were feeding on a bee captured by an ambush bug. And, of course, chemical attraction could also explain how flies other than milichiids that also visit spider prey, such as chloropids and phorids, home in on their targets. It is interesting in this connection that our cardboard traps baited with *trans*-2-hexenal drew phorid and chloropid flies as well, although in very small numbers.

■ **THE MORE I EXPERIMENTED** with orb weavers, the more I came to admire them. They were a prime example of evolutionary success. Unlike so many other arthropods that lost out in competition with insects and were forced to leave unchallenged the latter's conquest of the air, orb weavers literally invaded the airspace themselves, not to fly, but to prey on those that did. It was a marvelous evolutionary ploy and a whole field of biological exploration—insect-spider interaction—was its consequence. And best of all, the field was wide open. Spiders have countless ways of dealing with insects, and insects have complex ways of dealing with web entanglement, and most of these strategies remain to be discovered. What interested me in particular was how insects get out of webs. I was convinced that to discover new insectan escape strategies all one had to do is flip insects into orb webs and watch. I still adhere to that belief and feed orb weavers whenever I can. Indeed, when I was younger, and more terrified of the prospect of having my research support dry up—of having my National Institutes of Health grant turned down—I had always thought that in hard times I would fall back on research with orb weavers. There was much to be discovered with these animals with no more equipment than a notebook, a stopwatch, and time. Nowadays I worry about the shortage of time.

There were some fascinating studies done by others on how insects avoid capture by that other great enemy that threatens them in the night—bats. Most interesting was the discovery by Kenneth Roeder and Asher Treat that certain moths have ears sensitive to the echolocating pulses of bats. The moths can therefore hear when they are being chased by bats and take evasive action. Such ears have been found in other insects as well, including for instance praying mantids. I wondered how many insects might be protected against both orb weavers and bats. I thought green lacewings—the members of the family Chrysopidae—were likely candidates. Chrysopids have bat-detecting ears, and in addition have a pair of thoracic defensive glands that I was sure would protect them against spiders. But I was wrong. Chrysopids are quite acceptable to orb weavers despite their glands. But they are not as helpless as you might think, since there are circumstances under which they are able to work themselves out of webs.

Chrysopids can escape from webs by sliding out of them. Mitch Masters and I discovered this in the fall of 1987, at Archbold's. It was a banner year for chrysopids and they were to be had almost everywhere on the station grounds. We collected a number, put them in vials, and flipped them into webs in accord with our usual spider-catering service.

The first thing we noted is that the spider did not always respond to the offerings. Whether it did so depended on its "attack readiness." When the spider was busy feeding, or in the process of subduing another prey item, it could completely ignore our own offering. There is evidently only so much a spider can do, and the limitation can benefit the prey. With both *Nephila* and *Eriophora* we found that the spiders attacked the chrysopids only if they were free to do so. And when they did attack they usually ended up eating the chrysopid. There was no indication that the spiders were in any way discouraged by the chrysopid's defensive secretion, although we were certain, judging from the detectable stench, that the chrysopids were discharging. The German name for chrysopids is "Stink-fliegen," and the term is appropriate. Chrysopid secretion contains skatol, a compound with a distinct fecal odor.

Chrysopids that were ignored proceeded immediately to implement their escape strategy. They first concentrated on cutting away the sticky

An adult chrysopid making its escape from an *Argiope aurantia* web. In the beginning (left), the chrysopid is positioned ventral side down across four sticky strands. It immediately cuts the strand touching its forelegs, the ends of which snap back (arrows). The chrysopid begins to assume a cruciform posture as it topples head first through the hole formed in the web. It then proceeds to cut the last thread that remains attached to its body, with the result that it comes to hang below the plane of the web, from two sticky strands that run to its wing edges. Under the force of its own weight, it now slides free. Two cut and two coalesced sticky strands are all that remain as evidence of the chrysopid's entanglement.

strands entangling their heads, feet, and antennae. The long filamentous antennae could give them the most trouble. If they failed to free these by pulling on them, they performed what we called antenna climbing. They grasped an antenna in the forelegs and pulled it in the mouth. Holding the antenna in the mandibles they then slid the legs along it, pushing any attached sticky strands forward. Resuming their grasp on the antenna with the forelegs and releasing their grip with the jaws, they then pulled the legs to the mouth again, transferred their hold from legs to jaws, and repeated the cycle. As the mouth and forelegs were moved along the antenna, sticky strands were pushed toward the tip, at which point the chrysopid cut the antenna free.

With their bodies disentangled the chrysopids next became quiescent. They were now suspended by the wings and we wondered how they might proceed to detach these. We watched carefully, but nothing seemed to happen. But then we noticed that a very gradual process was taking place. The chrysopids were very slowly sliding downward under the pull of their own weight. While doing so they sometimes assumed a characteristic cruciform posture, so called because the wings would be turned edge outward at such angles as to form a cross in a head-on view. They maintained this posture rigidly for minutes on end as gravity did its job.

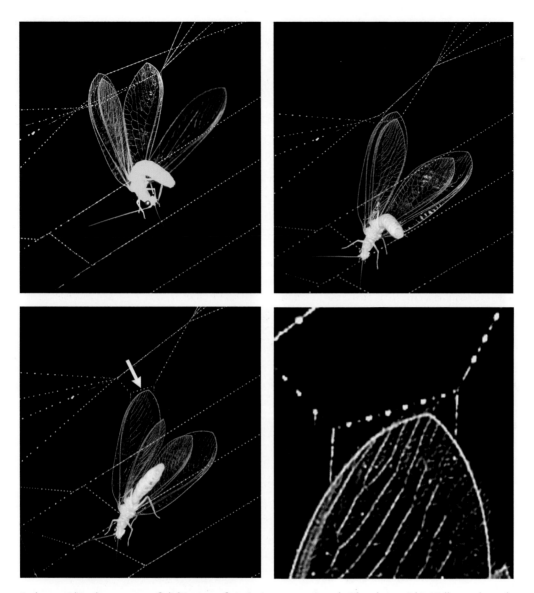

A chrysopid in the process of sliding out of an *Argiope aurantia* web. The chrysopid initially cut through strands running to its legs and antennae, and has here reached the point where it is suspended by four sticky strands running to its wing tips (top left). Note how the chrysopid slips downward (top right) as the threads slide to the edge of the wings. At the bottom left, in a photo taken an instant before the chrysopid fell free, the insect is held by a single strand. An enlargement of the area denoted by the arrow (bottom right) showed that this strand had already pulled loose, and that the chrysopid was hanging from no more than two stretched glue droplets.

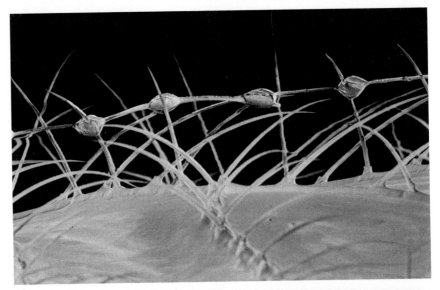

A scanning electron micrograph of a portion of a chrysopid wing, showing the dense covering of hairs. Note how the spider web strand is prevented by the hairs from making contact with the wing surface.

Left: A chrysopid, stuck in a spider web, performing the antenna climb. The insect is hanging by its antennae, both of which are caught by sticky strands. It has encircled one antenna with its forelegs and is about to draw that antenna to its mouth, the first move in the climbing maneuver. Right: A chrysopid in cruciform posture. The insect is suspended by one leg, and its body is nearly horizontal. The cruciform posture is frequently assumed by chrysopids as they wait to slide out of the web.

We wondered why the sliding tactic should work for insects whose average weight is only about 10 milligrams. We found the answer when we looked at the chrysopid's wings. Chrysopids can slide out of webs because the viscid threads never really make good contact with the wing surface. The wings are densely covered with short hairs, and it is with these that the strands make contact. The hold on the wing is thus intrinsically a loose one, to the benefit of the chrysopid, which can glide free as a result. Being in the cruciform posture greatly facilitates the sliding process since the wings are then so positioned as to shield the body from establishing renewed contact with the strands. When the chrysopids finally slide free, they may become trapped again, lower on the web, in which case they simply repeat the escape procedure. But we have also seen chrysopids in the cruciform posture literally tumble down the web when they became detached. Maintenance of the cruciform posture is therefore of help to the chrysopid throughout the escape procedure.

Mitch, forever the perfectionist, was not satisfied with the way we were introducing the chrysopids into the webs. He wanted to see the chrysopids fly into webs under their own power. So he released chrysopids into a flight cage, hung a spider web in the center of the cage, and watched the proceedings under red light, which most insects cannot see (he had obtained the web by the clothes hanger technique). Most web contacts that chrysopids made were momentary and did not result in entanglement. Only in about a third of the cases did the chrysopids get stuck. There is therefore considerable advantage that chrysopids derive merely from being light and slow on the wing. It is when they do get stuck, and the spider happens to be free for action, that they are likely to get into trouble.

■ **ANYONE** who has walked barefoot through clover in bloom knows that bee stings hurt. The stinger is the honey bee's defensive apparatus and it makes sense that a bee should be able to inflict pain with it. Wasp stings hurt as well, and so do scorpion stings, the stings of centipedes, and the venom injections of any number of animals, including snakes. For some of these animals the induction of pain may pose a problem, since they use their venom primarily for incapacitation of prey and the induced pain may exacerbate the prey's struggles, to the disadvantage of the predator. The reason such venoms are painful is that they are also used in defense and

An ambush bug, *Phymata fasciata*, feeding on a syrphid fly.

are therefore meant to be perceived instantly as being noxious. It is principally because they are painful that venoms can function in defense.

I had been interested in the question of pain for some time, and specifically in whether the so-called lower animals, the invertebrates, perceive pain. I had no doubts that they did, but how could I put the question to the test? I was not interested in whether invertebrates perceive pain consciously. Consciousness is a subjective notion applicable to humans but essentially untestable with animals. I was interested merely in whether they "felt" pain.

The question, it turned out, was also of interest to Scott Camazine, a physician and nature enthusiast who had joined my group in the early eighties. Could one not subject an invertebrate, we asked, to specific stimuli known to be painful to us, to see whether these stimuli induced a self-preserving response in the invertebrate as well? We were particularly ea-

TALES FROM THE WEBSITE

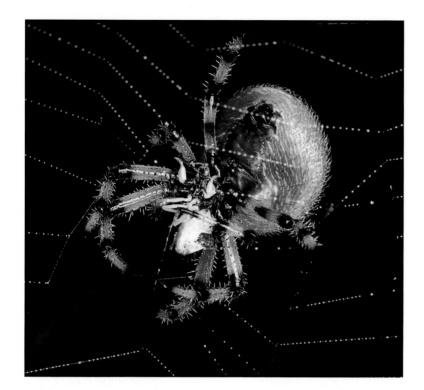

An araneid spider feeding on *Phymata fasciata.*

ger to test the notion in an experimental situation that made sense in terms of the real world—a situation where an invertebrate was exposed, in a natural context, to experiences known to be painful to humans.

I had made an observation some years earlier that put us on the right track. I had been web watching at the Huyck Preserve near Albany when I saw an ambush bug, *Phymata fasciata,* fly into the orb of *Argiope aurantia,* a relative of the Florida *Argiope* I knew so well. I expected the bug to be spun in but wondered whether there might be some trouble. Ambush bugs are just that—insects that lurk in hiding and prey on the incautious. They are little insects, no more than about a centimeter in length, and they kill their prey by injecting venom from a short beak. They also use the beak in defense and can administer a nasty sting with it, as I found out once when I handled one carelessly. They occur throughout the United States, but are not widely known because they are so well hidden in the flowers where they sit and wait for prey. They do have wings and fly about, and as a result may land in orbs.

What I wondered as I watched that encounter is whether the phymatid would use its venom against the spider, and it turned out that it did. The spider was just about to spin in the phymatid when the phymatid grasped

Argiope aurantia versus the ambush bug *Phymata fasciata*. Top photos and bottom left: The spider approaches the bug and is then stung in a leg, which it autotomizes. Giving up its leg saves its life. The encounter depicted at the bottom right had a different outcome. Here the spider failed to autotomize when its leg was stung, and it died as a consequence.

one of the spider's legs with its prehensile forelegs. It then proceeded to probe the spider's leg with its beak, causing the spider to become momentarily quiescent and then, abruptly, to autotomize the leg. The leg literally snapped off at its base, as if it had been amputated by invisible scissors. Did the venom do it, and was it in response to the pain-inducing factors in the venom that the spider had acted? Was it essential for the spider to autotomize? Was this its way of preventing the venom from seeping into its body and causing its death?

I released a quantity of phymatids into *Argiope* webs and witnessed sev-

TALES FROM THE WEBSITE

eral other instances of leg autotomy. One encounter was particularly telling because the spider did not autotomize when stung. That spider became quiescent and I wondered after 20 minutes of observation what
might be wrong. The phymatid during that period had maintained its prehensile hold. I finally lost patience and poked the spider, only to find that
it was dead. Failure to autotomize could evidently be fatal.

The first thing Scott and I decided to look into is whether it is in response to the sting itself that the spider autotomizes its leg. We were able
to demonstrate that neither the mere perforation of the leg nor the mere
grasp of the spider's leg by the phymatid induced autotomy. Venom had to
be injected. We watched with the microscope as we induced tethered
phymatids to sting tethered spiders in the leg, and could see that it was
only after beak penetration that autotomy occurred (we could actually see
through the spider's integument when the sharp inner stylets of the beak
made their penetration). We also were able to establish that failure to
autotomize in response to phymatid venom injection is indeed fatal to the
spider. Stings to the body of the spider are fatal as well. By shedding its
leg, therefore, the spider was saving its life.

Was it the pain-inducing factors that had triggered the autotomy? Little was known about the chemistry of phymatid venom, so we checked,
first of all, whether other venoms also induced leg autotomy, and then,
whether known pain-inducing factors from these venoms could in themselves elicit the response. Bee venom and wasp venom both proved active.
And so did histamine, serotonin, and phospholipase A, well-known pain
inducers, all commonly present in animal (and even plant) venoms.

What hurt us, evidently, caused spiders to react as if it hurt them as
well. In more formal terms one could say that the sensing mechanism by
which spiders detect injected harmful chemicals such as venoms may be
fundamentally similar to the one in humans that is responsible for the
perception of pain. The sensitivity of spiders does not exactly match our
own—the spiders proved not to autotomize in response to injection of
acetyl choline and bradykinin, both known to be pain-inducing in humans—but the overlap in sensitivity is substantial. Quite aside from the
issue of consciousness, we came to the conclusion that invertebrates perceive pain, and that their sensory basis for doing so may not be much different from our own. There is therefore good reason for treating invertebrates humanely.

7

The Circumventers

Fugu is Japanese for puffer fish. Valued for its exquisite taste, the fish is a cherished culinary item in Japan, sought out by thousands annually who are willing to risk consuming it. Eating puffers is dangerous indeed, and to indulge in it has been likened to playing the gastronomic equivalent of Russian roulette. The dangers are so great, in fact, that it is forbidden to serve puffer fish in any form to members of the imperial family. More than 50 people die every year from eating puffer fish, and the death is not a pleasant one.

Puffer fish contain tetrodotoxin, a lethal poison for which there is no antidote. The liver, ovaries, and to a lesser extent the skin of the fish contain the toxin, which acts by blocking cellular sodium channels, thereby preventing nerves from generating the propagated spikes that activate muscles. The consequence is paralysis and death by asphyxiation. The muscles of the puffer are devoid of tetrodotoxin and if kept uncontaminated during culinary preparation of the fish are perfectly edible. The puffer market is strictly regulated in Japan. Only special restaurants are allowed to serve fugu, and the chefs who prepare the fish must undergo extensive training before they are licensed. Yet mistakes do happen and the proof is in the statistics. The slightest cut into the tetrodotoxin-bearing organs of the fish during preparation may contaminate the edible parts, with potentially fatal consequences to the patron. Despite the risks, customers pay as much as $400 a serving for fugu. Importation of puffers into the United States is also regulated, but the rules have not prevented the occasional ill-prepared serving from ending up on someone's plate. Three cases of fugu poisoning, none lethal, were reported a few years ago from California.

In biological terms what we humans do when we partake of puffers is deal with poisonous food that has been depoisoned by special preparation. By laying open the puffer and selectively eating those parts that are harmless we are essentially circumventing the fish's natural defense. With a counterstrategy of our own, we circumvent the animal's defensive strategy, and by so doing add the species to our choice of edibles. Other animals resort to the same strategy. Prominent among these are insects, but the examples include vertebrates as well. Incredibly, there is a predator in the southern Appalachians, most likely a mammal, with the habit of deal-

Tetrodotoxin.

ing with a tetrodotoxin-producing newt by subjecting it, in fugu-chef fashion, to preingestive dissection.

We owe the discovery of this circumventing behavior to Donald J. Shure and his collaborators at Emory University. Working in the vicinity of Highlands, North Carolina, they found remnants of *Notophthalmus viridescens,* the red-spotted newt, strewn about on a logging trail. It had been raining and the newts had been active. The carcasses were fresh and had been mutilated. Some had been decapitated, others had been sliced open, and all were missing some of the internal organs. Whoever was feeding on these newts knew that the viscera were the harmless parts. Tetrodotoxin in the newt is stored in the skin. The spectacle of all those eviscerated corpses was pitiful—some had not quite stopped moving. Shure failed to catch the predator in the act, and to my knowledge the culprit has never been identified. Shure suggests that skunks might have been responsible.

■ **I HAVE MYSELF** come across evidence that there are predators able to deal with protected prey, and I too have been stymied at times by not being able to identify the predator. There is a wonderful site near Cornell, Taughannock State Park, which I love to visit because it includes a gorge that is a haven for millipedes. Hydrogen cyanide–producing polydesmoid millipedes abound there, as do benzoquinone-producing spiroboloids. The latter are represented by *Narceus annularis,* a large species, active at night, found in logs and leaf litter in the daytime. *Narceus* is easily provoked. Its glands are segmentally arranged along the sides of the body, and it tends to discharge even when mildly disturbed. The emission leaves the millipede virtually coated with the foul-smelling fluid.

The millipede's defense is imperfect in one respect. The segments im-

The red-spotted newt
(*Notophthalmus viridescens*).

mediately behind the head, like the head itself, lack glands. The front end of the millipede is therefore unprotected, which explains why the animal responds to disturbance by coiling its body with the head at the center of the coil.

Knowing that millipedes are best collected at night when they are active, I used to make it a point to scout around Taughannock Park after dark with my headlamp when I was in need of the animals. But there came a time when I had to be at the site in early dawn to collect insects, and it was then that I became aware of an animal I called Robespierre. Some predator at the park knew that the front end of *Narceus* was undefended and edible. At one location the evidence was all around me. Headless *Narceus*, some of them dead but others still writhing, were scattered over the leaf litter, in evidence of what must have been a feeding frenzy the night before. I suspected a rodent and set out traps at the scene, but neither the mice nor the shrews I caught proved able to deal with *Narceus*. I offered *Narceus* to each in experimental tests but none proved capable of subduing the millipede. All attacked *Narceus*, but they elicited discharges and were repelled by the secretion. To this day I remain ignorant of who Robespierre is. I still believe it is a rodent, but lack proof. I am also left wondering whether Robespierre knows from birth how to deal with *Narceus* or whether it acquires the ability through experience. I have visited the gorge again in recent years but failed to find evidence that Robespierre was still up to its tricks.

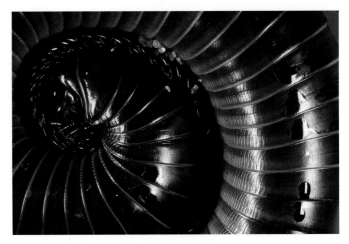

Top: A millipede coiled to defend itself, with its front end at the center. This posture ensures that the head and the first postcephalic segments, which have no defensive glands, are protected. Note that these glandless segments remain dry when secretion is oozing from the other segments. Bottom: Robespierre's depredations: Left: Headless *Narceus annularis* millipedes, victims of Robespierre's refined feeding habits. Right: The front end of one of Robespierre's victims.

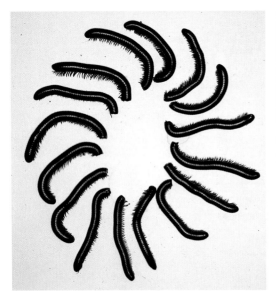

I had run into another Robespierre of sorts years earlier, in 1959, on my first trip to the Southwestern Research Station in Arizona, but on that occasion the detective work paid off and I was able to unmask the predator.

I had just made my acquaintance with the desert, and with the desert-dwelling beetles of the genus *Eleodes,* the same beetles that I was later to take to Ithaca to study in the laboratory (see Chapter 2). It wasn't long before I noticed that when these beetles were touched, they responded in a peculiar way—they assumed a headstand. They propped themselves up on their legs and raised the rear end. *Eleodes* are flightless and nocturnal.

They spend the night foraging for plant food and the day hiding, which means that at dawn and dusk they are found on the desert floor, walking about at leisure in search of food or shelter. Poke them with a finger when they are on the march, and they will halt at once and point their rears to the sky.

There is a passage in John Steinbeck's *Cannery Row* that alludes to the behavior:

> "I wonder why they got their asses up in the air for?" . . .
>
> Hazel turned one of the stink bugs over . . . and the shining black beetle strove madly with floundering legs to get upright again," . . . why do you think they do it?"
>
> "I think they're praying," said Doc.

The behavior, of course, is defensive. The animals undertake it when threatened, announcing thereby that they are ready to discharge their quinonoid spray from the abdominal tip. If further provoked, they do indeed spray, which causes them to take on a terrible stink. The glands responsible, it will be recalled, are two large sacs situated just inside the abdominal tip. Needless to say, the beetles derive advantage from the spray. But, as I was to find out, there is one enemy against which *Eleodes* has not a ghost of a chance.

There were several species of *Eleodes* at my desert site, all jet black, as is typical for the genus. The black color is part of *Eleodes*'s advertisement, since it makes the beetles extremely conspicuous against the light-colored desert soil, especially in the twilight, when they are active. Colors are barely discernible in twilight, and it is through contrast, therefore, that an animal can best achieve conspicuousness.

Eleodes were not the only black beetles at the site. Several different species also had that contrasting appearance and they were not all easy to tell apart. They differed somewhat in body shape, size, and shininess, but on the whole looked very much alike. Most belonged to the same family as *Eleodes*, the Tenebrionidae, whose members are appropriately known as darkling beetles. The family includes many desert dwellers, some from Africa, also similar in appearance to *Eleodes*.

Any insect that crawls about on the desert floor has ants to contend with, and I did experiments that showed *Eleodes* to be virtually invulnera-

The heads of kamikaze ants that died clamped onto the legs of the beetles they attacked. The beetles are, clockwise from the top left, a tenebrionid *(Eleodes longicollis)*, a dung beetle, and a blister beetle.

ble to ants. That was not surprising, given that the *Eleodes* secretion contains primarily benzoquinones. I released individual *Eleodes* near the entrance of ant colonies so they were sure to be attacked and watched as the secretion took effect. Whole swarms of ants sometimes materialized to take part in the assault, but the moment the beetle discharged, the swarms dispersed. Sometimes an individual ant clung to the beetle with its mandibles even as the beetle walked away. As I found out by experimenting at Cornell, ants that cling in this fashion may do so persistently until death, which indicates that they could have the ability, if engaged collectively in such behavior, to subdue even chemically protected prey. Such sacrificial behavior on the part of individual ants, if carried out close to the nest entrance, where masses of nestmates could rally to the assault, could pay off for the colony. With *Eleodes,* however, the ants did not cling in sufficient numbers for the kamikaze strategy to take effect. The beetles all managed to escape. I do have evidence, however, that the suicidal behavior occurs under natural circumstances. I have noted heads of ants, either bodiless or with the body still attached, clinging by their locked mandibles to one part or another of insects that I caught in the wild. To judge from the dried condition of the clinging parts, the ants had all died beforehand. Among the insects that I found to be thus encumbered was an individual *Eleodes.*

Evidence that some predator was feasting on *Eleodes* was readily at hand. Strewn over the desert floor, and distributed with some regularity, were little piles of beetle remains that were unmistakably those of *Eleodes.* They consisted typically of the wing covers (the elytra, which in *Eleodes* are fused to form a shield over the abdomen), plus the abdominal tip (still attached to this shield), and parts of the legs. Some of the remains, particularly the elytral shields, were sometimes found by themselves, an indication that they had been dispersed by the wind. The shields always bore jagged edges where they had become disconnected from the thorax, providing evidence that they had been chewed off, and implicating a rodent.

The station had a supply of rodent traps, and I set these out one night in an area where elytral remnants had been abundant. I was totally ignorant of rodents, but others at the station alerted me to the presence of grasshopper mice in the region, rodents with a ferocious appetite for insects and a terrifying name—*Onychomys torridus.* I did eventually trap grass-

Top: A grasshopper mouse (*Onychomys torridus*) eating an *Eleodes longicollis*. Bottom: Remnants of *Eleodes* found in the desert, in grasshopper mouse territory.

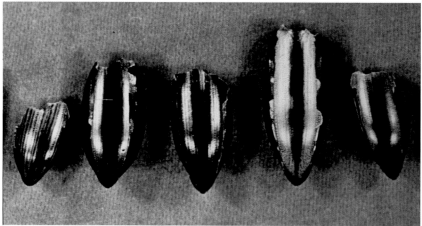

hopper mice and they not only lived up to their reputation but proved unmistakably that they were the *Eleodes* eaters I was after.

I released the mice in cages containing some desert soil and they were quick to adjust to the new premises. I then fed them a variety of insects other than *Eleodes* so they would get used to me. They eagerly consumed more than a dozen mid-sized insects per feeding session, which led me to wonder where they put it all. When I finally offered them *Eleodes,* the results were quite spectacular.

Eleodes longicollis (left) and its
mimic Moneilema appressum
(right).

The moment I introduced an *Eleodes,* the mouse grabbed it with the front feet, and while holding it upright, forced the beetle's rear into the soil. It then commenced feeding on the beetle's head, and onward into the thorax, undisturbed by the beetle's discharges, which were ineffectually ejected into the sand. I could hear the crunching sounds. Without ever releasing its grip or extricating the beetle's rear from the soil, the mouse then proceeded systematically to eat the abdomen, stopping only when it reached the abdominal tip. It seemed to know exactly what to do. When finished it retired to a corner of the cage, leaving behind a pile of remnants exactly like those I had been finding in the desert. I retrieved the elytral shield with the attached abdominal tip and by checking with a microscope was able to confirm that the mouse had stopped just short of chewing into the glands. These were almost empty, as expected, but they had not been torn open.

My camera at the time was a Leica with attached reflex housing that had kindly been lent to me by my student Ben Dane. I had the camera at the ready when I did these tests and still remember the mouse staring obligingly into the lens as I photographed it eating *Eleodes.*

I fed *Eleodes* to several *Onychomys* and the results were always the same. The attacks proceed so quickly that the *Eleodes* hardly ever had time to assume the headstand. I was fascinated by the headstand and had photographed a number of *Eleodes* in the field, including individuals of the largest species, *Eleodes longicollis,* while they were performing the trick. Predators other than grasshopper mice, I reasoned, must heed the warning implicit in the headstand or the headstand would not be such a standard feature of *Eleodes*'s behavior. The headstand could be deterrent to other rodents as well, even to skunks and birds. There was indirect evidence that beetles benefit from pointing their rears upward. I had noted at the time that among the black beetles that roamed the desert floor with *Eleodes* there were some that assumed the headstand even though they lacked defensive glands. These beetles are evidently mimics of *Eleodes* that benefit from being misdiagnosed, when headstanding, as being chemically armed like *Eleodes.* The most spectacular of these fakers is a flightless, cactus-eating beetle of the family Cerambycidae, *Moneilema appressum,* whose headstand is uncannily imitative of that of *Eleodes.* Cerambycids all have long antennae and *Moneilema* is no exception. The long antennae could blow *Moneilema's* cover, but predators may not notice it.

Formic acid.

■ **ANTS,** one would imagine, by virtue of availability alone, should be many a predator's first dietary choice. Yet ants have relatively few enemies. The reason is that most are chemically protected, and that it may take special skill to subdue them.

Take, for instance, the ants of the large and successful subfamily Formicinae, called the formicine ants. These ant species all possess the formic acid gland, a large sac in the gaster (the posterior abdominal portion of the ant's body) from which they discharge jets of fluid, either for defensive purposes or to incapacitate prey. With a formic acid content that may exceed 50 percent, the fluid is an irritant of the first order.

The individual formicine ant, as it goes about its business, is thus to be viewed as a spray gun on legs. But it is really more than that. It is also a foraging machine. The ants that one sees ordinarily scurrying about outside a nest are worker ants, members of the principal caste of the colony, whose task it is to gather food and provide protection. As a rule worker ants are sterile, having either no reproductive organs or only vestiges thereof. Instead of the organs they have the formic acid sac and a highly distensible chamber, the crop, that serves as the ant's stomach. The crop can accommodate far more food than the worker ant requires for itself. In fact, returning foragers quickly regurgitate most of their crop contents to fellow workers, which in turn share it with other workers, with the result that the incoming food is equitably shared by the colony as a whole. The worker crops, taken collectively, therefore amount to a social stomach, containing at any one time a substantial fraction of the food reserves of the colony. And none of this food is in solid form. These ants swallow only liquids; the workers have a sievelike mechanism at the entrance to the mouth that excludes all but the tiniest particles, with the result that the crop contains fluid only.

To a predator, therefore, a formicine worker is indeed an ambulatory spray gun, but one that is worth subduing because it comes with nutrients, including a potentially large load of crop fluid. What a predator that feeds on a formicine ant must do is somehow gain access to the nutrients, including the crop contents, without risking exposure to the formic acid. As it turns out, there is a predator that does just that. It is none other than the antlion, that familiar pit-building hunter sometimes called the doodlebug.

Antlions make up a separate family, the Myrmeleontidae, within the insectan order Neuroptera. The adults are nocturnal, slow-flying, and soft-bodied, and therefore potentially vulnerable to predation, but too little is known about the biology of the adults to speculate on the secrets of their survival. The "lions" are the larvae of myrmelontids, and it is they that are best known. An early account of their biology was given by August Johann Roesel von Rosenhof, one of the greatest insect illustrators of all time, whose 1774 essay on how antlions construct their funnel traps and catch their prey is noteworthy for its accuracy and charm.

Antlions are sit-and-wait predators. They typically construct funnel-shaped pits in sandy soil and lie in wait at the bottom with only the head protruding. Any small arthropod that oversteps the pit margin runs the risk of sliding down the pit's slope into the jaws of the predator. Roesel describes in detail how the antlion builds its pit by scooping up sand with its front end and flipping it away over its back, repeating the action time and again until it has fashioned a pit of sufficient depth. He also describes how the antlion resorts to this flipping behavior to shower the prey with sand as the prey struggles to avoid sliding into the pit. The antlion always orients itself in the pit in such a fashion that it can direct the flips with accu-

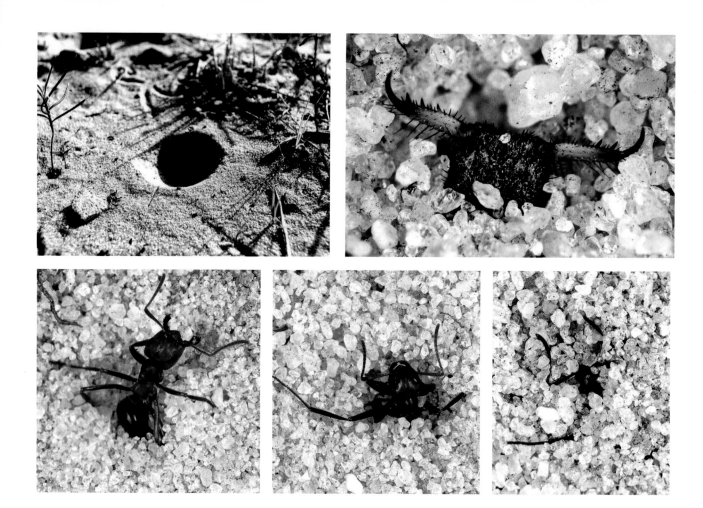

Top left: An antlion pit. Top right: The head of an antlion, lying in wait with jaws spread at the bottom of the pit. Bottom: An ant being pulled into the sand by an antlion.

racy. Pelted by sand, the prey finds itself swept down the slope and into the jaws of the larva.

There are many species of antlions worldwide but the larvae don't differ much in appearance. Stout-bodied and usually drab in color, they are easily recognized by the ferocious-looking jaws that project from their head. The jaws are curved at the tip and sharply pointed, as if fashioned for piercing, which is indeed how they are used. Insects that slide to the bottom of the pit are instantly impaled upon the jaws and slowly killed by salivary injection. The soft insides of the prey are then liquefied under action of the injected enzymes and the ensuing "soup" is sucked up by the larva through the hollow core of its jaws. Ants, without question, are the principal insects to fall victim to this strategy.

I first became interested in antlions on the trip I took cross-country

THE CIRCUMVENTERS

with Ed Wilson in 1952. We both had been collecting by a roadside in Arizona and the site was pockmarked with larval pits. The craters ranged in size from less than 1 to several centimeters in diameter, and were so close together in some places that we wondered how ants could even navigate the terrain. But the ants were there and I happened to catch sight of one just as it slid into a pit. I watched as the ant struggled to crawl up the incline, only to be swept farther and farther down the slope by the sand flipped at it by the larva. The flippings were effective not only because they hit the ant itself, but because they generated a downward avalanche that kept the ant from getting a purchase on the slope. Sand that slid to the bottom of the pit was flipped up the slope again, where it triggered further slides. The ant fighting to make its way up the slope was thus caught in a losing battle with a shifting substrate that carried it inexorably to its doom.

I experimented that day by catching ants and dropping them into pits and found that the flipping behavior is very much a part of the attack ritual of the larvae. It was uncanny to watch how accurately the larvae delivered their flippings in the direction of the struggling prey. They seemed to know exactly where the ant was on the incline. It was not that they could see the ant. Antlions have poor eyesight. It was more likely that they were getting their fix from the direction in which sand was caused to trickle down the incline by the struggling ant. I remember being able to confirm this. Using a small stick, I stirred the sand along the edge of a crater, causing some grains to slide all the way down the slope. The larva instantly positioned itself with its rear toward the incoming sand and commenced flipping in the direction of the stick. If I next stirred another area along the edge of the crater, the larva quickly reoriented itself and resumed flipping in the new direction. It was as if I was controlling a marionette. No matter where I stirred the crater the larva retaliated with accuracy.

I found out later that there was nothing new about these observations. Others had made them years earlier. There is a saying that "5 minutes in the library can save you weeks in the laboratory," which has considerable merit. I prefer the naturalist's version, which says that "weeks in the field can save you minutes in the library." But be that as it may, it does pay to check the literature before yielding to the assumption that a given observation is a genuine discovery. This exercise does not detract from the sheer joy you experience when you learn something from nature that you didn't know before. Finding out later that your "discovery" was no discovery at all in no way diminishes the pleasure of the recollection. There are countless observations that I made that turned out later to be "old hat." What I remember about these observations is the pleasure of experiencing them, not how I reacted when I discovered the published accounts. The naturalist explorer need not always be a pioneer. To discover for yourself what is already known can still be a source of wonder—which is why the study of nature can never disappoint.

The truth is, though, that what we do know about nature is dwarfed by the immensity of what we don't know. It was not until 30 years after the trip with Ed Wilson that I linked up with antlions again, and this time I found out something about them that was not already in books.

As I knew from my first trip to the Archbold Station, Florida is antlion country. There are places along the sandy lanes of the station where the craters are so dense that the terrain looks as though a Lilliputian air force had used it for bombing practice. I had also learned that antlions make good pets. All you have to do is release some larvae into a large box filled with sand. Chances are they will construct their pits overnight, from which point on you can feed them by presenting them with prey, which they will treat exactly as they do in nature. Eventually they will pupate, and when they emerge as adults they can be set free. The only condition is that they not be overcrowded. If you put too many larvae in the box you will discover eventually that some have gone missing while others have grown unduly large. Antlions have no qualms about eating each other.

It was when I put my nose to work while feeding formicine ants to captive antlions that I got the first clue that the larvae were subduing the ants without causing them to spray. The antlions quickly caught the ants as I released them into the pits, flipping sand at them and seizing them the moment they slid within reach. I could detect no odor of formic acid when this was happening, nor any later when the antlions pulled the struggling ants into the sand. I literally stuck my nose into the crater to do the sniffing, but the telltale odor of formic acid was not to be detected. Typically, when an antlion has sucked out its prey it disposes of the carcass by flipping it out of the pit. I waited, and when the carcasses were thus discarded I dissected them to check on the condition of the formic acid sac. To my amazement the sac was intact and full. The rest of the ant's insides, including the contents of the crop, had pretty well vanished.

I forsook the antlions for a number of years again, and then tackled the problem in earnest when two of my graduate students expressed interest in collaborating. Jeffrey Conner was a Harvard graduate, a talented behavioral ecologist who would eventually join the faculty of Michigan State University, a quiet self-reliant researcher, clever, articulate, and productive. The other was Ian Baldwin, a chemistry major from Dartmouth and a born naturalist, a visionary researcher now making his mark as director of the new Max Planck Institute of Chemical Ecology in Jena, Germany. Both were superb naturalists and great company.

We set up antlions in sandboxes and transferred a colony of *Camponotus floridanus,* a pugnacious local formicine ant, to the laboratory. We also had access to a good balance. We knew we would be doing a lot of weighing.

For starters we wanted to know quantitatively how much the antlion gets from an ant meal. So we put the balance to work. Here is how we described the procedure: "25 pre-weighed ants were individually fed to repletion on honey solution and re-weighed; 24 of these were then offered to pre-weighed ant lions, which were re-weighed after they fed on the ants. The discarded ant carcasses were weighed." (We don't say what happened to the twenty-fifth ant, but the fact is it escaped.)

We sat down with a calculator and played with the numbers and came up with some interesting facts. The ants on average gained about a third in body weight by gorging on honey solution and then lost about half their body weight (about 20 milligrams) through predation. The antlion gained only about 9 milligrams by feeding on the ant, so we presume that some of the ant's body fluids are lost to evaporation or through leakage during feeding.

We were able to confirm that the ant does not spray when grabbed by the antlion. We held strips of red phenolphthalein indicator paper millimeters away from the ant while it struggled in the antlion's hold, and then again when it was being sucked out by the larva, but the paper failed to turn white, as it typically does in the presence of acid vapor.

We fed some more ants, this time on honey solution that we had dyed red with commercial food coloring, and offered these to antlions, and after the meal dissected the ant carcasses to check on the condition of the crop and acid sac. We also dissected the antlions to check on the condi-

Left: The formicine ant *Camponotus floridanus*. Right: The same, showing the condition of the crop (cr), and the acid sac (ac) before and after consumption by an antlion. The crop (replete with stained sugar solution) is totally sucked out by the antlion while the acid sac survives intact.

tion of their crops. Antlions have no anus. Their gut ends blindly and consists primarily of the capacious crop. They store their wastes elsewhere in the gut and don't void them until they emerge from the pupa, a behavior that saves them the task of having to defecate in the pit where they lie in wait.

When we dissected the ant carcasses we found that the crops were reduced to a shriveled mass while the acid sac was consistently intact. The antlion crop was replete with red fluid, a sure indication that the animal had imbibed the ant's crop contents. Again we made use of the balance and could express our findings in numbers. The crop of an ant that had gorged on honey solution weighed about 10 milligrams. Crop remnants retrieved from carcasses discarded by antlions weighed little more than 1 milligram. The difference is roughly equivalent to the total weight gained by the antlion as a consequence of the meal. The ant's crop contents are thus very much a part of that meal. This is not to say that the antlion feeds only on the ant's crop. It also imbibes much of the musculature and other

THE CIRCUMVENTERS

Left: An antlion larva (Myrmeleon carolinus). Right: The head of an antlion larva, with its crop attached, showing the crop's condition before and after consumption of an ant. The red color of the replete crop indicates that the crop contents of the ant have been imbibed.

noncuticular components, but the crop's contents may be the largest single item that an antlion gains from the ant.

Inexplicable to me to this day is how the antlion manages not to pierce the acid sac of the ant. It obviously perforates and drains the ant's crop. But how it avoids puncturing the acid sac when it probes the ant's innards with its needle-sharp jaws remains a mystery. Does it sense the sac's contents through the sac's membranous wall? We did show that antlions are repelled by formic acid. If a droplet of acid is applied directly to an antlion's body while it has an ant in its grasp, the antlion releases its hold and scurries into the sand.

I think we have an explanation of why the ant does not spray when caught by the antlion. Most formicine ants do not spray unless they manage to bite at the same time. As becomes apparent when you dig up a formicine nest, the ants do not spray at random but usually only when biting. Biting seems to be their first priority. Once their mandibles have secured a grasp they are quick to bend the gaster forward beneath the body

The defensive behavior of *Camponotus floridanus*. Typically, the ant combines biting with spraying. In the top photo, it is seen bearing down with its jaws while bending its gaster forward to direct the spray toward the bitten site. At the bottom left, an ant that has been seized by forceps, but is being kept from biting, is withholding its spray. At the bottom right, an ant has sunk its jaws into a rubber cube presented to it, and has sprayed.

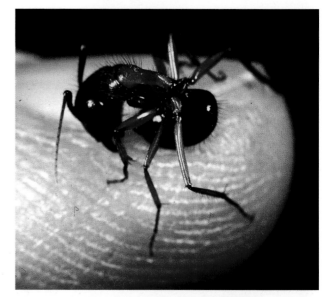

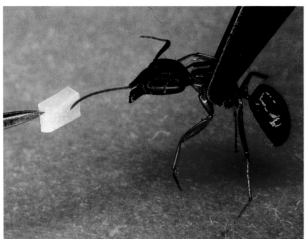

and spray. We could demonstrate this experimentally. When we grasped individual ants by the thorax with forceps in such fashion that they could not bite, and held them over indicator paper, we could tell from the failure of the paper to change color that the ants were not spraying. But if we then gave the ants a small rubber cube (cut from a rubber band) into which they could sink their mandibles, they sprayed at once. I think ants are programmed to spray into wounds or abraded integumental sites because formic acid is a slow penetrant. An open wound is vastly more sensitive to the irritant qualities of formic acid than an undamaged surface, and it is by biting and inflicting wounds that the ants ensure that their spray will take maximum effect.

We concluded that in each case the ant had refrained from spraying because it had not succeeded in biting the antlion. With its body drawn partly into the sand, and the antlion virtually buried from the outset of the assault, the ant may have had difficulty sensing the precise position of the antlion and, as a result, withholds firing. The very behavioral program that provides formicines with a positive edge in countless other hostile encounters—the ant's predisposition to combine spraying with biting—may be the cause of their undoing in their interaction with antlions.

ROBESPIERRE was a very wasteful feeder. Unable to deal with the millipede's chemical weaponry, it was forced to eat only the glandless anterior portion of the millipede and to discard the remainder of the millipede's body. A more effective strategy by far would be for a predator to consume the entire millipede after somehow inactivating the millipede's defenses. Could there be a predator able to deal with millipedes as antlions deal with ants, by killing the prey without inducing the prey to discharge?

Imagine how a predator might implement such a strategy. And imagine specifically how it might implement it with quinone-secreting millipedes. Recall that the glands of these millipedes (Chapter 2) are spherical sacs arranged linearly along the sides of the body, two per segment, each equipped with an exit duct that leads to the outside by way of a small pore. Only the first few body segments and the last lack glands.

The glands operate in a simple way. They are ordinarily leak-proof because the exit duct, just inside the gland opening, is kept occluded by a spring-like inflection of the duct wall. There is a special muscle that opens this valve. It inserts on the inflected portion of the duct and attaches to the body wall. The sacs themselves lack compressor muscles, so they must be squeezed indirectly in some way. The most likely possibility is that they are compressed by a rise in pressure of the body fluid that surrounds them. All the millipede would need to do to bring about such a pressure increase is contract its intersegmental muscles momentarily so as to cause the body segments to telescope. Discharges could be effected by coupling such telescoping with the contraction of the muscles that open the valves. Moreover, the glands are lined internally by cuticle, as is generally the case in arthropods, so that the millipede itself is ordinarily shielded from being exposed internally to the stored secretion.

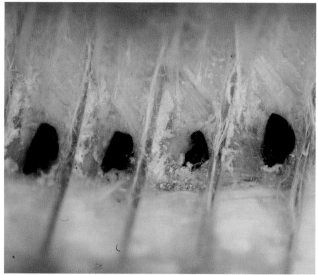

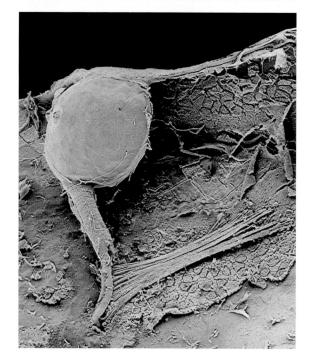

Top left: The millipede *Floridobolus penneri* coiled in its defensive posture, discharging secretion. Top right: An inside view of the millipede's body wall, showing the defensive glands, replete with dark secretion, embedded in the body musculature. Bottom: An individual defensive gland, exposed by dissecting away the surrounding body wall muscles. The gland consists of a sac, an exit duct, and an opener muscle that inserts on the inflected terminal portion of the duct. Contraction of that muscle clears the path for emission of the secretion.

For a predator to subdue a millipede, therefore, it must inactivate the prey's musculature. It must somehow knock out the millipede's body wall muscles as well as the muscles that operate the glandular valves. Doing so would trap the secretion in the glands, and all the predator would need to do next is consume the millipede's insides without piercing the glands. That seems like a tall order, but there is a group of beetles, the Phengodidae, whose larvae feed only on millipedes and deal with these in precisely that way. Phengodids are of widespread distribution, yet rarely abundant, and have therefore been largely neglected by entomologists. They are extremely interesting insects. Their larvae are luminescent, and when aglow, with their rows of lateral light organs turned on, look like miniature railroad cars. The females are flightless and lure the males with a pheromone. The males bear enormous branched antennae, presumably for detection and localization of the females. And, of course, the larvae have that unique feeding habit.

I first learned about phengodids from an interesting paper by D. L. Tiemann, a naturalist who had made it his specialty to study these insects and who provided the first detailed account of their millipede-hunting habits. The paper was accompanied by excellent color photos and it whetted my appetite. Somehow, I would have to study these larvae.

Tiemann worked with a California species, *Zarhipis integripennis,* but that did not matter. It seemed that his account would apply to other phengodid larvae as well since they all appeared to feed on millipedes. My turn to witness the larval behavior in its full course came at the Archbold Station, where the larvae of one phengodid species, *Phengodes laticollis,* are occasionally taken in pit traps. Such traps are fashioned simply by burying soup cans or other cylindrical containers in the soil, flush with the surface, so that ambulatory arthropods will tumble into them. Pit traps are particularly useful for capturing nocturnal species that might otherwise escape detection. Researchers usually set these traps in numbers and check them as a matter of routine in the early morning hours, before the heat of the sun can take its toll on the catch. Making the rounds each day to check on the night's take can be exciting, particularly when you are set on trapping something rare. The Florida phengodid larva was in that category, but with a number of us checking traps, including Mark Deyrup and one of his assistants, Paige Martin, we eventually came up with eight individuals, which turned out to be enough for our purposes.

Phengodes shares the habitat of the millipede *Floridobolus penneri,* and it seemed logical that it should prey on that millipede. I therefore confined the larvae individually in large plastic cages with soil from their habitat, and after giving them a period of days in which to adjust to the new conditions, offered individual *Floridobolus* to several of them. What followed conformed to Tiemann's script.

The larvae, which had remained quiescent in their cages during the period of adjustment, came to life when I introduced the millipedes. I have no idea how it is that they become aware of the nearby presence of their intended prey but suspect that chemical cues are responsible. Although I cannot be absolutely certain that the larvae never touched the millipede before springing into action, contact appeared not to be necessary to turn them on. Whatever the sensing mechanism, each larva began to crawl quickly once the *Floridobolus* was in its midst with the result that each eventually located the millipede. When it did, it promptly mounted it and threw a coil around the millipede's front end. The millipede coiled in response to this "embrace," but it offered no resistance, and in what appeared to be a matter of seconds, suddenly went limp. It was apparent that the larva had been prodding the millipede's neck membrane with its mouthparts, ventrally just in front of the millipede's first pair of legs, but I could not see clearly what it was achieving by that action. I judged "going limp" to be coincident with the moment of the millipede's death. Indeed, once limp, the millipede could easily be uncoiled by hand, which indicates that its intersegmental muscles had lost their tone. When alive, millipedes maintain tension in these muscles and are therefore difficult to straighten out when coiled. Tiemann thought that the larvae managed somehow to sever the millipede's nerve chord, something that could indeed account for the abrupt incapacitation of the animal.

Even though limp, the millipede did not become totally motionless immediately. For minutes it continued to show some wiggling of its legs, but that movement was feeble and uncoordinated. Most remarkable, though, was that throughout the attack there was virtually no secretory emission from the millipede's glands. Tiny quantities of secretion sometimes oozed from a few of the glandular pores, but there was never the massive emission of fluid that I knew *Floridobolus* was capable of delivering in response to assault.

The larva maintained its coil around the millipede for some minutes, during which it kept its head immovably appressed against the millipede's neck membrane. It was then so thoroughly devoted to its task that I could prod around its head with forceps without causing it to desist. I was thus able to determine that the larva actually penetrates the millipede's neck

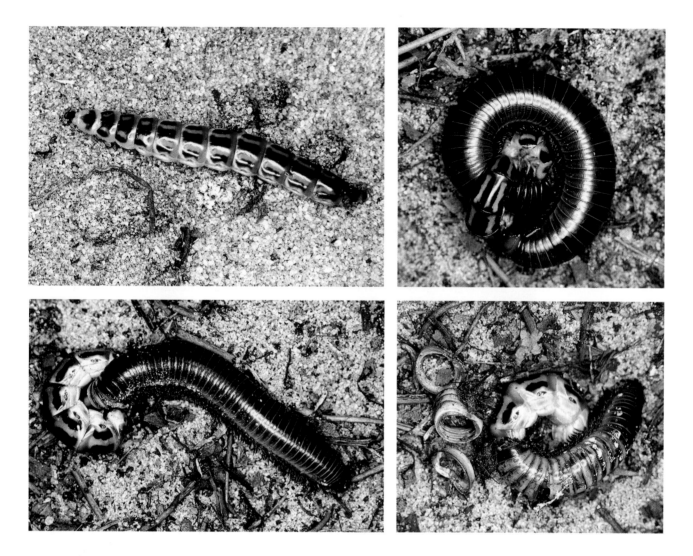

membrane with its mouthparts for the entire period that it maintains its coil.

When the larva eventually uncoils, it does an odd thing. It crawls a few centimeters away and then buries itself in the sand. It remains thus concealed for an hour or so, at which point it surfaces, crawls back to the millipede, and commences eating. Why the interruption? The only explanation I can hazard is that by digging in the larva avoids exposure to whatever quinonoid traces might be emanating from the incapacitated millipede.

The larvae consume only the innards of the millipede. They suck out the head capsule first, and then proceed posteriorly to imbibe the contents

Stages in the attack of *Phengodes laticollis* (top left) on the millipede *Floridobolus penneri.*

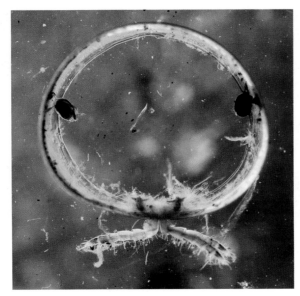

Left: The remains of a *Floridobolus penneri* segment that was sucked out by a phengodid larva. Notice the intact glands, replete with secretion, still attached to the skeletal wall of the segment. Right: A close-up view of one such gland, showing how the inflected portion of the duct keeps the secretion from leaking out.

of one segment after another. By checking on a half-eaten millipede I could determine that its uneaten inner portions were liquefied, which indicates that the larva probably predigests its meal by injecting enzymatic fluids (the millipede, incidentally, was taken back when I returned it to the larva after the experiment, and completely eaten). Consumption of a millipede can take a larva well over a day.

As the larvae imbibe the contents of the individual segments, they reduce these to skeletal rings, eventually leaving no more than a group of these rings in a disorderly array. After having fed, a larva can survive for weeks before needing to eat again. Returned to its cage, it will take up a quiescent life style until stirred into action by the presence of another millipede.

I took some of the segmental rings from the pile of leftovers, put them in a dish with water, and examined them under a microscope. There was not a trace of any soft body tissues on them. Gone were the wall of the gut, all organs ordinarily present between the gut and the body wall, and the body wall muscles themselves. Preserved intact, however, and replete with secretion, were the glandular sacs. These were reduced to their cuticular linings, all cellular components having been digested away, and it was clear that they had not been ruptured. No trace remained of the muscle ordinarily associated with the valve of the glands, and the valve itself was clamped shut in its passive state, so that the glands were essentially leakproof. I collected some of these sacs and asked Athula Attygalle to analyze

THE CIRCUMVENTERS

 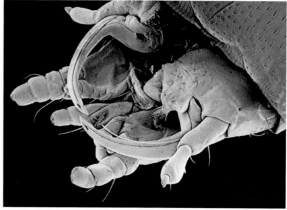

them, and it turned out that they contained full loads of the chemicals, no less than the amounts present in the glands of undisturbed *Floridobolus*. This finding confirmed what was already apparent from watching the encounters, that the phengodid makes its kill without causing the millipede to discharge.

But how does it make its kill? I dissected one millipede that I retrieved from an attacking phengodid larva immediately after the larva had thrown a coil around it and caused it to go limp, and found that the millipede's nerve cord was intact. Severance of the nerve chord, therefore, was not the means of execution.

I took a careful look at the mouthparts of the larva and found that the mandibles were ill suited for cutting. They were piercing devices. They were curved and needle sharp and had an opening at the tip that suggested they were hollow. I concluded that it was with the mandibles that the larva sucked up the millipede's body contents. But then I thought that the mandibles might serve also for the delivery of fluid and that they might be the "hypodermic needles" by which the larva administers a lethal injection to the prey. It seemed to me that there had to be some sort of venom glands associated with the mandibles and I proceeded to dissect the head of a larva to see if I could find these.

I was wrong. With Maria, I made some preparations of the larva that we examined with the scanning electron microscope and it was clear that there were no venom glands. The mandibles were indeed hollow and they opened directly into the mouth cavity so there was no question they could function for food uptake. But the idea of special mandibular venom glands had to be abandoned.

It then occurred to me that the larva, by virtue of its contained gut fluid,

The head of *Phengodes laticollis*. The hollow sickle-shaped jaws have perforations at the tip and at the base. The basal opening is visible on the right, where half the frontal plate has been dissected away.

was itself a bag of deliverable venom and that it might be by injection of that very fluid that it killed its prey. I had thought all along that the larva secured the enzymatic digestion of the prey's insides by injection of its enteric fluid, but it did no occur to me initially that it might be by use of that fluid that the larva made the millipede go limp. I dissected two larvae and drained the fluid from the midsection of their guts, where the fluid is stored, and by use of a small syringe injected the fluid through the neck membrane into individual *Floridobolus*. Two millipedes that received undiluted quantities of the fluid went instantaneously limp, as if phengodids themselves had delivered the fluid. They gave off some secretion when I handled them and prepared them for injection, but once limp they no longer emitted secretion, even when tapped or otherwise stimulated.

I concluded that the enteric fluid is both the larva's killing agent and its means for liquefying the millipede's contents. Its action is so quick that the millipede is rendered defenseless literally from the moment it is infused. I am still somewhat mystified by why the millipede fails to eject secretion during the initial phases of the assault, when the larva first throws its coil around it, or when it then probes the millipede's neck membrane before injecting fluid, since I know from experience that it does not take much disturbance to cause a *Floridobolus* to discharge. *Phengodes* evidently has some means of avoiding exposure to the prey's defense from the very outset of its assault.

I have very limited experience with other phengodid species, although it is known that phengodids do not feed exclusively on quinone-producing millipedes. That first phengodid encountered, back in Texas, it will be remembered, was in the process of consuming a hydrogen cyanide–producing polydesmoid millipede.

MOST ANIMALS feed on plants rather than other animals, so it should come as no surprise that plants are, on the whole, well protected against animal assault. Plants have spines, thorns, and countless chemical defenses, clear evidence that their evolution has been a long struggle against herbivory. The resin that oozes from a cut in a pine tree, the terpenoids that give odor to a mint plant, and the nicotine that imparts bitterness to the tobacco leaf are all manifestations of plant defenses. To any one animal, the consequence of the widespread possession of defenses by plants is that it can't eat everything. Individual animals, as a rule, feed only on a limited number of plants, in part doubtless because they have been unable to evolve the capacity to cope with more than a few plant defenses. There are of course "generalist" herbivores that feed on a broad range of plants, but most herbivores restrict their feeding to a few species. We ourselves are specialists of sorts. Think of how few plants, given the thousands that inhabit our planet, actually figure on our menus. And think of the trouble we take to avoid exposure to the noxious chemicals contained in some of the plants we do eat. We feed selectively on the leaf stems of rhubarb, and discard the leaves themselves, since it is in the leaves that the plant stores the bulk of its defensive oxalates. We peel citrus fruit to avoid the bitter alkaloids and concentrated oils placed in the rind for protective purposes. And we resort to cooking, as no other animals do, to leach out or inactivate undesirable chemicals from our plant foods. We also resort to cultivation, to selective breeding, as a means of increasing not just the yield of an edible plant product, but its acceptability. Some of the plants that we eat have been bred to be free, or relatively free, of noxious chemicals present in the wild strains from which they were derived.

Animals do not cook, nor do they engage in the selective breeding of their food plants. They do, however, sometimes engage in complex behaviors by which they manage to circumvent the defenses of the plants they eat. One example involves insects that feed on laticiferous plants, plants that produce latex for defense, a viscous, milky juice that they emit when injured.

Most of us are familiar with laticiferous plants. They include species of several families, of which the Asclepiadaceae, the common milkweeds, are the best known. Laticiferous plants are also found in the families Apocynaceae, Euphorbiaceae, Asteraceae, Caricaceae, and Moraceae. In most laticiferous plants the latex is stored under pressure in elongate cellular tubes called laticifers. These tubes follow the venational system of the leaves, the standard phloem and xylem vessels that conduct nutrients and water in the plant. When you cut into a milkweed leaf, therefore, latex appears to ooze from these veins, when in fact it is emerging from the accompanying laticifers.

It was by no means always accepted that latex is a defensive fluid in plants. One view, for instance, was that latex represented stored nutrients,

The experiment everyone can do. The dogbane leaf on the left *(Apocynum cannabinum)* has been cut in half and latex is oozing from the severed veins. The leaf on the right has been similarly transected, but after severance of the midvein. Latex oozes from the ruptured midvein, but not from the cut edge of the leaf.

white on account of contained starch. But eventually, as knowledge spread that latex is typically foul-tasting and sticky, the idea took hold that the fluid is protective. Convincing evidence was provided by an ingenious experiment carried out in 1905 by the German investigator H. Kniep. Kniep was fascinated by the fact that slugs refused to eat certain laticiferous plants of the family Euphorbiaceae, so he decided to see whether it was on account of the latex. He found that he could drain the latex from the euphorbs by repeatedly puncturing the leaves, so he proceeded to treat one group of the plants in this fashion, while keeping another group untreated as controls. To what must have been his great satisfaction, he found that while the slugs spared the controls, they left little trace of the treated counterparts.

Kniep's experiment, because of its clear-cut results, stands as a classic in the annals of chemical ecology. What Kniep did not realize, however, but would doubtless have enjoyed knowing, is that his experiment is performed as a matter of dietary routine by a number of insects that feed on milkweeds and other laticiferous plants. This was beautifully documented by David Dussourd, a graduate student in my laboratory, in studies that became a part of his doctoral thesis and that he is continuing to this day as he pursues his career on the faculty of the University of Central Arkansas.

A simple experiment, which I urge the reader to try, will illustrate how easily a laticiferous system can be inactivated. The test can be done for instance with one of the well-known milkweeds, such as *Asclepias syriaca,* or

Top: A droplet of latex oozing from a severed vein of a milkweed leaf. Bottom left: The milkweed beetle *Tetraopes tetrophthalmus* with a fresh droplet of latex applied to its mouth. Upon drying (bottom right) the droplet hardens into a glue that gums up the beetle's mandibles.

with an apocynaceous species, such as the dogbane *Apocynum cannabinum*. All that it requires is a pair of scissors. To begin, select a leaf of the plant and cut transversely across its center, so that the distal half falls off. Droplets of latex will immediately appear along the cut edge of the leaf, wherever a leaf vein, and therefore a laticifer, was severed. Now select a second leaf, and at about a third of the way from its base, cut across the midvein only. Latex will ooze in quantity from the severed vein. Now treat the second leaf as you did the first, by cutting off the tip, making sure the transection is distal to the point of severance of the midvein. The cut edge of the leaf will now remain latex-free. By cutting through the midvein and causing a localized emission of latex, you can block the plant's ability to convey latex to parts of the leaf beyond the cut, with the result that such parts are rendered defenseless. The laticiferous system is thus subject to

THE CIRCUMVENTERS

easy circumvention. It was inevitable that insects would evolve ways of exploiting this weakness.

In the northeastern United States, if you come upon a stand of the milkweed *Asclepias syriaca* in midsummer, chances are you will find some of the leaves have been eaten away at the tip, so that they appear terminally notched. Examination of the midvein of such leaves shows that they have been transected, sometimes more than once, just proximal to the notch. The insect responsible for inflicting such injury is a long-horned beetle (family Cerambycidae) that goes by the charming name of *Tetraopes tetrophthalmus*. *Tetraopes* pretreats leaves essentially as we pretreated the second leaf in our experiment, by cutting through the midvein. It usually does so repeatedly, and thereafter feeds on the part of the leaf distal to the transections. As it chews away, hardly a trace of latex emerges from where it inflicts its bites. Precutting the midvein evidently works for the beetle just as it worked for us. Latex does emerge in quantity from where it cuts the midvein, but the beetle somehow manages, probably by intermittently wiping its mouth against the leaf, to prevent the fluid from building up on its mouth. You can easily demonstrate that the secretion has the potential to incapacitate the beetle. If a droplet of latex is placed on the mouthparts of a *Tetraopes* and the beetle is tethered so it cannot clean itself, the droplet gradually hardens and becomes sticky, with the result that the beetle is literally muzzled.

Another beetle, *Labidomera clivicollis,* a member of the family Chrysomelidae, uses a technique similar to that employed by *Tetraopes tetrophthalmus,* except that it tends to cut through a series of branches of the midvein preparatory to feeding, rather than through the midvein itself. The effect is the same. The beetle induces latex emission initially, when it bites through the veins, but not afterward, when it consumes leaf tissue distal to the incisions.

Labidomera is interesting in that the larvae feed on milkweeds as well and engage in the same vein-cutting technique as the adults. Some in-

THE CIRCUMVENTERS

The chrysomelid beetle *Labidomera clivicollis* cutting veins of a milkweed leaf (top) and then feeding on the leaf distal to the cuts (bottom left). Bottom right: A larva of *Labidomera clivicollis* biting into a milkweed leaf vein in preparation for feeding on the leaf.

triguing questions remain unanswered about the larval behavior. It takes a larva a considerable amount of time to cut the veins and ready a leaf for consumption. Would it not make sense, therefore, for larvae to avoid undertaking such effort by "muscling in" on larvae that have already completed the task? I am tempted to predict that such pushiness should occur in *Labidomera* larvae, and that it might be exercised more intensely toward non-kin than toward kin. Indeed, I can envision the adult *Labidomera,* and for that matter vein cutters generally, competing for pre-

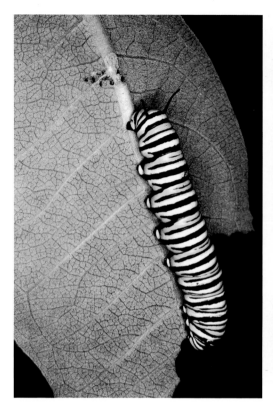

treated leaf areas, although admittedly there is so far no evidence to that effect.

Vein cutting is performed by a number of other insects, including caterpillars of several families of moths and butterflies. Even the monarch butterfly *(Danaus plexippus)* and the queen butterfly *(Danaus gilippus),* which both derive defensive steroids (cardenolides) from the milkweeds they eat as larvae, cut the veins of these plants. They are evidently adapted to tolerate and utilize the steroidal weaponry of their food plants while still needing to circumvent the plants' laticiferous system.

Not all laticiferous plants are as easily disarmed as milkweeds. In milkweeds the venational system is laid out as a midvein with linearly diverging branches, so that severance of the midvein alone suffices to block latex flow to leaf areas beyond the cut. There are laticiferous plants, however, such as *Carica papaya,* in which the leaves have a venational network laid out in loops, such that the latex can bypass severed veins by flowing through neighboring veins. Such leaves cannot be rendered edible by transection of the midvein alone. But the leaf can be disarmed if a trench

Left: A larva of the monarch butterfly, *Danaus plexippus,* feeding on a milkweed leaf after having repeatedly punctured the midvein with the mandibles. Right: Larvae of the queen butterfly, *Danaus gilippus,* feeding on leaves of a milkweed plant. The upper larva is chewing into the midvein of a leaf; the lower larva is feeding on a leaf after having severed the midvein. In both cases the severance of the midvein has caused the leaf to droop.

A geometrid caterpillar *(Trichoplusia ni)* cutting a trench (left) into a leaf of willow-leaved lettuce *(Lactuca saligna)*, and then (right) feeding distal to the trench.

is cut clear across it, so that the veins are severed over the entire width of the blade. David Dussourd showed that there is a caterpillar, *Erinnyis alope,* that feeds on *C. papaya* and cuts just such trenches into the leaves, as a way to block latex flow to the feeding sites. Similar feeding behavior is carried out by the caterpillar of *Trichoplusia ni,* a typical inchworm, on willow-leaved lettuce *(Lactuca saligna).*

The interaction of insects and their food plants is extraordinarily complex and provides countless examples of evolutionary adjustment of insect to plant, and of plant to insect—of insect-plant coevolution. Insect herbivory forced the evolution of plant defenses, and plant defenses in turn forced the evolution of insectan countermeasures. The presence of laticiferous systems in plants and the capacity of insects to circumvent these systems are both the result of coevolutionary interaction. The further occurrence of looping laticiferous networks, and of trenching behavior for dealing with these, provides evidence that coevolution in some of these cases has progressed beyond the initial reciprocal adjustments.

Plant defenses are not always visually apparent, like laticiferous sys-

THE CIRCUMVENTERS

tems. Most plant defenses are covert, in the sense that they consist of toxic or distasteful factors that the plants maintain invisibly hidden in their tissues. To counter such defenses animals resort to biochemical processes that are themselves covert. By use of special enzymes, animals transform noxious compounds into derivatives that are either harmless or more readily excretable. One group of such enzymes, the P450 enzymes, has been particularly well studied. Many insects that feed on poisonous plants are richly endowed with P450 enzymes, upon which they depend to counter the plants' poisons. The P450 enzymes have also given insects an edge in their interaction with humans. It is in part due to their possession of P450 enzymes, and the many ways they put these to use, that insects are so difficult to control by chemical means.

In providing examples of how animals cope with the defenses of organisms they eat, I deliberately picked cases where coping meant having the means, or the behavioral capacity, to circumvent the defenses. There is another option, open to animals that are able to tolerate noxious chemicals. Such animals, having no need to circumvent or to inactivate the substances, may instead put them to their own use. Example after example is coming to light of insects that in some ingenious way manage to appropriate exogenous chemicals, or materials, for one purpose or another. Insects are particularly apt to use acquired substances for defense, since defenses are "expensive" in the sense that they often require high concentrations of chemical to be effective. But they may use acquired substances for other purposes as well. In fact, insects are extremely adept in reaching out to nature for what they need. What it is that they use from their surroundings, and how they put it to use, is the topic of the next chapter.

8

The Opportunists

I know that wasps don't carry flowers around, so when one day in Arizona I saw what I took to be an *Ammophila* with a flower in its grasp I was baffled. *Ammophila* are the thread-wasted wasps (subfamily Sphecinae of the family Sphecidae) and I knew them to be caterpillar hunters. They paralyze caterpillars with a sting and carry them to their burrows, where the caterpillars are fed upon by the wasps' larvae. I had photographed *Ammophila* in the process of carrying paralyzed prey and also had noted how they "slept" at night, clamped to vegetation by their mandibles. What could an *Ammophila* possibly do with a flower? I grabbed my camera and photographed the wasp as it was on the ground (see the facing page), homing in on its burrow. In close-up view, through the viewfinder, I realized my mistake. The "flower" was not a flower at all, but a caterpillar masquerading as a blossom. I had made my first acquaintance with *Synchlora,* a caterpillar with the remarkable habit of decorating itself with real petals.

Others had studied *Synchlora* so the habit was known, but to me the observation was new. I decided I had to find that caterpillar. It was probably dressing up for camouflage and would be found on flowers, but I had no idea on which ones. I ended up spending hours scrutinizing every yellow flower in the area and the search paid off. I came upon several *Synchlora* larvae, each on a floral head, adorned with pieces of petals and well concealed. Not the kind of dietary item I would want to depend on, I thought. Too difficult to find.

Eventually I learned that there are a number of *Synchlora* species, all with the same larval habit. I also learned that they occur not only in Arizona, the site of that first encounter, but in Florida and on my own four acres in Ithaca. I have become rather adept at finding them. I make it a habit of taking them indoors, clipping the floral branches on which they are found, and keeping these fresh in water. I can thus watch them "get dressed" and follow their development into pupae and adults. Over the years I have kept a photographic record of their habits.

Synchlora larvae have a predilection for flowers of the aster family (Asteraceae). They are found on disk flowers, such as fleabane *(Erigeron philadelphicus),* Spanish needles *(Bidens pilosa),* and black-eyed Susans *(Rudbeckia hirta),* but are best camouflaged on more irregular inflores-

Left: *Ammophila* "asleep," clinging to vegetation with its mandibles. Right: *Ammophila* in the process of transporting a freshly paralyzed caterpillar to its burrow.

cences such as those of goldenrod (*Solidago* species) or camphorweed (*Heterotheca subaxillaris*). On *Solidago* and *Heterotheca* especially, thanks to their petal covering, they may be almost impossible to detect. The petal pieces can be pulled off with forceps and the larvae then look like ordinary inchworms (family Geometridae), which is what they are. They don't stay naked long. Within minutes they begin fashioning a new cover, which they do quite systematically by cutting off pieces of petal with the mandibles, and fastening these onto tiny clusters of spines on their backs, using strands produced by their silk glands. It takes time for them to dress up, but they give the activity priority. They usually do not resume feeding until they are fully cloaked. Since they don't have miniature vases on their back, the petal pieces wilt, but they add new pieces periodically, and thereby keep up appearances. When they molt, they shed the entire investiture with the skin, but they quickly undertake the task of dressing up again.

When the time comes to pupate, the larvae crawl to a site away from the flower to construct a cocoon. My laboratory specimens usually pupated on the floral stalk itself. They fashioned a lose silken enclosure for the purpose, using strands derived from the same glands that supply the threads used to attach the petal pieces in the larva. They retain these pieces when they pupate, weaving them into the fabric of the cocoon. Depending on where it is located, the cocoon can itself be well camou-

Synchlora larvae. Top: A fully "dressed" larva (arrow) on a flower. Bottom left: A naked larva from which the petal cover has been plucked away. Bottom right: A larva with a partially rebuilt petal cover.

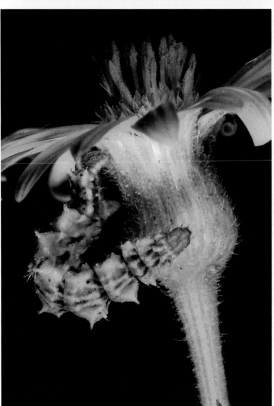

Top left: A mature *Synchlora* larva in search of a pupation site. Note the tiny hitchhiker (a meloid beetle larva). Top right: A *Synchlora* cocoon. Bottom: A chalcidoid wasp emerging from the *Synchlora* larva it had parasitized.

An adult *Synchlora*.

flaged. The adult that eventually emerges is a beautiful green moth, itself
capable of escaping detection in a world where green is the dominant
color.

I don't know who is fooled by *Synchlora*'s larval disguise, but would
imagine that birds and lizards are prime candidates. As usual, the defense
is not impregnable. *Ammophila* is evidently not taken in by the camouflage, and neither are certain parasitoids. I have had chalcidoid wasps
emerge from larvae in the laboratory and could imagine these parasitoids
being major natural enemies of *Synchlora*. I was amazed to find that
among those fooled could be the tiny larvae (triangulins) of blister beetles,
which are often found on flowers. In one of my photos one such trian-
gulin is seen perched on *Synchlora*'s petal packet, seemingly at home in
what is obviously not its home.

Floral imitators are rare among insects, but *Synchlora* is not unique.
The Malaysian praying mantid, *Hymenopus coronatus,* for instance, which
sits in ambush on pink orchid flowers, is stunningly built to match the
flower in both structure and coloration. Similarly, the African membracid
Etyraea nigrocincta imitates flowers by aggregating in clusters at the end of
branches. But *Synchlora* is definitely unusual in that it achieves its floral
image by using actual floral parts.

Many organisms, like the Malaysian mantid, derive protection by blending in with a specific plant background. Evolution has forged the imita-

The caterpillar of the geometrid moth *Nemoria outina* occurs in two forms, a twig-mimicking form (left) and a leaf-mimicking form (right).

tion, and the organisms may be rigorously programmed to take advantage of the resemblance. The bark mimic *Brochymena quadripustulata*, when at rest in the daytime, is typically found on tree trunks, while the moth *Schinia gloriosa* spends the day snuggled amidst the flowers of *Palafoxia feayi*, whose pink coloration it matches exactly. Over the years I have kept a photographic record of such imitators, which may be quite difficult to spot. The mantid *Gonatista grisea*, when lurking in ambush beside the lichens it imitates, is extremely easy to overlook. So are the caterpillars of *Nemoria outina*, which feed exclusively on *Ceratiola ericoides* (family Empetraceae), and which occur in two forms, a summer form that imitates the leaves and a winter form that imitates the branches.

The most extraordinary case known to me of imitative dimorphism is that of a caterpillar discovered by Erik Greene, at the time a graduate stu-

---

Facing page: Camouflage. Clockwise from the top left: A praying mantis (*Gonatista grisea*) among paramelioid lichens; a moth *(Schinia gloriosa)* on flowers of *Palafoxia feayi;* an unidentified caterpillar on the disk of a flower; and a stink bug (*Brochymena quadripustulata*) on a tree trunk.

dent at Princeton, now on the faculty of the University of Wyoming. Greene found that in the geometrid moth *Nemoria arizonaria,* the caterpillars resemble either the catkins or the twigs of the oak tree they inhabit. Remarkably, whether they develop into one morph or the other depends on what they eat. When they feed on catkins they end up taking on the appearance of catkins, and when they feed on leaves they end up looking like twigs.

■ **SPINES** are a common feature of plants, meant to keep animals at bay. But they also provide opportunities. Animals may seek temporary safety among the spines of a plant, or they may go so far as to establish residence on thorny plants. The cactus wren *(Campylorhynchus lorunneicapillus),* for instance, builds its nest on cacti, thereby securing protection for both itself and its brood. I have a particularly fond memory of an *Anolis carolinensis* lizard that I encountered regularly over a period of days on my exploratory walks at the Archbold Station. I would find it basking in the sun during the early morning hours, always on the same *Opuntia* cactus, shielded by spines. *Anolis* do not as a matter of routine rest on *Opuntia.* But that particular individual appeared to have developed the habit of doing so, to what I would imagine was considerable protective advantage.

At the Archbold Station I made the acquaintance of a squash bug, *Chelinidea vittiger* (family Coreidae), that literally spends its life on *Opuntia.* The adults have wings so they must be in the air at times, as when they are dispersing or in search of a mate, but the bugs feed on *Opuntia* and are present in all developmental stages on the cactus. It is virtually impossible to pick *Chelinidea* off a cactus pad by hand. The spines get in the way. The female even lays her eggs on the spines, and it is not uncommon to find individuals holding on to spines while molting. The adults themselves, while resting, may choose to cling to a spine. I haven't had occasion to determine whether *Chelinidea* inflicts much damage on its host. But heavily infested *Opuntia* pads have a senescent appearance and bear signs of having been partially sucked out by the bugs.

Recently, in Arizona, Maria and I became interested in a spiny plant that is a veritable insect killer. We thought initially that the plant had no

The nest of a cactus wren *(Campylorhynchus lorunneicapillus)* on a cholla cactus.

An *Anolis carolinensis* lizard resting amidst spines of *Opuntia* cactus.

insect enemy, but then found that it did, and that this enemy actually benefits from the plant's possession of spines.

The plant, *Mentzelia pumila,* a member of the daisy family (Asteraceae), is a multibranched herb, which we found growing in the desert in Arizona, in our usual haunts, near Portal. What first drew our attention to the plant is that it seemed invariably to have numbers of dead insects stuck to its surface. The insects were fastened to leaves and stems, and came in all sizes and kinds. What was holding them in place were the tiny surface elaborations called trichomes, structures that are commonly present in plants in the form of minute hairs or spines, but that in *Mentzelia* are of a particularly lethal design. *Mentzelia*'s trichomes are barely visible to the naked eye, but at higher magnification stand revealed as the devilish little

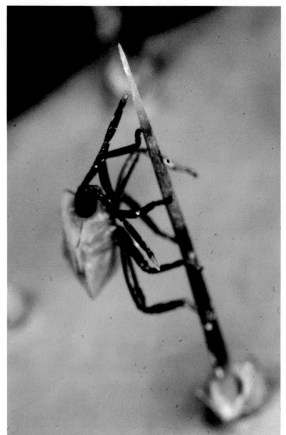

Left: A squash bug *(Chelinidea vittiger)* on *Opuntia* cactus. The bugs often rest, and lay their eggs, on the host's spines. Right: The eggs of the bug hatching.

piercing and trapping devices that they are. The trichomes come in three forms. One type, the most numerous, bears a characteristic crown of re-curved barbs at the tip, as well as occasional barbs along the shaft. Another is a somewhat stouter, more densely barbed version of the first, and the third—which we dubbed the Christmas-tree type—is tapered, barbed, and sharply pointed. The trichomes are rigid in consistency and not easily broken off. They give the entire plant a sand paper–like feel.

The trapped insects included beetles of various families, as well as moths, mayflies, ants, bees, flies, leafhoppers, and others. Most had probably become stuck by chance since they appeared to have no behavioral ties to the plant. They might simply have made a casual landing upon the plant or collided with it in flight. Maria and I took counts of the number of insects entrapped per plant, and for a sample of 12 plants came up with an average figure of 72. The count did not include ladybird beetles, which—for reasons that will become apparent—we tallied separately.

The plant *Mentzelia pumila* and its trichomes. Top left: The entire plant. Top right: The edge of a leaf. Bottom: The three primary types of trichomes.

Insects trapped on the surface of *Mentzelia pumila* plants. Clockwise from the top left: a bee, a leaf-hopper, a fly, and a mayfly.

We took some leaves bearing freshly caught insects and with the microscope watched how these struggled against the restraining action of the trichomes. Just a few trichomes sufficed to do the job. They hooked onto the insects' legs, mouthparts, antennae, and wings, often piercing the integument and causing the insects to bleed. Only very rarely did an insect that was well stuck succeed in freeing itself.

We had a hard time imagining how a plant with such a nasty integument could have any insect enemies, but we thought we would check, and when we did found that *Mentzelia* was quite consistently infected with an aphid. *Macrosyphum mentzeliae* was its name and it was known only from

Mentzelia plants. It was evidently a specialist, fashioned by evolution to be able to cope with Mentzelia.

We looked at Macrosyphum carefully, expecting it to be "Teflon-footed" or in some other fancy way protected against entrapment, but instead found that it had no obvious physical adaptation for coping with Mentzelia. Its legs had very slender tips, which fit neatly between the trichomes, and it moved with extreme caution on the rare occasion that it did any walking. But aside from that there was nothing about its structure or behavior that was anomalous. Macrosyphum was atypical only in that it grazed, so to speak, on "turf" that was out of bounds to others.

But the relationship of Macrosyphum to Mentzelia turned out to be more

Top: Mentzelia's enemy, the aphid Macrosyphum mentzeliae. Note the aphid's slender legs. Bottom: Dead coccinellid beetles trapped on Mentzelia. As enemies of Macrosyphum, the coccinellids are potentially beneficial to the plant. Their death reduces the predation pressure on the aphids, to the detriment of the plant.

A tarflower (*Befaria racemosa*). The outer surface of the bud is protected by a viscous glue, which oozes from the margins of the calyx. The glue, which can be drawn into strands, clings tenaciously to solid surfaces. Insects by the hundreds, including sizable ones, get stuck on the plant.

complex. The aphid not only avoids the trichomes but benefits from them. Aphids are ordinarily heavily preyed upon by ladybird beetles, which in a sense are the plant's allies. On *Mentzelia,* however, the ladybirds tend to get stuck, with the result that *Macrosyphum* is spared the full brunt of their assault. It is as if the aphid puts its host's defense to use.

The principal ladybird beetle we found on *Mentzelia* was *Hippodamia convergens.* Many of these beetles, including both larvae and adults, were trapped on the plant and were already dead, but others were free and walking about. The adults did occasionally make it to aphid colonies, and they also laid eggs, but these too were sometimes pierced and killed by the trichomes. Larvae here and there also found their way to aphid colonies, but

THE OPPORTUNISTS

Top left: The chrysopid larva *Ceraeochrysa lineaticornis* walking on a sycamore leaf, shielded by its packet of sycamore leaf trichomes. Top right: The packet is protective, as against the predaceous reduviid bug shown here. The reduviid attempts to impale the larva on its proboscis, but if the packet is thick enough, it may be thwarted. Bottom: The chrysopid builds the packet by scooping up trichomes with its jaws, and piling load upon load on its back.

The tingid *Corythucha confraterna*, a pest on sycamore trees, and the favored prey of *Ceraeochrysa lineaticornis*.

the fraction that didn't, of both larvae and adults, was substantial. Eighteen of the 100 adult beetles that we counted on 23 *Mentzelia* plants were stuck. *Macrosyphum* is thus a "winner" both for being unaffected by the trichomes and for deriving indirect benefit from them. For *Mentzelia*, vis-à-vis one type of enemy, the defense, to a degree, has backfired.

And what about the many insects that are incidentally killed by the plant? Could *Mentzelia* derive benefit from the killing? Insects trapped on the plant, once dead, inevitably must decay, and I could imagine the nitrogenous products of such decay leaching into the ground and becoming available to the plant as fertilizer. The nutritional supplement might be meager but perhaps under poor soil conditions not insignificant. There is a plant, endemic to the coastal plain of Florida and Georgia, *Befaria racemosa*, a member of the blueberry family (Ericaceae), that is also lethal to insects and that could also derive nutrients from its victims. The buds, the abaxial (rear), surface of the petals, and the developing fruit of the

THE OPPORTUNISTS

plant are coated with a viscous adhesive in which insects, including robust species such as honey bees, become trapped in large numbers. The glue probably serves primarily to protect *Befaria* against herbivores and nectar thieves, but it could also enable the plant to be an occasional insectivore.

In Arizona, back in the sixties, on sycamore trees growing in Cave Creek Canyon, near Portal, I discovered a chrysopid larva of the trash-carrying variety (see Chapter 4) that constructs its packet out of trichomes. The species, *Ceraeochrysa lineaticornis,* was previously known, but only as an adult. Sycamore leaves are densely covered with trichomes, particularly on the underside. The trichomes are not of the lethal kind. They take the form of multibranched hairs, hinged at the base, which as anti-insectan structures could act primarily as impediments to locomotion and feeding. They could also serve to trap a layer of "dead air" directly adjacent to the leaf surface, providing thereby for retardation of evaporative water loss from the leaves. The chrysopid, it turns out, plucks the trichomes from the leaves, tearing them off at the hinge, and loading them upon its back. The finished packet eventually shields it from view so that when it walks about it looks like a little wad of cotton. The packet protects the larva against ants and reduviid bugs. Constructing the packet takes time and requires the equivalent of two leaf-undersides' worth of trichomes. By seizing the sycamore's trichomes, the larva is essentially usurping the plant's defense, but it does not practice the thievery without reciprocation. The larva, like all chrysopid larvae, is insectivorous, and among its prey is a tingid (family Tingidae of the order Hemiptera) that is one of the sycamore's main enemies. The larva, therefore, pays for the plant's services. In exchange for free trichomes it provides pest control.

■ **GRASSHOPPERS** have a nasty habit. If you pick them up, they vomit. It hardly ever fails to happen. They may spit up a mere droplet at first, but if you handle them for any length of time they often bring forth enough fluid to mess up your entire hand. In tobacco-chewing country kids call the fluid tobacco juice.

Years ago at the Archbold Station I did some experiments with Fotis Kafatos to test the hypothesis that by vomiting grasshoppers are putting to defensive use repellent chemicals that they obtain from their diet. We knew that plants contained the most diverse kinds of secondary metabolites, distasteful, malodorous, and poisonous substances by which they se-

The lubber grasshopper *Romalea guttata* regurgitating.

cured protection, and we knew that this chemical arsenal was not fail-safe. There were always herbivores that breached the defenses, animals that, having evolved the ability to tolerate a plant's protective chemicals, had taken up the habit of feeding on the plant. Why couldn't herbivores have taken matters one step further? If they are tolerant of a plant's noxious metabolites could they not themselves use such metabolites for defense? And isn't that exactly what grasshoppers might be doing?

If we were correct, the grasshopper's oral effluent should be not only offensive but variably offensive to predators, depending on the plants eaten. *Romalea guttata*, the large lubber grasshopper, was available to us in Florida, and we decided we'd experiment with it. We caused individual

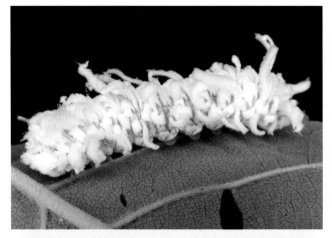

Sawfly larvae. Top left: The wax-covered *Eriocampa ovata*. Top right: The slime-covered *Caliroa cerasi*. Bottom: A gland-possessing *Nematus* species larva; *Nematus* everts its serial abdominal glands when disturbed, as shown, while simultaneously arching its rear upward.

Romalea to vomit by handling them and picked up the fluid with micropipettes, then added small quantities of the liquid to dishes with sugar solution that we presented to ants. The ants wouldn't have anything to do with the samples bearing the additive, although they eagerly guzzled up pure sugar solution that we offered as a control. This experiment gave us proof that the regurgitated fluid has a protective capacity. We then took two groups of *Romalea* and fed one group on a natural food plant, *Eupatorium capillifolium,* and the other on commercial iceberg lettuce. We had tested extracts of both plants as additives to sugar solution and found that only the *Eupatorium* extract was noxious to the ants. Accordingly, we predicted that the oral effluent from the *Eupatorium*-fed *Romalea* should deter the ants, which turned out to be the case. We also predicted that the vomit from lettuce-fed *Romalea* would be innocuous, and that too proved true. It did indeed matter whether the grasshopper ate noxious plants, and it did appear that, by vomiting, it was reusing ingested noxious chemicals. *Eupatorium,* unlike lettuce, is laden with defensive metabolites, including pyrrolizidine alkaloids, so the findings made sense. But since we never got around to doing actual analyses of the vomited fluid, we decided not to publish the results. Some years later, however, we looked into the role of defensive vomiting in an entirely different insect, a conifer-feeding sawfly larva, and then we did get chemical data.

Sawflies are members of the Hymenoptera, the insect order that also includes the bees, wasps, and ants. They are primitive Hymenoptera. The adults are relatively vulnerable in that they are solitary and do not sting. The larvae, on the other hand, often have means of defense. Some bear an integumental covering of dense waxy "fluff," others have a bodily covering of slime, and still others have eversible abdominal glands that they "present" to the enemy by special body posturing. The sawfly larvae that caught my interest were conifer feeders that all had the habit of responding to disturbance by vomiting. It seemed to me that in one species, *Neodiprion sertifer,* which was readily available to me in Ithaca, the fluid regurgitated by the larva was the actual resin of the food plant. If so, and if the larva truly had a way of separating the resin from the ingested plant tissue and regurgitating the resin only, it would be very interesting. *Neodiprion's* vomit certainly had the sticky consistency and typical fragrance of conifer resin.

Neodiprion sertifer is an introduced species in the United States, as is its preferred host, the Scotch pine *(Pinus sylvestris).* Its larvae are gregarious and feed routinely on the needles of the tree. They show a characteristic startle response when disturbed, involving typically the rearing of the front end and the appearance at the mouth of the resin-like fluid. If pinched with forceps, they revolve the front end and attempt to deposit the droplet directly on the forceps. The response is accurate and quick and the larva can reach any part of its back or flanks with the mouth. Other investigators had shown the oral fluid to be a deterrent to birds and parasitoids. Simple tests that I did with two of my graduate students, Judy Johnessee and Jim Carrel, showed that the fluid is protective against ants and wolf spiders. In both these arthropod predators, direct contamination with the fluid evoked prolonged cleansing activities.

Lawrence B. Hendry, a student of Jerry's, became our collaborating

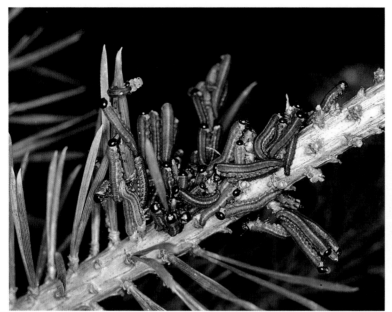

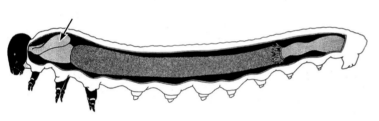

The sawfly *Neodiprion sertifer*. Top left: A larval aggregation on Scotch pine. Top right: A larva responding to disturbance by arching its front and rear upward while regurgitating a drop of resin. Bottom left: Diagram of the digestive tract of the larva. The dark middle section is the midgut. The foregut, which is lined with cuticle, is the light-colored anterior portion of the gut. The foregut bears the esophageal pouches in which the resin is stored. The left pouch is denoted by an arrow. Bottom right: A larva, under attack by an ant, revolving its front end while depositing a droplet of resin directly on the attacker.

chemist. We looked into the composition of the oral fluid and it was indeed, for all intents and purposes, pine resin. It contained the characteristic carboxylic acids (resin acids) responsible for the resin's sticky consistency, and the simple isoprenoids, such as α-pinene and β-pinene, that give resin its odor. The composition of the fluid appeared to indicate that the larvae ate both needle resin and branch resin. The two resins differ

The isoprenoids α-pinene and β-pinene (left and middle) and abietic acid (right).

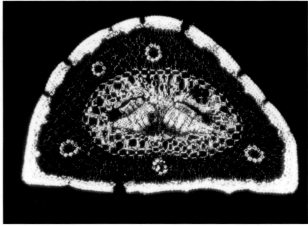

Left: A transected pine needle, oozing resin from the severed resin canals. Right: Cross-section of a pine needle, showing the five peripheral resin canals.

somewhat in composition and the larvae appeared to regurgitate a mixture of both. We paid close attention to the larvae's feeding habits and noted that they did indeed eat not only the shaft of the needles, which contain needle resin, but the needles' basal fascicled portions, which are penetrated by branch resin canals and are therefore laden with branch resin.

We dissected larvae and found that the resin is stored in two pouches connected to the esophagus. The pouches are quite large, and could easily be dissected out and extracted with solvents. All the compounds that we found in needle and branch resin were present in the pouches as well. The pouches were also notably free of solids, which indicated that only the resin, and none of the accompanying plant matter, had been diverted into them. Analysis of the plant matter retrieved from the digestive tract itself, and of the fecal pellets, revealed no trace of pinenes or resin acids. Whatever the mechanism that shunts the ingested resin into the pouches, it must operate with great efficiency.

The pouches are part of the foregut of the larva, a portion of the digestive tract that arises in the embryo by invagination of the body wall and

THE OPPORTUNISTS

therefore has cuticle-secreting capacity. Indeed, the foregut is lined with a cuticular membrane in all insects, and for *Neodiprion* this is of special importance. It is thanks to the foregut lining, which extends into the pouches as well, that *Neodiprion* is able to cope with the resin. It never really swallows the fluid. It sequesters it from the ingested matter and stores it in the insulated confines of the pouches, without ever exposing its own potentially sensitive tissues to the hazardous liquid. The larva mobilizes the stored resin for defensive purposes only. Starved individuals may have totally depleted guts but their pouch contents remain undiminished.

Neodiprion's opportunistic use of resin extends beyond the larval stage. In midsummer the larvae crawl to the ground to spin their cocoons. Within the enclosure they retain the larval condition for several weeks without pupating, and during that time respond to tapping or scraping of the cocoon by revolving the body so as to bring the mouth toward the site of the disturbance, and by regurgitating if the cocoon is pierced. Only when they shed the last larval skin do they finally relinquish the resin. The fluid is then neatly extricated from the body, packaged within the shed cuticular lining of the foregut.

In Australia there are other sawflies that have evolved a comparable defensive strategy using eucalyptus oil. They have only one esophageal pouch, which indicates that they may have evolved their defense independently from *Neodiprion* and its relatives, but they use the plant oil in much the same way as the conifer feeders use resin. As they ingest the eucalyptus leaves they shunt the leaf oil into the pouch, and they vomit the oil when disturbed. The larvae are called spitfires in Australia, and the term is appropriate. The oil has the characteristic odor of Vick's VapoRub, and although it deters predators, I wouldn't be surprised if it could be used as a decongestive inhalant. The partitioning of ingested oil into the oil sac occurs with extraordinary efficiency in the Australian sawflies. Gas chromatograms of sac contents and of eucalyptus leaf oil are virtually superimposable.

The Australian sawflies, which I studied with two collaborators, Thomas Bellas and Patrice Morrow, during my 1972–73 Australian stay, are gregarious as larvae, particularly in the daytime when they are not feeding. The members of a cluster may respond collectively by regurgitating when disturbed, and in some species the aggregated individuals show

Australian sawflies. Left: A cluster of recently emerged *Pseudoperga* species larvae, guarded by the mother. The mother remains in attendance for as long as it takes the larvae to build up defensive oil reserves. Right: Aggregated larvae of *Pseudoperga guerini,* responding collectively by regurgitating droplets of eucalyptus oil.

Left: A mature larva of *Perga affinis* beside a replete esophageal sac. Right: A gas chromatogram of a eucalyptus leaf oil extract (top) and of an extract of oral fluid from a sawfly that fed on that eucalyptus tree. The concordance of peaks of the two chromatograms is virtually absolute, which indicates that the samples are of the same chemical composition.

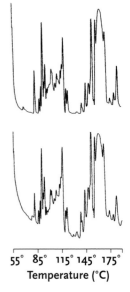

55° 85° 115° 145° 175°
Temperature (°C)

a synchronous tail-flicking behavior that results in the regurgitant being smeared over their own backs. It takes the newly hatched larvae several days to fill the esophageal pouch with oil. In some species the mother remains with the offspring during this early developmental period, providing protection by biting and leg-kicking when the brood is molested. The

THE OPPORTUNISTS

oil sacs in the Australian sawflies may be enormous. In fully grown larvae of one species, *Perga affinis,* we found the sac accounted for nearly 20 percent of the body weight.

The Australian sawflies also put their stored oil to use when they spin their cocoons, but not in the same way as *Neodiprion.* They incorporate the oil, together with feces, into the silken wall of the cocoon, with the result that the pupae are chemically shielded. Initially, the cocoon reeks of eucalyptus oil. Later, as the cocoon ages, the oil provides additional protection by hardening.

■ ONE OF THE EARLIEST ACCOUNTS—if not *the* earliest account—of an insect making use of plant resin is that by the eighteenth-century Swedish entomologist Karl De Geer. De Geer is best remembered for his many insect descriptions. A casual glance at Mortimer Demarest Leonard's 1928 book, *A List of the Insects of New York,* shows De Geer's name appended to that of a number of cockroaches, katydids, and crickets, many described originally by De Geer in his seven-volume treatise on insects, published in 1776. In these volumes, which contain mostly descriptions and commentary, De Geer also provides accounts of actual experiments, which were rarely done in his time. The experiments were often spur of the moment in design, but they are presented lucidly, conceived rationally, and were carefully controlled. In volume 1 of the set he provides a delightful account of a small moth that as a larva lives on pine, within a gall made of resin from the host. The moth is almost certainly *Petrova* (formerly *Evitria*) *resinella,* a member of the Tortricidae.

De Geer describes in some detail how the larva feeds by carving out a groove along the length of a small branch of its host, and how the surrounding gall, which takes the form of an igloo-like enclosure, is formed by the resin that oozes from the plant wound. He notes that the larva lives protectively confined within the gall and that it encrusts the enclosure from within with some of its feces. Not surprisingly, given that the difference between the vascular and resin systems of plants had not been clarified at the time, he errs in his interpretation of the nature of resin, which he takes to be part of the plant's "nutrient juices." He does, however, evince awareness of the special chemical peculiarities of resin, and it is this awareness that prompts him to wonder about the insect's seeming

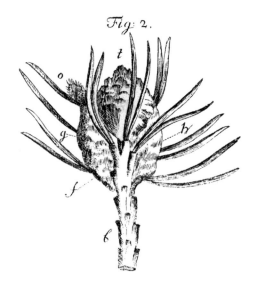

De Geer's drawing of the larval gall of *Petrova resinella* above photographs of the intact and opened gall. Notice the larva in the opened gall.

ability to tolerate the substance. Conifer resin, he states, like the odorous turpentine oil derived from it, is known to be ill tolerated by insects. How, then, can *Petrova* withstand living so close to resin?

As a start, he sought reassurance that the gall was constituted of resin. He found that it was soluble in alcohol, smelled of turpentine oil, and emitted upon ignition the odor of burned mastic, a well-known resin product, commonly used as a varnish ingredient.

Then he experimented. He took a *Petrova* larva and placed it on a piece of paper soaked in turpentine oil. The larva crawled about and became visibly wetted with oil, but remained alive. A control arctiid caterpillar, similarly placed, promptly became restless and had to be herded by hand to be kept on the paper. It quickly weakened and within 4 minutes "gave up the ghost."

A further three *Petrova* larvae, placed in a closed jar with pieces of turpentine-soaked paper, fared well initially, but drowned overnight in excess oil. Proper precautions to prevent drowning showed yet another larva to emerge alive from 2 days of such confinement. Prolonged tolerance also was shown by a larva that was placed in a jar while still within the hollow of its partly opened gall, and therefore exposed to the vapors of the oil only. Two "smooth green caterpillars" confined as controls died in 2 minutes; a fly lasted only half an hour.

De Geer concludes succinctly that neither turpentine vapors nor turpentine oil is fatal to *Petrova*. The larva, he reasons, must be "structured differently" from other insects, in some fashion that he presumes to involve the respiratory system. He expresses regret that the larva's "respiratory orifices and tubes" are too small for him to examine in detail. He does not elaborate beyond this point, nor does he look for parallels with other insects. But he does raise the notion of food specialization and of adaptive tolerance by an herbivore of the noxious chemistry of its host, and by so doing brings up a topic that was not to be addressed again for a century. Although De Geer is remembered mainly as a describer of insects, he was evidently also a pioneer experimenter.

■ SINCE TIME IMMEMORIAL, humans have had a visual craving for colors, and they have displayed great ingenuity in discovering the dyes and pigments with which to satisfy that craving. Colors figure prominently in art, in spiritual adoration, and in self-decoration, as well as the world over in fabric design. In early times, before the advent of synthetic dyes, coloring matter came from nature. There were "earthen" materials, such as iron and copper oxides, and countless colored products from plants. Animal dyes were rare, but they were highly valued because they were so often red. As the color of blood, of fire, and of the sun, red is often assigned preeminent status among colors and adopted as a cultural symbol of majesty. The animal sources of red dyestuffs, in consequence, especially if of exotic origin, often became important items of commerce, cultivated and shipped around the world in what can only be described as astounding quantity. *Dactylopius coccus,* the cochineal bug, for nearly three centuries the main source of red coloring matter, is a case in point.

Cochineals are scale insects (family Coccidae; order Hemiptera). Their host plants are prickly pear cacti (*Opuntia* species), on which they typically form aggregations. They are cloaked in a fluffy investiture of waxy powder and silken threads, which gives them a woolly appearance and makes the aggregations easy to spot. The "wool" is absent from newly born cochineals and males. The adult females are larger than the males, and unlike males are wingless. They are typically grapelike in shape, about half the size of a blueberry. The red dye—or cochineal red as it is called—is present in the body of cochineals, as is clearly apparent if one of the bugs is accidentally squashed.

Cochineals were cultivated for their dye in the New World long before the arrival of the Spaniards. When Hernán Cortés entered Mexico in 1519 he was much taken by the widespread use of red in Aztec culture, and was downright dazzled by the magnificence of Montezuma's robe. He managed to obtain some sacs of coloring matter that the Aztecs were deriving from cochineals, and he sent these on to Spain. The dye was an instant success and it spawned an industry. By the year 1600, cochineal had already become one of Mexico's prime exports, second only to silver and gold in value on a pound-per-pound basis. Production spread, and it continued on a vast scale throughout the seventeenth and eighteenth centuries. From the region of Oaxaca alone, for the period 1758 to 1780, the production of cochineal amounted on average to about 1 million pounds per year.

As a colonial enterprise, cochineal farming was unusual in that it was carried out almost entirely by the indigenous population. That meant cultivation on small landholdings and a genuine sharing of the wealth. Growing cochineals involved farming the *Opuntia* cactus, "inoculating" it with young cochineals, and eventually harvesting the globular females. These were then sun dried or otherwise desiccated, and reduced to the shriveled pellets that even to this day constitute the cochineals of com-

THE OPPORTUNISTS

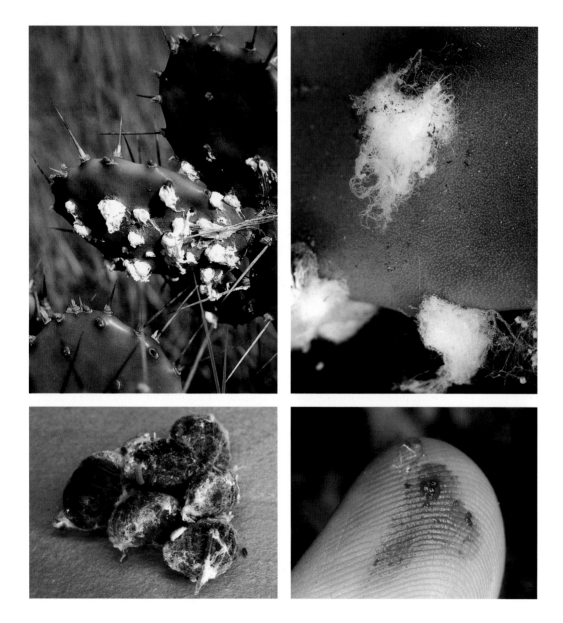

merce. The Indians grew the cochineals with great skill. They fertilized the cacti with refuse and wood ash, and kept the land well weeded. It was a family business in which the women helped with the brushing by which the cochineals were swept from the cactus pads.

Spain tried hard to maintain a monopoly on cochineal production by keeping the details of the farming procedure secret. The authorities enforced an embargo on the export of live cochineals and tried even to con-

Cochineal bugs (*Dactylopius confusus*). Top left: A colony on *Opuntia* cactus. Top right: A female concealed beneath her coating of wax and silk. Bottom left: A cluster of females, divested of their coating. Bottom right: A squashed female cochineal.

Harvesting cochineal in Algiers; from L. Figuier, *The Insect World*, London, 1872.

ceal the fact that cochineals were insects. Word had it in some European quarters that cochineals were derived from plants, and it was not until Antonin van Leeuwenhoek himself settled the matter by use of his microscope that the true nature of cochineals became generally known.

Eventually, after persistent efforts, the French too managed to establish cochineal farms in their territories, and by the mid 1800s were successfully producing cochineals in Algiers. Cochineals were also introduced into the Canary Islands, which in very short order became major exporters of the dye. In 1831 the entire production of cochineal for the Canary Islands was 4 kilograms. By 1850 the annual production had risen to 400,000 kilograms.

Chemical work on the nature of cochineal red was slow to proceed. We now know the compound is an anthraquinone called carminic acid. Although the compound was isolated in 1818, its definitive structure was not elucidated until 1959.

The mid-1800s saw a turning point in the use of cochineal. It was at that time that the first aniline dye was synthesized, a development that was nothing short of sensational, since it meant that pigments of diverse hues could now be made in the laboratory. The science of organic chemistry had arrived, and it was off to a colorful start. The effect on the cochineal industry was disastrous. Within the span of a few years it underwent a near collapse worldwide, and although cochineals are still grown nowa-

Carminic acid.

days in a few areas for dyeing and food-coloring purposes, the overall demand for the insect is trivial in comparison to what it was in its heyday.

The history of cochineal, of how it was discovered, cultivated, marketed, and put to use, is rich in detail relevant even to our own North American history. Cochineal probably provided the color for the red stripes on the very first American flag, as well as for the red coats of the British soldiers who confronted Americans in the Revolutionary War. In fact, cochineal may have played a role in initiating that war. Merchants objected to paying escalated prices for cochineal from British middlemen, when they could themselves buy the dye directly from Spain or its colonies. So it was not just the issue of stamp duties or tariffs on tea. Even then the color red had political connotations.

What interested me, when I first learned about cochineals, is the function of their pigment. What was carminic acid for? As an anthraquinone, I thought, the compound had to be defensive, but the literature had nothing to offer in support of that view. It seemed strange that what was undoubtedly one of the most widely used natural products of all times should have no proven function.

Carminic acid was commercially available, and I bought a few milligrams so I could test its effect on ants. Steve Nowicki, with whom it had been so much fun to work on spider stabilimenta, joined in the experiments. We were at the Archbold Station, where we were experienced in baiting ants, so we lured worker ants from a colony to a sugar source and put them to the test. The object was to see whether carminic acid, when added to a sugar solution, would cause the ants to reject the solution. We presented the ants with a rectangular plastic plate bearing eight conical feeding depressions filled to the brim with test solution. Four depressions (the controls) were filled with sucrose solution, and the other four, which alternated with the controls, were filled with sucrose solution plus carminic acid. We then kept score of the number of ants that were visiting the control and the experimental depressions, and from the difference in visitation rates calculated the deterrency rating of carminic acid. We tested

THE OPPORTUNISTS

Ants feeding on an offering of sugar solution while ignoring an offering of sugar solution plus carminic acid. The same results were obtained in darkness, which proves that the avoidance of the carminic acid–containing sample was not due to its color

the compound at various concentrations and found it to be absolutely deterrent—in other words, to bar visitation entirely—at a concentration of 10^{-1} M, which corresponds closely to the level of carminic acid in the cochineals. We repeated the experiments in darkness to check whether the red color of the treated samples could have played a role in the deterrency, but found that it did not. We concluded that carminic acid is an antiinsectan substance. Evidence that we got from an entirely unexpected quarter bolstered the conclusion.

Florida is *Opuntia* country, and we found the cactus, complete with occasional infesting cochineal colonies, growing wild at the Archbold Station. In the course of examining such colonies we found that they were themselves sometimes infested by a caterpillar that makes it a habit to feed on them. The year was 1979 and the caterpillar was that of a moth described exactly 100 years earlier by the renowned Cornell entomologist John Henry Comstock. The moth fed exclusively on scale insects and Comstock had called it *Laetilia coccidivora*. It was an interesting moth. Having evolved the ability to feed on cochineals, it was evidently insensitive to carminic acid. But it also used the carminic acid for its own defense, thereby demonstrating both that it is an evolutionary opportunist and that carminic acid is indeed a protective substance.

Laetilia uses carminic acid by disgorging its crop contents. It never goes hungry because it lives in the cochineal aggregations and feeds more or less continuously except when molting. Its crop, which is a sizable sac, is

THE OPPORTUNISTS

Laetilia coccidivora. Top: An adult moth. Bottom left: A larva vomiting its crop contents in response to being pinched with forceps. Bottom right: An ant leaving a red streak in its wake, from fluid regurgitated on it by a *Laetilia* larva that the ant had attacked.

therefore always filled or partly filled with a sort of cochineal *au jus,* a red viscous goo rich in carminic acid.

Laetilia can administer droplets of crop contents upon demand and with accuracy, just as larval *Neodiprion* do with the pine resin. We pinched *Laetilia* with forceps, and the larvae responded instantly by flexing the front end and vomiting upon the instrument. Even more impressive were the results with ants. Released near the trail of ants, the larvae were soon encountered and attacked, but they always dispersed the ants with their vomit. No sooner were the ants wetted than they fled, wiping their body against the substrate and leaving a conspicuous red streak in their wake.

Thanks to Michael Goetz, our collaborator in the firefly work, we were able to obtain data on the carminic acid concentration in *Laetilia* vomit. At

Two additional larvae that prey on cochineals: a chamaemyiid fly larva (*Leucopis* species) (left) and the coccinellid beetle larva *Hyperaspis trifurcata* (right). Note the red Malpighian tubules (execretory organs) showing through the body wall of *Leucopis*.

a level of 2.7 percent, that concentration proved slightly higher even than that of carminic acid in the cochineals themselves. *Laetilia* does not incorporate carminic acid into its body. It uses the compound strictly as part of the larval gut fluid and appears not to transmit it to the pupa or the adult. Pupation occurs within the cochineal colonies themselves, where the pupae derive some benefit from being visually concealed. The adult *Laetilia* is a typical pyralid moth, delicate and beautiful.

Thanks to another collaborator, Rolf Ziegler, who provided some cochineal colonies from Arizona, and J. L. McCormick and E. R. Hoebecke from Cornell, we were able to work on two additional opportunists that also feed on cochineals, a coccinellid beetle larva *(Hyperaspis trifurcata)* and a chamaemyiid fly larva (an unidentified species of *Leucopis*). Both also used acquired carminic acid for defense as larvae, but they did not do so by vomiting.

The *Hyperaspis* larva emits droplets of blood when disturbed. It is a typical reflex bleeder. The droplets are an intense red, and upon analysis proved to contain carminic acid. The concentration of the compound in the blood varied somewhat but was on average a hefty 1.7 percent, about as high as the concentration in the female cochineals themselves. The *Hyperaspis* larva evidently absorbs the carminic acid enterically, probably through the wall of the midsection of the gut, a portion known in insects as the midgut, where most of the digestion and absorption take place. Blood in insects is the fluid that fills the body cavity. The absorbed carminic acid therefore inevitably passes into the blood, there to be held in readiness until needed for defense. Reflex bleeding can occur from virtually anywhere on the larva's surface. Larvae that I exposed to ants all reflex-bled when assaulted and survived uninjured.

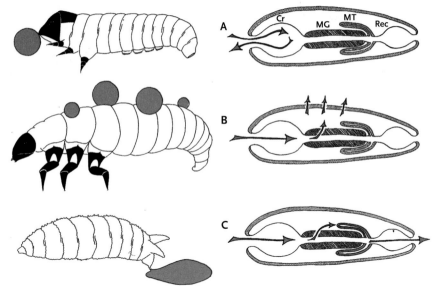

Defense strategies of the three larval predators of cochineals. Mechanisms of fluid emission are shown on the left, while the postulated pathways of the ingested carminic acid are shown on the right. A, *Laetilia coccidivora,* B, *Hyperaspis trifurcata,* C, *Leucopis* species. Cr = crop; MG = midgut; MT = Malpighian tubules; Rec = rectum.

The *Leucopis* larvae also survived ant assault, but they did so by defecating defensively. *Leucopis* emit drops of red fluid from the rear when disturbed and the emission is clearly seen to stem from the anus. Dissection of larvae showed them to have a large rectal pouch in which the defensive fluid is stored. The carminic acid is apparently not directly routed into this pouch, but is first absorbed in the midgut and then passed back into the gut by the larva's tubular excretory organs, the Malpighian tubules. These tubules are always an intense red in *Leucopis* larvae, which indicates that they might also serve for the storage of carminic acid.

Laetilia, Hyperaspis, and *Leucopis* illustrate nicely how opportunistic strategies can differ from insect to insect. Evolutionarily all three have achieved the same thing. Through specialization they have "crashed" through the defensive chemical barrier of their host, and have seized the opportunity of appropriating the host's weaponry for protective purposes of their own. All three use carminic acid for defense, but each does so in its own way. One expels the compound orally, another does so from the rear, and the third deploys it by bleeding.

9

The Love Potion

In the year 1893, in an obscure French medical journal, the *Archives de Médecine et de Pharmacie Militaires,* there appeared a remarkable paper. The author, J. Meynier, was a physician who years earlier, in 1869, while attached to French military forces in northern Algeria, had witnessed a most peculiar medical incident. It was the month of May, and he had been assigned to a contingent of *chasseurs d'Afrique,* under orders to march from Sidi-bel-Abbès to a mountain site some days away to help establish a new post. It seems that during one of the stopovers a group of men appeared at his quarters with medical complaints. Meynier does not specify how large a group, saying only that it was "grand," and that the men all shared the same symptoms: abdominal pain, dryness of mouth, pronounced thirst, frequent and painful urination, general weakness, depressed pulse rate, reduced arterial pressure, lowered body temperature, nausea, and anxiety. Such a constellation of symptoms was suggestive of any number of ailments, but there was an additional complaint that did point to a cause. The men had, in Meynier's words, *érections douloureuses et prolongées*—painful, long-lasting erections—something that under regimental circumstances, I would imagine, could be embarrassing, but that to an enlightened French physician was grounds for diagnosis. The 1800s were pre-Viagra years, so there had to be another cause. Known to the physicians of the time was a compound called cantharidin, which induced erections and was for that reason occasionally taken by men for remedial purposes. Cantharidin was also known as Spanishfly, although it was of neither dipteran nor exclusively Iberian origin. It did however stem from insects, from beetles of the family Meloidae, and that fact was known to Doctor Meynier. He put two and two together and, suspecting that his patients had somehow ingested cantharidin, asked them whether they had eaten anything unusual. It turned out they had. *Grenouilles* was the answer. In the best French tradition they had eaten frog legs. Eager probably for a change from army chow, they had gone to the local river and helped themselves to what they expected would be a gourmet delight.

Cantharidin poisoning from eating frogs? Doctor Meynier decided he'd investigate. He went to the river and found that the frogs were abundant and that they were gorging themselves on a kind of beetle that was also

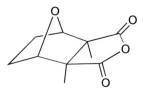

Cantharidin.

abundant. And sure enough, the beetles were meloids. The frogs, he concluded, had incorporated cantharidin from the meloids they ate and as a result become poisonous themselves. Remarkably, according to Meynier, cantharidin poisoning through frog ingestion may have occurred with some frequency at that time among French military personnel in Algeria. Meynier also comments that his own patients all recovered in due course.

Cantharidin has a long history of being used by humans, often for unsavory purposes. The compound is remarkably toxic. As little as 100 milligrams is said to be lethal to humans. Individual meloids may contain milligram quantities of the compound, meaning that ingestion of no more than a few individuals can be fatal. The peculiar erectile consequences of cantharidin ingestion are responsible for the compound's reputation as an aphrodisiac and for its frequent misuse in one context or another. Cantharidin is not used as a love potion any more but it was in many places in times past, including France, where cantharidin pills were known as *pastilles galantes*. Cases of poisoning may have been fairly frequent among users. In ancient times, Lucretius is said to have been among those who died of cantharidin poisoning. There are also gruesome accounts of how the Marquis de Sade experimented with cantharidin, supposedly driving some prostitutes in Marseilles to suicide under influence of the compound. Cantharidin was used extensively for medicinal purposes in the past. Hippocrates prescribed it in ancient Greece and in the times of Frederick the Great it was used for treatment of tuberculosis and rabies. The compound was isolated in 1810, but its stereochemistry was not elucidated until 1941, and it was not synthesized until 1953.

Cantharidin is also topically active. Applied to the skin it induces large fluid-filled blisters. In the past many an ailment was treated by induction of blisters with cantharidin. Medical claims are still made for cantharidin in several places in the world, where the compound is sometimes still available in the form in which it was originally marketed, that is as Spanishfly, the dried pulverized remains of meloid beetles, also called blister beetles. On the practical side, cantharidin today poses a real hazard to horses, which may be poisoned by feeding on meloid-infested hay. Since race horses worth millions are at risk, it should come as no surprise that substantial sums are being expended in research aimed at shielding the equine elite from cantharidin poisoning.

Both Arizona and Florida are meloid country and I became acquainted

THE LOVE POTION

with the blister beetles in both states. I had handled meloids without getting blisters and had concluded I might be topically insensitive to cantharidin. The alternative explanation was that I was insensitive on the fingers only, where the skin is relatively thick. I had read that cantharidin was stored in the blood of the beetles, and had found out on my own that meloids reflex-bleed when disturbed. So I decided I'd take a meloid and

Blister beetles. Clockwise from the top left: *Nemognatha* species, *Meloe levis*, *Zonitis* species, and *Megeta optata*.

 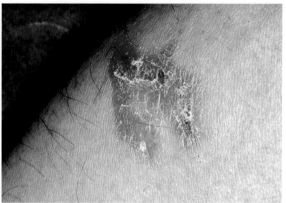

Blistering induced by blister beetle blood. On the left, after 24 hours; on the right, after 2 weeks.

cause it to bleed directly on the inner surface of my forearm, to see whether on tender skin I might be sensitive after all. It was the last time I deliberately applied meloid blood to myself. I did indeed blister, and the nasty wound that developed on the site took weeks to heal.

Was it by induction of blisters or its other delayed effects that cantharidin fulfilled its defensive role? It occurred to me that there might be more to the story. Could cantharidin be distasteful or in some other immediate way offensive? At first I thought I would taste meloids myself but decided not to because of reluctance to risk *érections douloureuses*. There is evidence that some of the systemic effects induced by cantharidin, such as abdominal discomfort, are of quick onset, quick enough perhaps for a predator readily to associate the ill effect with the cause—for the predator to blame the symptoms on the beetle. But does this mean that the beetle has to be swallowed in order for the predator to develop an aversion to that type of prey item? Blister beetles do commonly live in aggregations, so it is not impossible that the beneficiaries of the predator's acquired aversion are the relatives of the beetle ingested. But to my knowledge it has not been established that meloid aggregations are made up of close relatives. And if they are not, I am hard put to imagine how defensive suicide could have evolved in meloids. The whole question of the defensive role of cantharidin against vertebrates remains a bit of a mystery. For one thing, it is clear that not all vertebrates are affected by cantharidin. The Algerian frogs are a case in point.

WE'RE ON MORE SOLID FOOTING explaining how cantharidin protects meloids against arthropods. When I noted that blister beetles reflex-bleed,

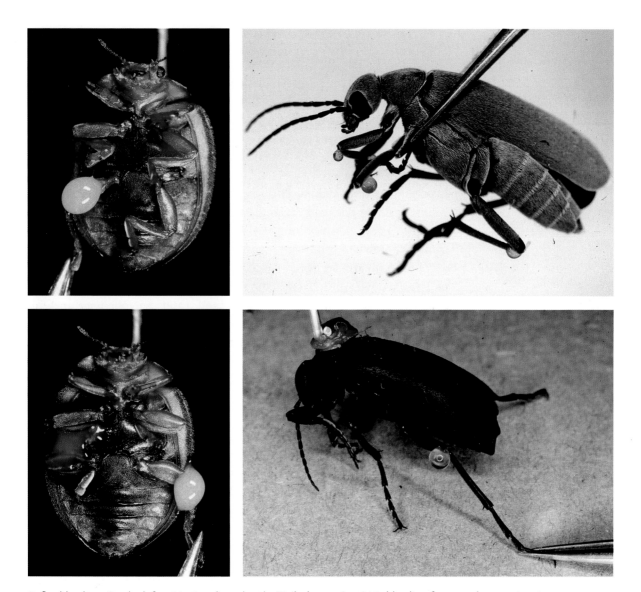

Reflex bleeding. On the left, a Mexican bean beetle *(Epilachna varivestis)* is bleeding from one leg at a time in response to the pinching of individual legs. On the right at the top, a blister beetle (*Epicauta* species) is bleeding from all legs in response to pinching of the thorax. In the bottom right photo, *Epicauta immaculata* is being pinched in one leg and is bleeding from that leg only.

I was already interested in the general phenomenon of defensive bleeding, because it occurred in other insects as well. I had published a paper with one of my first graduate students, George Happ, on reflex bleeding in the Mexican bean beetle, *Epilachna varivestis,* a member of the family Coccinellidae, which includes also the familiar ladybugs. *Epilachna* reflex-bleeds from the knee joints, but it does not necessarily bleed from all legs at once. It exercises control over the emission and emits blood from only the leg or legs stimulated. In fact, it is not necessary for a leg to be directly stimulated in order for that leg to bleed. All that is needed is that a body site close to the leg be stimulated. *Epilachna* is therefore programmed to respond in a measured way, which could indicate that it is plagued primarily by little predators, such as would inflict localized assaults, and which could be deterred by no more than a single well-placed droplet of blood. In short, the bleeding mechanism seemed to be adapted to serve against insects. As usual, ants came to mind, and the tests George Happ and I did with ants met our expectations. The beetles bled when attacked, and they tended to bleed only from the legs closest to where they had been bitten. The blood was instantly deterrent. Later, in collaboration with Jerry's group, we isolated a tropane alkaloid from *Epilachna*, a compound called euphococcinine, which accounts for the noxiousness of the blood. But the blood also acted mechanically, by hardening upon clotting and gumming up the mouthparts of the ants.

Given these facts about *Epilachna*, I thought I'd check whether blister beetles also reflex-bleed from individual legs, and I found that they did. Interestingly, the site of emission in their case is also the knee joint. They responded exactly like *Epilachna*. If I seized them bodily, with broad-tipped forceps, they tended to bleed from all legs at once (and even also from the neck region), but if I pinched a single leg, they bled from that leg only. They too seemed adapted to deal with localized assaults. I did some tests with ants and found that they rejected meloids and were deterred by cantharidin. But my tests were preliminary. More refined experiments were in order. Jim Carrel, an excellent experimentalist, joined the project and as part of his thesis research decided to take a good look at cantharidin. Nowadays, as an award-winning teacher at the University of Missouri, he is a world authority on Spanishfly.

The test that Jim designed is one in which we offered ants sugar solutions containing cantharidin at various concentrations, and measured the rate at which the ants consumed the various samples. The ants were from a colony we had in captivity and they were fed on a platform adjacent to the colony where they routinely received their food. We presented the liquid samples to the ants in open-ended glass capillary tubes, maintained at

A bioassay for cantharidin. The carabid beetle *Calosoma prominens* is responding to application of a cantharidin solution to its mouthparts by wiping them in the sand.

a tilt, so that downward flow automatically replenished the amounts withdrawn by the ants. We used calibrated tubes so that we could directly determine the rates of fluid inhibition. The ants proved to be extraordinarily sensitive to cantharidin. Even at the minimal concentration tested, 10^{-5} Molar, the compound had a significant depressant effect on the drinking rate. The concentration of cantharidin in the blood of meloids is, on average, about 100 times higher.

We showed cantharidin to be deterrent also to a carabid beetle, *Calosoma prominens*. We had earlier worked with that beetle and noted that it had a very specific way of cleaning itself when its mouthparts were contaminated with a noxious chemical. It dragged its head in the soil. Carabids are predaceous and likely therefore to have their mouths contaminated occasionally with defensive chemicals discharged by prey. It made sense, therefore, that they should use the soil as a napkin. We took individual *Calosoma* and stimulated them orally by placing a brush, dipped in cantharidin solution, between their mandibles. The beetle reacted by biting into the brush, upon which we released the beetle on sand and kept track of whether it performed the cleaning response. Again, we found that cantharidin was effective even at a concentration of 10^{-5} Molar.

I later did experiments with other collaborators and other predators and

THE LOVE POTION

learned more about the vulnerability of meloids. We found, for example, that three species of spiders, the orb weavers *Nephila clavipes* and *Gasteracantha cancriformis* and the lynx spider *Peucetia viridans,* reject meloids, while another spider, the orb weaver *Argiope florida,* accepts the beetles. With *Nephila* we also showed that cantharidin itself, when added to the surface of an edible item (mealworm), renders that item inedible.

One year I had the good fortune of collaborating with the Israeli ornithologist Reuven Yosef, who had been studying loggerhead shrikes *(Lanius ludovicianus)* at the Archbold Station and had a number of these birds in captivity. The shrikes, which all came from an area where they could have come to know meloids, consistently ignored all meloids they were offered.

At one point I also thought I would check into what we had come to refer to in the lab as the grenouilles syndrome. We were by no means ready to feast on meloid-fed frogs, but we thought we'd check chemically to see whether cantharidin is retained in the body of frogs that gain access to the chemical.

Feeding cantharidin to the frogs proved relatively easy. We used leopard frogs *(Rana pipiens)* that we obtained from a commercial source and that adapted quickly to laboratory conditions. We offered some of the frogs meloid beetles, which they took without hesitation, and others mealworms bearing a surface coating of crystalline cantharidin, which they also took. We then collected slime samples over a period of days from the body surface of some of the frogs, as well as the frogs' feces, and analyzed these samples for cantharidin. We also sacrificed some of the cantharidin-fed frogs and analyzed their body parts for cantharidin. The data told a pretty straightforward story. First, it was clear that the frogs tolerated cantharidin without ill effects. And second, they did indeed absorb cantharidin into the body. They did not simply void the compound with the feces. After ingestion of cantharidin, the compound made its appearance in the internal body parts, as well as in the skin, which left little doubt that meloid-fed frogs could be poisonous to humans. But they would not be poisonous for long. To judge from the slime samples in particular, it was evident that the retention of cantharidin was time-limited. The frogs rid themselves of the compound in a matter of days and they appeared to do so largely by excreting the chemical through the skin. So, if you *must* eat frogs that coexisted with meloids you better not cook them until after they were kept in isolation for some days. But why not forgo eating frogs altogether? They seem to have a hard enough time these days surviving in nature.

We thought we would also look into the question of whether frogs can derive benefits from retention of cantharidin, and obtained results that suggested that they do not. We exposed cantharidin-fed frogs to leeches and found that these sucked about as much blood from such treated frogs as from frogs that were cantharidin-free. Thanks to a close friend, the well-known naturalist and reptilian expert Carl Gans, at that time on the faculty of the University of Michigan, we were able to check also into the acceptability of cantharidin-fed frogs to the broad-banded water snake, *Nerodia sipedon*. Four such snakes that Carl had caught showed no discrimination against cantharidin-fed frogs and seemed none the worse for having eaten them when released at their capture site 10 days later.

Some years ago, through the courtesy of a German colleague, Michael Boppré, I learned that the grenouilles syndrome manifests itself at times in visitors to North Benin, on the Niger, as a sequel to eating spur-winged geese *(Plectropterus gambiensis)* rather than frogs. Some species of meloids appear to occur there in enormous numbers at times and the geese apparently feed on these.

Clearly, cantharidin, from the perspective of the meloid beetles that produce it, is a compound of mixed defensive merit. Active against some predators, but not against others, it provides what is undoubtedly an important measure of protection but not absolute protection. Cantharidin is therefore, like other defensive agents, imperfect. Evolution, more often than not, provides partial solutions to problems.

But there is more to the cantharidin story. It has been known for some time that while cantharidin is present in both sexes in meloids it is not necessarily synthesized by both sexes. Indeed, it is the males alone that synthesize the compound in some species, and it is they that supply the females, by transferring the compound to the female with the sperm package at mating. Swiss investigators were the first to demonstrate this. Cantharidin is synthesized from a small molecular building block called mevalonic acid. Males produce radio-labeled cantharidin if given radio-labeled mevalonic acid, but females do not. But if you mate a female to a male loaded with radio-labeled cantharidin, you subsequently find radio-labeled cantharidin in the female. Also interesting, as Jim Carrel showed while he was still in my lab, is that female meloids bestow some of the cantharidin upon the eggs, thereby providing them with protection.

And finally there is the remarkable phenomenon called cantharidiphilia, a term that means, quite literally, love for cantharidin. Noxious as cantharidin is to some insects, it is attractant to others. The phenomenon has long been known, but has only recently come under scrutiny. Cantharidin, it turns out, the very compound wrongly purported to have an excitatory aphrodisiac effect on humans, may in fact have such an effect on the cantharidiphilic insects that crave the compound.

■ CANTHARIDIPHILIA is now well documented. Put some crystalline cantharidin in a glass jar outdoors and chances are you will attract some insects. If the jar is fashioned with a one-way entrance, the visitors will become trapped and you can then keep an accurate count of their number and kind. The visitors are diverse and may vary somewhat depending on where in the world you do the trapping, but they are known to include representatives of four major orders—Coleoptera, Diptera, Hemiptera, and Hymenoptera.

My interest in cantharidiphilia was spurred by a 1937 paper, in German, entitled *Cantharidin als Gift und Anlockungsmittel für Insekten* ("Cantharidin as Poison and Attractant to Insects"). It was an intriguing paper, and full of detail. The author, Karl Görnitz, an employee of a major industrial concern, Schering-Kahlbaum A.G., had been engaged in the search for new insect-control agents when he discovered that cantharidin, which was available to him in crystalline form, had insecticidal potential. The compound, he found, was lethal to certain insects, including some well-known caterpillar pests. This was in itself worth reporting. But what he hadn't counted on and found compelling was the astonishing attractancy of cantharidin.

His first observations were in a greenhouse. He had sprinkled some cantharidin on *Tradescantia* plants wishing to check into the protective action of the toxin, but found instead that the compound had luring potential. Both an anthomyiid fly and an anthicid beetle were attracted to the plants, as well as to dishes with crystalline cantharidin that he set up on subsequent days to see whether the attractancy was due to the cantharidin itself. The dishes also attracted a species of midge and a braconid wasp.

Görnitz set out dishes baited with cantharidin in diverse habitats at

THE LOVE POTION

many locations in Germany and found that he could always lure the same four species. He was fascinated. In commenting on the "enormous attractancy" of cantharidin, he recalls automobile trips that he took in the summer of 1934 to the lake district south of Berlin, where he went swimming, and where he always took the time to set up baited dishes. He found that the attractancy was transferred to items that he had packed in the suitcase together with the cantharidin bottle. His bathing suit and blanket thus became attractive, with the result that the insects landed on him when he rested. Even the unopened car, if he had made a stop on the way, sometimes drew numbers of beetles and flies. If he put the cork-stoppered cantharidin bottle on the ground, anthicid beetles would in short order materialize to gather, like clustered grapes, around the cork.

Another investigator, F. Fey, wanting to determine the distance over which cantharidin can effect attraction, put baited dishes on floating buoys at distances upwind of up to 500 meters from the shore of a Berlin lake. Anthomyiid flies were attracted in numbers to as far as 200 meters from shore; one fly was even drawn to the farthest site. Considering that the individual baits consisted of no more than 0.4 milligrams, these results were astonishing.

What I found most fascinating about Görnitz's paper is what he says about the sex ratio of the insects drawn to the cantharidin. Of a sample of 693 anthicid beetles that he collected at baited dishes in 14 separate locations, 642 were males. Görnitz found this result puzzling—a "Rätsel," he called it—and I agreed. He found that the male anthicids actually ate the cantharidin upon arrival at the baits, but he was at a loss to explain what need they might be satisfying by doing so. I thought it might be worthwhile to look into these questions, so I set out to lure cantharidiphiles myself. I thought anthicids would be the most logical cantharidiphiles to study, because they were abundant and I would have no difficulty finding them at the Archbold Station. But I was a bit leery because anthicids are tiny. At any rate, I bought some crystalline cantharidin and set up some traps that I baited with the compound. I was too impatient to wait for a Florida trip, so I placed the traps outdoors in Ithaca, and to my great satisfaction found that there were cantharidiphiles in my neck of the woods. But best of all, one of these cantharidiphiles was large, large enough for experimental work, and it too was a cantharidin eater. Moreover, only its males came to the trap so it seemed suitable for study of the same questions I had hoped the anthicids would answer.

The beetle was *Neopyrochroa flabellata*, a member of the Pyrochroidae, which I thought had not previously been taken in a cantharidin trap. But I was wrong. Daniel Young, now an entomologist at the University of Wisconsin, had while a graduate student at the University of Michigan found that *Neopyrochroa* was a cantharidiphile, and he was just as curious about why the male of this species should have a love for cantharidin, so we decided to collaborate. His cantharidin traps drew many more *Neopyrochroa* than my Ithaca traps, which in itself proved helpful, but most important, Dan knew how to find the larvae, which made it possible to do courtship studies with "home-raised" adult virgins of both sexes. And why would we want to do courtship studies? Because, simply put, we had a notion that cantharidin might play a key role in the sex life of this beetle.

■ **THE *NEOPYROCHROA* MALES** that came to the traps, as well as those that emerged from the pupae, were cantharidin-free, and so were the newly emerged females. We determined this by chemical analysis, and it told us that *Neopyrochroa* is not itself capable of synthesizing the compound. But we found out early on why the *Neopyrochroa* males come to the traps. They have a compelling hunger for cantharidin. If a male was placed in a petri dish with a few micrograms of cantharidin, he would promptly wave his antennae, walk toward the sample, and the moment he came in contact with it devour it to the last crystal. It took quite a bit of cantharidin to satisfy the hunger of a male. Samples of 50 micrograms were typically eaten all at once, but over a period of days a beetle could consume several times that quantity. The total amount ingested could come close to 1 percent of male body weight. And what a critical acquisition it was, not because the male derived nutrients from it, but because through its possession he gained access to the female.

Courtship in *Neopyrochroa* can be easily staged. The sexes are utterly shameless and usually "go at it" the moment a pair is introduced into a petri dish. The male typically takes the initiative. Sensing the presence of the female, he is quick to walk toward her and upon finding her, orients face-to-face toward her as if he were presenting himself. The two then do a remarkable thing. They both raise the front end, and while the male places his forelegs and midlegs on the female's flanks, the female gets a grip on the male's head. The male's head is especially fashioned for the

Neopyrochroa flabellata (male). Top left: Feeding on crystalline cantharidin. Top right: A close-up of the head showing the glandular cleft. Bottom left: Secretion being extracted from the glandular cleft with a fine pin. Bottom right: Cantharidin crystals in the secretion of a cantharidin-fed male (photographed in polarized light).

THE LOVE POTION

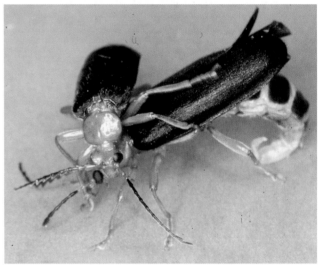

Events in the courtship of *Neopyrochroa flabellata*. Top left: The male (left) presents his gland. Top right: The female (right) samples the gland. Bottom left: Copulation. Bottom right: The female coiling her abdomen and assuming the blocks-intromission posture in response to attempted mounting by the male.

THE LOVE POTION

purpose. Clear across his forehead is a deep cleft that looks as though he had been hit with an ax. The cleft has flared ends and it is into these that the female inserts her mandibles. The pair remain locked head-to-head for a protracted period, during which the female's mouthparts can be seen to be kept in motion, as if she were feeding from the cleft. Eventually the female relinquishes her hold, upon which—if things are going "well"— the male mounts the female and the pair proceed to mate. The female may remain completely passive and offer no resistance to the male's copulatory attempts. Mating typically lasts for several minutes, during which the pair remain quiescent, with the male astride the female. The male eventually dismounts and becomes flaccid, whereupon the female begins walking and the pair uncouples.

Things did not always go well, and we soon found out why. If males had not fed on cantharidin, they tended to be rejected. The females walked away from them when they presented their heads, or after briefly having locked heads with them, and if the males nonetheless persisted, the females actively fought them off by flexing their rear ends and making themselves inaccessible. And how did the females "know" whether the males had eaten cantharidin? It turned out that they can tell by inspecting the male's cleft. The cleft is glandular and it secretes cantharidin when the males have had access to the compound. The female's mouthparts are in motion when she grasps the male because she is then feeding from the cleft. If she is thereby able to detect cantharidin, she accepts the male. If he doesn't have it, he's out of luck.

We videotaped the courtships and the graphic plots we constructed, based on the videotape analysis, told the story. The plots were in the form of behavioral flow charts, in which arrows denote the sequence of occurrence of behavioral events, and the width of the arrows denotes the frequency with which behaviors occurred in the sequences indicated. To get an idea from such a chart of the prevailing flow of events one simply follows the sequence denoted by the thickest arrows. Thus as can be seen from the chart on page 342, when females were paired with cantharidin-fed males, the events proceeded with little variation from male-female encounter, to male presentation of the head, to female feeding from the cleft (gland sampling), to mounting and copulation. The males were accepted in 20 of 21 such pairings.

In contrast, only 3 of 24 cantharidin-free males succeeded in copulating

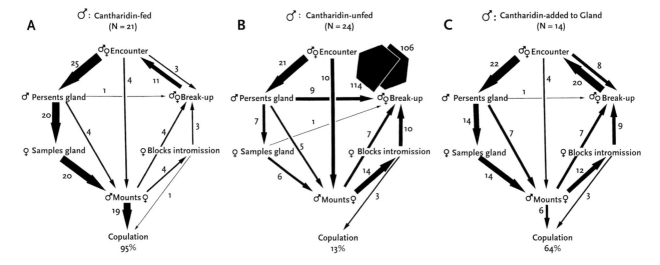

A ♂: Cantharidin-fed
(N = 21)

♂♀Encounter

25 4 11 3

♂ Persents gland ———1———→ ♂♀Break-up

20 4 4 3

♀ Samples gland ♀ Blocks intromission

20 4

♂Mounts♀ / 1

19

Copulation
95%

B ♂: Cantharidin-unfed
(N = 24)

♂♀Encounter 106

21 10 114

9

♂ Persents gland ———————→ ♂♀Break-up

7 1 7 10

5

♀ Samples gland ♀ Blocks intromission

6 14

♂Mounts♀ 3

Copulation
13%

C ♂: Cantharidin-added to Gland
(N = 14)

♂♀Encounter

22 4 8 20

♂ Persents gland ———1———→ ♂♀Break-up

14 7 7 9

♀ Samples gland ♀ Blocks intromission

14 12

♂Mounts♀ 3

6

Copulation
64%

The flow of events in courtship trials with three categories of *Neopyrochroa flabellata* males.

(see the chart above). In the cramped quarters of the petri dish it was inevitable that male and female encountered each other frequently, but such encounters mostly led to break-ups. Break-ups were often abrupt, after no more than brief antennal contact. The males sometimes did succeed in mounting but in no case did mounting lead directly to copulation. In the 3 cases where copulation took place, it occurred *despite* the female's attempt to block intromission. Gland samplings were relatively rare since the female tended not to proceed to this behavior after the male presented his head. Such presentations, mostly, led to break-up of the pair. The gland samplings, when they occurred, were also of short duration. Whereas with cantharidin-bearing males the females spent on average 25 seconds feeding from the cleft, they spent less than 5 seconds locked head-to-head with cantharidin-free males.

The secretion in the male's cleft is semisolid and sticky. It can be pulled out with a pin and therefore easily collected for analysis. Its cantharidin content, in cantharidin-fed males, turned out to be 17 percent, a value so high as to suggest that the secretion might be saturated with the compound. We looked at secretion with a microscope under polarized light and, sure enough, it was chock full of crystals. Despite the concentration, the total amount of cantharidin in the male's cleft available to the female in courtship is only about 1.5 micrograms, far less than the quantity we knew the males were able to take in. Yet that token amount appeared to be the male's ticket to reproductive success.

To check whether it was indeed the possession of cephalic cantharidin that made the difference, we put some virgin males to the test that we had kept away from cantharidin, but "loaded" by introducing cantharidin into

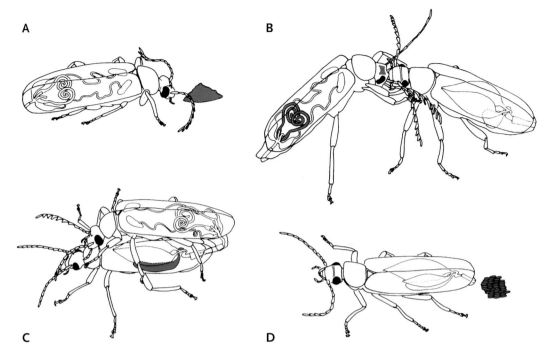

The use of cantharidin by *Neopyrochroa flabellata*. The male procures the chemical (A) and, after feeding on it, stores a little of it in the cephalic gland and the remainder in the large accessory gland of the reproductive system. The female samples the secretion from the male's cephalic gland during courtship (B) and then yields to the male's copulatory attempt. The male inseminates the female and in the process transmits cantharidin from the accessory glands to the female's spermatheca, her sperm receptacle (C). The female in turn bestows the acquired cantharidin on the eggs (D).

their cleft. To load the males we simply picked up some cantharidin crystals with a fine pin and stuck these to the gummy contents of the cleft. Offered to females, these males were treated as if they had fed on cantharidin (see the chart on page 342). Encounters tended to lead directly to male head presentations, which in turn led to cleft feeding, mounting, and copulation. Fully 9 of 14 males passed muster. Gland sampling lasted on average 18 seconds with these males, significantly longer than with cantharidin-free males, and nearly as long as with cantharidin-fed males.

But what does the male do with the bulk of his cantharidin? The amount he offers to the female in the cleft is, after all, only a small fraction of what he took in. It turns out that he actually holds a much larger cantharidin offering in store for the female, a gift that he bestows only when he is assured access to her eggs. The cantharidin in the cleft is but a promissory note, a way of announcing that, given a certain condition, there is more to come. "Accept me," the male says by way of his glandular message, "and you will get a lot more." "But you will need to take me on, accommodate my sperm, and give me the opportunity to father our offspring." It is only following intromission, as a part of the sperm package, that the male bestows that additional gift, which exceeds by far the amount made available in his gland. By favoring males with laden clefts the female is essentially providing proof that she is capable of heeding the message conveyed by the male's cephalic offering. "Accept only males with loaded clefts" is the rule by which she abides, "or remain forever impoverished."

By dissecting cantharidin-fed males and analyzing their body parts for cantharidin content, and doing the same with another group of such males that had mated, we could determine how the chemical is internally distributed after feeding and from where it is missing after mating. We found that the compound, after feeding, is shunted in part to the head, and in larger measure to a set of glands, the accessory glands, of the reproductive system. Passage to the head was anticipated given what we knew about the role of the cephalic gland. And presence in the reproductive system confirmed our suspicion that cantharidin was transmitted to the female at mating.

The male reproductive system of *Neopyrochroa* is structured as such systems typically are in beetles. The sperm are produced in a pair of testes, stored in a pair of seminal vesicles, and expelled at mating through the ejaculatory duct. The two pairs of accessory glands, a large pair and a small pair, connect to the base of the ejaculatory duct. Cantharidin is stored primarily in the large accessory glands. Before mating these glands are chock full of material. After mating they are translucent and devoid of much of their cantharidin.

The female, which as a virgin is cantharidin-free and has an empty sperm receptacle (spermatheca), has a full receptacle after mating, containing cantharidin in an amount roughly equivalent to what the male loses from the accessory glands at mating.

Dozens of analyses for cantharidin content, of both individual beetles and of their body parts, were required to put the story together. We were extremely lucky, as usual, to have a chemical collaborator from Jerry's lab willing to join in the venture. Brady Roach slaved for days on end to provide the necessary numbers. Now a researcher with an industrial laboratory and a veritable wizard with instrumentation, he resides in Ithaca and remains a friend to this day.

THE LOVE POTION

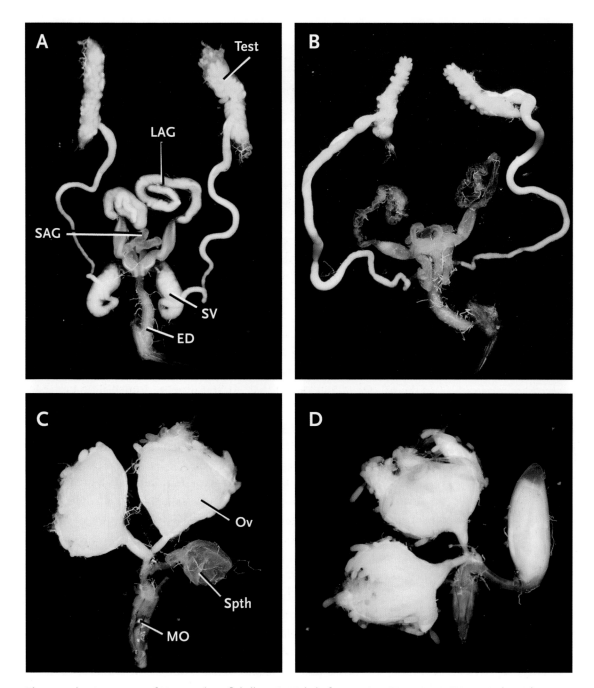

The reproductive system of *Neopyrochroa flabellata*. A: Male before mating (Test, testes; SV, seminal vesicles; SAG, small accessory gland; LAG, large accessory gland; ED, ejaculatory duct). B: Male after mating; note that the seminal vesicles and large accessory glands are empty. C: Female before mating (Ov, ovary; MO, median oviduct; Spth, spermatheca). D: Female after mating; note that the spermatheca is full.

A dead blister beetle (*Nemognatha* species) being visited by anthicid beetles (*Notoxus* species). The anthicids are sequestering cantharidin by licking the body surface of the corpse.

By analyzing the eggs of the female *Neopyrochroa,* we showed that like the female meloid, she invests part of her cantharidin in the eggs. Moreover, by subjecting the eggs to predation tests with coccinellid beetles, we were able to show that the chemical endowment conveys a significant degree of protection. Eggs sired by cantharidin-fed males proved much less vulnerable than those fathered by cantharidin-free males.

The results could be expressed in numbers. By feeding on cantharidin, the male *Neopyrochroa* takes in some 70 micrograms of the chemical. He transmits about 40 micrograms of this load to the female, who in turn bestows the gift upon the eggs, to the tune of about 0.02 micrograms per egg. The female's total cantharidin output by way of the eggs (about 13 micrograms) amounts to about 33 percent of the cantharidin she receives from her mate. The overall "flow" of cantharidin in *Neopyrochroa,* from male to female to eggs, proceeds therefore with considerable efficiency.

There is a mystery remaining that I would love to see solved. Where do male *Neopyrochroa,* or for that matter cantharidiphiles generally, get their cantharidin? The assumption, of course, is that they get it from meloid beetles, or from the related beetles of the family Oedomeridae, both of which produce cantharidin. But are there really enough meloids and oedomerids in the world to meet the needs of cantharidiphiles? Meloids and oedomerids at all stages of development, including the egg stage, could serve as cantharidin sources, but are they in fact the only sources of the compound? Cantharidiphiles themselves could certainly also act as sources once they acquired cantharidin, but that does not provide for an increased net availability of the compound. Whatever the answer, there is evidence that some cantharidiphiles do make contact with cantharidin producers in nature. Anthicids, for example, have on numerous occasions been spotted sitting on meloid beetles. As early as 1827 Thomas Say, a great pioneering entomologist, noted a pyrochroid beetle of the genus *Pedilus* perched on a blister beetle. "I found this species . . . attached to the side of a *Meloe angusticollis* which was perfectly at rest upon the ground, not appearing to be in the slightest degree incommoded by the weight of its temporary parasite."

There is also evidence that in other cantharidiphiles males transfer acquired cantharidin to the eggs by way of the female. Konrad Dettner and his coworkers in Germany, who had the courage to work on the tiny anthicids that I spurned, showed that these beetles use cantharidin essentially in the same manner as *Neopyrochroa*. But ultimately, there is a broader picture here. There are other male insects that also procure exogenous chemicals for protection of their offspring, and they too may use this commodity as a bargaining chip in courtship. The next chapter deals with this theme.

10

The Sweet Smell
of Success

Detachable scales prevent moths from sticking tightly to spider webs, but they do not provide moths with absolute protection. Spiders may react so quickly to moths in their webs that they may make their catch before the moths have had a chance to flutter free. And besides, there are specialist spiders that feed habitually on moths. *Scoloderus cordatus,* for instance, constructs a ladder web, a long, upright, bandlike structure that apparently acts as a chute to convey fluttering moths downward to where the spider sits in wait. Even more remarkable are the bolas spiders, evolutionary descendants of orb weavers that do entirely without orbs. Their name stems from the device they use to catch their prey, a short silken chord with a drop of viscous glue at the tip, which ordinarily hangs motionless from a leg. When a moth comes within range they rapidly twirl the bolas and if they make the catch they quickly hurl it in to the fangs for the kill. It had long been a mystery how bolas spiders managed to "fish" with such a short line, until it was noticed that they caught only male moths of certain species. This finding raised the possibility that they might be masquerading as female moths and luring the males by emitting replicas or near replicas of the female moths' sex pheromones. That such is indeed the case has now been established beyond doubt by, among others, two leading arachnologists, William Eberhard and Mark Stowe, and the pioneer chemical ecologist James Tumlinson.

Scales being imperfect defenses against spiders, and in some cases even useless, suggested that there had to be moths "out there" with improved defensive capacities, protected in ways that might be totally unimagined. After all, nature often provides alternative solutions to problems. I became aware years ago that for moth defense the alternative was distastefulness. Moths could be chemically unacceptable to spiders and for that reason safe from attack. I learned this from a single observation, from seeing a moth lie still after flying into a web and being set free by the spider, although I know now that I could have made the same observation with a variety of moths. But it so happens that it was that very first observation, with the moth *Utetheisa ornatrix,* that was to lead my collaborators and me down a path of inquiry that changed forever the way we thought about insect survival. Distastefulness, it turns out, and the defensive benefit it conveys, can be at issue in all fundamental aspects of a moth's life, in

Left: The ladder web of *Scoloderus cordatus*. Right: The bolas spider *Mastophora bisaccata*.

the dialogue of courtship, in the achievement of status and genetic eminence, and in the quantification of parental commitment. Survival in *Utetheisa* means adherence to a script, to a set of rules that appear all to be written in chemical terms. We have had *Utetheisa* in culture for some 30 years now, long enough to have deciphered some of its rules, but by no means all of them.

That first observation with *Utetheisa* dates back to 1966, to Florida, to an occasion when I was doing some general collecting. I was familiar with *Utetheisa*. The moth was stunningly colored, pink and white with touches of black and orange, and it had the habit of flying in the daytime as well as at night. If I had been asked, I would have guessed *Utetheisa* to be distasteful. How else to explain its gaudiness and ability to risk diurnal exposure? But being witness to that act of rejection by a spider was a compelling experience nonetheless. The *Utetheisa* had flown into the web and, quite atypically for moths, did not attempt to flutter free, but became instantly motionless. The spider darted toward it and inspected it briefly but then,

THE SWEET SMELL OF SUCCESS

An attack by a spider *(Nephila clavipes)* on *Utetheisa ornatrix*. Initially, the moth remains motionless in the web (top left). The spider then inspects the moth (top right), and having decided upon its unacceptability, cuts it from the web (bottom left). Eventually the moth falls free (bottom right).

THE SWEET SMELL OF SUCCESS

instead of following through with a bite, proceeded to cut the moth loose. Systematically, by use of the fangs, and with the help of legs and palps (the first set of appendages after the fangs), it snipped one after the other the strands that were restraining the moth, until the moth fell free. Before even striking the ground the moth activated its wings and flew off.

To ensure that it was not for lack of appetite that the spider had rejected the moth, I offered the spider an edible scarab beetle, which it promptly took. I then collected about a dozen or so *Utetheisa* and flipped these alive, one after the other, into individual *Nephila* webs. Without exception, whether male or female, the moths were freed. The spiders all accepted edible insects that I offered as alternatives, so there was no question that they were hungry. None of the spiders rejected *Utetheisa* on near-contact. They had always touched the moths before deciding to free them. Odor, therefore, was not at issue in the rejections.

Utetheisa showed a peculiar behavior when disturbed, which in itself suggested the moths were protected. When mildly squeezed, or even when so much as touched, both male and female moths typically emitted two bubbling masses of froth from the anterior margin of the thorax, behind the head. The fluid oozed forth abruptly, sometimes from one side at a time. I examined the froth under the microscope and found that it contained the same cells as the body fluids of the moth—as the moth's "blood." By frothing, the moths were evidently externalizing their inner fluids, indicating that they might be distasteful throughout, on the inside as well as the outside.

In the tests with spiders, frothing had occurred, but inconsistently. Contact with the moth's blood was therefore not essential for appreciation of the moth's noxiousness. Mere palpation of the moth could suffice. In fact, I had noticed that some spiders cut the moth free after contacting no more than its wings. This prompted me to design what I called transvestite experiments, in which I took some edible moths that I outfitted with *Utetheisa* wings and some *Utetheisa* that I outfitted with the wings of edible moths, and offered these items to *Nephila*. The results were revealing. Moths that bore *Utetheisa* wings—edible moths from the families Noctuidae or Notodontidae whose wings I had replaced with *Utetheisa* wings (I had simply glued the new wings to the basal stubs of the original wings, using droplets of Elmer's Glue)—were almost all cut from the web as soon as the spider made contact with the replacement wings. The other moths were also cut out, this time without exception, and always after the

Top: *Utetheisa ornatrix,* female (left) and male (right). Bottom left: *Utetheisa* on pods of the larval food plant, *Crotalaria mucronata.* Middle right: *Utetheisa* emitting froth in response to disturbance. Bottom right: Cells such as those shown here, from the blood of *Utetheisa,* are present also in the froth, which lends support to the notion that froth is blood plus air.

Pyrrolizidine alkaloids: monocrotaline (left) and usaramine (right).

spider made contact with the *Utetheisa* bodies. It appeared that all parts of *Utetheisa* were unacceptable to *Nephila*, including the wings. The few transvestites of the first category that were retained by the spiders were treated as composite morsels—their bodies were eaten, but the attached *Utetheisa* wings were ignored and eventually cut from the web.

We all knew, from the classic studies of the monarch butterfly, that insects can acquire their defensive substances from the diet. In the case of the monarch it had been shown, notably by Lincoln Brower and his collaborators, that the toxic steroids (cardiac glycosides) that protect that butterfly against birds are derived from the milkweed plants eaten by the caterpillar. It occurred to me that *Utetheisa* might itself be a monarch of sorts in that it also obtained its defensive chemicals from its food. There was logic to the assumption. As a caterpillar, *Utetheisa* feeds on plants of the genus *Crotalaria,* members of a family (the legume family, Fabaceae) that includes, among others, the peas and beans upon which we ourselves depend. *Crotalaria* species are not on our menu for a very good reason. The plants contain certain alkaloids called pyrrolizidine alkaloids that make them extremely poisonous. Cattle sometimes browse on *Crotalaria,* and when they do may not survive. Not surprisingly, research funds have been expended on the study of pyrrolizidine alkaloids, and the basis of their toxicity to mammals is now well understood.

Could pyrrolizidine alkaloids be responsible for the unacceptability of *Utetheisa* to spiders? Could *Utetheisa,* for some reason, be insensitive to the alkaloids and actually incorporate the compounds from its larval food plants? And does it retain the compounds through metamorphosis, as the monarch does with its defensive steroids, so that they are present in the adult?

Analyzing for pyrrolizidine alkaloid content is time-consuming but can be done with great precision. It was therefore possible to establish unambiguously, thanks to collaborators from Jerry's group, notably Robert K.

Vander Meer, Karel Ubik, James Resch, and the late Carl Harvis, that the *Utetheisa* adult contains pyrrolizidine alkaloids. In field-collected adults the amounts were variable, but substantial. Single individuals contained on average 0.7 milligrams of the toxins, or about 0.4 percent of body mass.

But how to prove that it is the pyrrolizidine alkaloids that provide the protection? Could we somehow come up with *Utetheisa* that were alkaloid-free to check whether these might be palatable? Could we rear *Utetheisa* on an alkaloid-free diet? I was pessimistic because I thought it would take forever to formulate an artificial diet that would be an adequate substitute for the real thing. But I was wrong. Investigators elsewhere had developed a diet for caterpillars other than *Utetheisa,* which *Utetheisa* not only accepted, but on which it flourished. And best of all, the diet, which was based on pinto beans rather than *Crotalaria* beans, was entirely free of pyrrolizidine alkaloids. We called that diet the pinto bean diet, or (−) diet for short, and proceeded to raise *Utetheisa* on it. We also established a culture of *Utetheisa* on a second diet, intended to be a substitute for the natural diet. We called that diet the (+) diet, because it contained the seeds of *Crotalaria spectabilis,* one of *Utetheisa*'s natural food plants. The (+) diet was in fact of the exact same composition as the (−) diet except that it contained *Crotalaria* seeds in replacement of 10 percent of its pinto bean content.

Together with Bill Conner and Karen Hicks, both new to our group at the time, we tested the moths of our two dietary lineages with *Nephila.* Jerry's lab had carried out analyses beforehand that showed the (−) moths—those raised on the (−) diet—to be alkaloid free. The (+) moths in contrast, the products of the (+) diet, contained pyrrolizidine alkaloid in an amount averaging 0.6 milligrams. Our (+) moths, intended to be laboratory versions of natural *Utetheisa,* were evidently a close chemical match of the real thing. They were also a visual match. (+) *Utetheisa,* (−) *Utetheisa,* and field-collected *Utetheisa* all look alike.

The results exceeded all expectations. The (+) moths were almost all rejected by the spiders, but the (−) moths were consistently eaten. The (−) moths seemed "unaware" of their chemical deficiency. They remained passive when inspected by the spiders and did not even struggle when the latter bore down to bite. Some did emit froth when inspected or bitten, but the spiders appeared in no way to be affected by the fluid.

Left: *Utetheisa ornatrix* moths that were offered to the spider *Nephila clavipes*. The specimen on the left, rejected intact, was raised on its normal, pyrrolizidine alkaloid–containing food plant *(Crotalaria mucronata)*. The specimen on the right, raised on the (−) diet, was eaten, and reduced to a small packet of solid remains. Right: *Utetheisa ornatrix* larvae feeding on the (−) diet.

To determine whether it was specifically because of the pyrrolizidine alkaloids that the spiders were being "turned off," we did one more test. We offered the *Nephila* mealworms, which we knew they liked, and offered these both untreated and treated by addition of pyrrolizidine alkaloid. We had in our possession one pyrrolizidine alkaloid in crystalline form, a compound called monocrotaline, that we knew to be present in species of *Crotalaria*. We made a monocrotaline solution, trickled some of it on mealworms, and then, after allowing the solvent to evaporate, offered the mealworms to *Nephila*. The treated mealworms proved far less palatable than the controls. A much larger portion of the bodies of treated mealworms was left uneaten than of the control mealworms, which had received a coating of the solvent only.

The relationship of *Utetheisa ornatrix* to its food plants is an obligatory one. The larvae, in nature, are found only on *Crotalaria,* and the adult female, so far as is known, does not lay eggs on other plants. At the Archbold Station and its surroundings *Utetheisa* occur on two primary food plants, *Crotalaria mucronata* and *Crotalaria spectabilis.* The two plants differ chemically in that their primary pyrrolizidine alkaloids are slightly different. In *C. spectabilis* the primary alkaloid is monocrotaline, and in *C. mucronata* it is usaramine. Field-collected *Utetheisa* at the Archbold Station contain primarily usaramine, which indicates that the moths develop predominantly on *C. mucronata*. Field observation con-

THE SWEET SMELL OF SUCCESS

Crotalaria mucronata. Clockwise from the top left: a mature plant, an inflorescence, ripening pods, and mature pods.

Utetheisa ornatrix. Top left: Adults mating. Top right: An egg cluster. Bottom left: A larva in the process of chewing its way out of the egg. Bottom right: The same larva, moments later, emerging from the egg.

firms this. *Utetheisa* larvae are found mostly on *C. mucronata,* which happens also to be the more abundant of the two *Crotalaria* species at the station.

Utetheisa lay eggs in clusters ranging broadly in size. The average egg count per cluster is 20. The largest cluster I ever recorded had exactly 100 eggs, but I have also noted an occasional egg that had been laid singly. The larvae hatch 4 to 5 days after the eggs are laid and they emerge from the eggs in close synchrony.

At first the larvae feed on the leaves of the plant, but as they grow they shift their priority to the seeds. *Crotalaria,* like other legumes, encloses its seeds in pods, which the larvae breech by chewing a circular hole through the pod wall. The holes give the larvae away and I have made it a practice to collect larvae by checking perforated pods for their contents. The technique is not very efficient. In most cases, the pods turn out to be seedless and empty, in evidence of previous occupancy. *Utetheisa* populations vary greatly in density. When they are at a peak, the larvae may be so numerous as to cut significantly into the seed production of the plant. At such times I can imagine the larval search for seeds to be a competitive endeavor—essentially a quest for a limited resource. There could be winners and losers in that competition, the latter being forced to subsist on leaves alone. Bill Conner showed that the larvae can subsist on leaves, but that they then accrue smaller amounts of alkaloid. Losers in the competition for seeds therefore pay the penalty of being potentially more vulnerable to predation, which could put them at a serious disadvantage. Pod occupancy appears to be limited to one larva per pod, even when the occupant takes up no more than a fraction of the pod's interior. Competition can also come from another source. A pyralid moth with the delightful name of *Etiella zinckenella,* a species known to feed on many leguminous plants and to be a pest of some of our crops, also feeds on *Crotalaria* and takes up residence in its pods. *Etiella* can achieve high densities and may then occupy a large proportion of *Crotalaria's* pods. Double occupancy of pods by *Etiella* and *Utetheisa* appears to be rare, so I can imagine *Etiella* being a serious contender for the seeds. *Etiella,* incidentally, does not sequester pyrrolizidine alkaloids. It is entirely alkaloid-free, and as a result fully palatable to *Nephila.*

Utetheisa larvae migrate away from their food plant to pupate. At the Archbold Station they may crawl to nearby pine trees and pupate between flakes of bark or, alternatively, on shrubs or herbaceous vegetation.

The *Crotalaria* plants themselves are shrubby and grow in clusters. Shoulder high at maturity, the plants senesce after a few years and entire stands may then vanish, only to reappear at some later time. *C. mucronata* characteristically make their reappearance when an area has undergone a natural burn.

Utetheisa moths spend much of their adult life among the *Crotalaria* plants. They even court amidst *Crotalaria.* If you want to collect *Utetheisa,*

Utetheisa ornatrix. Top left: A half-grown larva. Top right: A nearly mature larva, chewing its way into a pod of *Crotalaria mucronata*. Bottom left: A larva inside a pod, feeding on seeds. Bottom right: A pupa on the trunk of a pine tree *(Pinus eliottii)*, exposed by the removal of bark flakes.

the recommended procedure in Florida is to drive along the back roads and pull to a stop when you see a *Crotalaria* stand. The chances of *Utetheisa* being absent from the surroundings are low. In the spring, after a frosty winter, you may not be so lucky. *Utetheisa* are sensitive to cold and may not survive prolonged freezes. But depopulated areas do not take long to be recolonized, which suggests that the moths are prone to disperse as adults.

Tests that Maria and I did with *Utetheisa* larvae left no doubt that these too derive protection from the pyrrolizidine alkaloids. We showed this in tests with wolf spiders. Larvae collected on *Crotalaria* in the wild, as well as those raised on the (+) diet, were consistently rejected uninjured by the spiders, while larvae raised on the (−) diet were consistently eaten. We did some of the tests with captive spiders, which we maintained on sand in plastic cages, but did others with spiders outdoors. Walking about at night on the sandy fire lanes of the Archbold Station with our headlamps, we would locate lycosids by their reflective eyeshine. They were there by the hundreds, poised in wait for prey, and could be tested simply by dropping larvae directly in front of them. They pounced on the offerings at once, releasing within seconds those that contained alkaloid. We also did tests with adult *Utetheisa* and showed that lycosids rejected these as well, but only if they contained alkaloid.

We had earlier checked whether the eggs of *Utetheisa* also contained pyrrolizidine alkaloid and found that they did. The mother moth evidently is a provider and the question was whether the provision was adequate. Jim Hare, a Canadian biologist with a magical talent for experimentation and a marvelously warm personality, showed that ants *(Leptothorax longispinosus)* are deterred by pyrrolizidine alkaloids and that they discriminate against *Utetheisa* eggs if these contain the toxins. Again, we were able to take advantage of the fact that we could use alkaloid-free items as controls. Our (−) *Utetheisa* produced alkaloid-free eggs, and these proved acceptable to the ants. Jim also showed that the ants developed a long-term aversion to *Utetheisa* eggs once they had been exposed to (+) eggs. They then were likely to ignore *Utetheisa* eggs, whether these contained alkaloid or not. Even as late as a month after the negative experience the ants still showed an aversion to the eggs. The mental capacity of ants is evidently not to be underestimated.

I had often seen chrysopid larvae scurrying about on *Crotalaria* plants and it seemed reasonable to assume that these versatile little hunters were natural enemies of *Utetheisa* eggs. The larvae were easy to maintain, and they were ferocious consumers. I confined some in petri dishes and found that individually, if kept unfed for a day or two, they could eat over 30 *Utetheisa* eggs in succession. They evidently had the capability of dispatching entire clusters of *Utetheisa* eggs.

We collected numbers of *Ceraeochrysa cubana* larvae and watched how they disposed of the *Utetheisa* eggs we offered them. Eggs that they accepted they impaled on their hollow, sickle-shaped jaws and sucked out. Eggs that they rejected they also impaled, but only briefly. Occasional eggs were rejected on the basis of mere prodding, but most were lanced first. It seemed that it was the inside of the egg that had to be tasted by the larva in order for an egg to be judged unacceptable.

We offered individual larvae a choice of 10 (−) and 10 (+) eggs and they made it very clear that they could tell the two types apart. They ate each of the 10 (−) eggs and not a single one of the (+) eggs. They always skewered some of the (+) eggs before deciding to forgo a given (+) cluster, but they never sucked out such "sampled" eggs. What was interesting is that they always sampled about the same number of eggs—2.4 eggs on average per cluster to be exact—before rejecting the cluster. Sampled eggs died, but they were but a fraction of the batch. On average, between 6 and 7 of the eggs in the (+) batches went on to hatch.

We also presented chrysopid larvae with natural *Utetheisa* egg clusters that we had collected on *Crotalaria* plants. These clusters contained anywhere from 1 to 54 eggs. Of the 24 clusters tested, 3 proved edible and were entirely or almost entirely eaten. The remaining 21 clusters were rejected, and rejection was again on the basis of egg sampling. The average number of eggs skewered per cluster was 2.3, almost the same as the number sampled in the choice tests. Chrysopid larvae evidently gauge the quality of a cluster on the basis of an assessment procedure that does not vary with cluster size. Large clusters lose the same number of eggs to sampling as small clusters. In terms of the net number of eggs surviving chrysopid assaults, large clusters are therefore at an advantage.

You would predict, judging from the chrysopid larva's strategy of rating clusters by a randomized subsampling of its eggs, that the eggs in a cluster are evenly endowed with pyrrolizidine alkaloid. If the endowment was variable, it would make more sense for the larva to sample persistently

Top left: A scrub jay *(Aphelocoma coerulescens)* in the wild, feeding from a dish in which it was offered a choice of *Utetheisa ornatrix* and other items. The dish is partly buried in the sand. The bird has in its bill a mealworm, which it favored over *Utetheisa*. Top right: A lycosid spider feeding on a (−) *Utetheisa* larva. Bottom left: A coccinellid beetle *(Coleomegilla maculata)* preying on *Utetheisa* eggs. Bottom right: A chrysopid larva *(Ceraeochrysa cubana)* feeding on *Utetheisa* eggs. The larva has just skewered one egg with its hollow sickle-shaped mandibles, and is imbibing its contents. Two eggs, already sucked out, are seen to its right.

and try out the entire cluster for taste. Thanks to Eva Benedict in Jerry's lab, who took on the demanding task of analyzing individual eggs, we learned that within clusters there is indeed little variation in egg alkaloid content. Eva's data did show, however, that egg alkaloid content varies substantially from cluster to cluster. Values she obtained for 15 natural egg clusters ranged from a high of 1.5 micrograms per egg to a low of zero. It is not surprising, therefore, that some of the natural egg clusters that we offered to chrysopids had proved palatable.

We also staked out, on *Crotalaria* plants outdoors, (+) and (−) egg clusters from our laboratory cultures of *Utetheisa*. As might have been predicted, the (+) clusters proved less vulnerable than the (−) eggs. Fully 38 percent of the 26 (−) clusters showed evidence of having been dispatched by chrysopids. The eggs in such clusters were hollowed out and bore the circular puncture marks indicative of jaw penetration. Of the 26 (+) clusters only one (4 percent) had fallen victim to a chrysopid.

Utetheisa appears also to be unacceptable to birds. I did some tests at the Archbold Station in which I offered adult *Utetheisa* to scrub jays and found that the birds would avoid the moths on sight. I did the tests outdoors, where the birds could be easily trained to feed from plastic dishes. I dug these dishes into the sand, so that they would be flush with the soil surface, and baited them with three items at a time; an *Utetheisa* adult, a mealworm, and a peanut (cut into several pieces). To keep the moth from flying away I had cut into the leading edge of one of its front wings. The birds ate the mealworm and the peanut pieces but in most cases did not even peck at the *Utetheisa*. Only young female jays actually seized the moths in the bill, but they did not eat them. The jays were being used in experiments by other investigators at the station and they were individually recognizable by colored bands that had been put on their feet. Hence my ability to recognize the young females. The birds, incidentally, discriminated also against (−) *Utetheisa,* so I cannot be certain that it was on account of the alkaloid that they were averse to the moths. In the absence of more extensive testing I cannot claim, therefore, that the jays had had previous experience with *Utetheisa,* that they found the moths bad-tasting because of the alkaloid, and that they had learned to ignore them henceforth.

Some predators appear to be insensitive to the alkaloids. Captive toads *(Bufo americanus),* for instance, took *Utetheisa* unhesitatingly, without suffering noticeable ill effects as a consequence.

An unexpected potential enemy of *Utetheisa* is *Utetheisa* itself. *Utetheisa* larvae have an avidity for pyrrolizidine alkaloid. They eagerly eat any chewable item, even filter paper or agar, so long as it bears alkaloid. But their hunger for alkaloid is manifest only if they themselves lack alkaloid—only if they have been raised on the (−) diet. Nice experiments demonstrating this were done by an undergraduate research student, Jack Pressman, and a graduate student and expert parasitologist, Curt Blankespoor.

A postdoctoral associate from Germany, Franz Bogner, went on to show that in the laboratory the appetite for pyrrolizidine alkaloid can be responsible for inducing cannibalism in (−) *Utetheisa* larvae. *Utetheisa* larvae, for as long as they lack the alkaloid, will eat both *Utetheisa* eggs and pupae, provided these bear the alkaloid. Alkaloid-free eggs and pupae are not in danger of being cannibalized. Chemical data showed that cannibalism paid off for the cannibal. The cannibal does indeed acquire the alkaloid of the victim.

The question was whether cannibalism is practiced by *Utetheisa* larvae in nature. Cannibalism, after all, could be a neat strategy by which alkaloid-deficient larvae—losers in the competition for seeds—made up for the chemical shortage. Maria and I, with Bogner's help, staked out a total of 137 (−) and (+) *Utetheisa* egg clusters on *Crotalaria* plants and by checking periodically came upon four instances where a larva was in the process of eating a (+) cluster. We concluded that egg cannibalism was probably of real significance in the life of *Utetheisa*.

We also staked out pupae on *Crotalaria* and found instances where (+) pupae were under cannibalistic attack by larvae, but the experiments were not true to nature because *Utetheisa* do not normally pupate on their food plant. Perhaps *Utetheisa*'s habit of pupating away from *Crotalaria* can be viewed as a tactic for escaping cannibalistic threat.

Jim Hare and I wondered whether *Utetheisa* larvae might be selective in their egg cannibalism and feed preferentially on the eggs of nonrelatives. We found that they exercised no such discrimination. I was particularly interested in whether newly emerged larvae ever attacked unhatched eggs of their own cluster, but they do not. For one thing they have little opportunity to do so, since the eggs of a cluster tend to hatch in near synchrony. And besides, at emergence the larvae seem intrinsically reluctant to attack eggs.

There are two types of enemies against which *Utetheisa* eggs appear to be defenseless: pathogenic fungi and parasitoid wasps. Gregory Storey, a collaborator from the University of Florida, found *Utetheisa* eggs to be no better off with than without the alkaloid in fighting off such disease-causing fungi as *Beauveria* and *Paeceliomyces*. And as regards parasitoids, I have had many a field-collected *Utetheisa* egg cluster give rise to tiny

Top left: Parasitoid (*Tolenemus* species) emerging from *Utetheisa ornatrix* egg. Top right: A *Utetheisa* larva, killed by microbial infection. Middle left: A (−) *Utetheisa* larva, feeding on (+) *Utetheisa* eggs. Middle right: A (−) *Utetheisa* larva feeding on a (+) *Utetheisa* pupa. Bottom left: A larva of *Etiella zinckenella* feeding inside a pod of *Crotalaria mucronata*. Bottom right: An adult *Etiella zinckenella*.

THE SWEET SMELL OF SUCCESS

brachonid wasps instead of the expected caterpillars. I have also seen parasitoid wasps in the act of laying eggs on *Utetheisa* eggs in the field. They were so intent on their task that I could transport them to the laboratory, together with the leaf and the egg cluster, for a photographic session, without scaring them off. The pupae of *Utetheisa* are also subject to parasitism. Among the insects that we had emerge from pupae were four species of tachinid flies, a chalcidid wasp, and an ichneumonid wasp of the genus *Corsoncus* that turned out to be an undescribed species.

◼ **HAVING *UTETHEISA*** in culture meant that we could broaden our inquiries about the moth. The whole subject of acquired defense was tantalizing. Questions in that domain were being asked about the monarch butterfly, but our little moth was much easier to work with because it could be maintained on an artificial diet and raised to be either alkaloid-laden or alkaloid-free. I was intrigued by how in their quest for alkaloid the larvae might compete for food plant seeds, but not everyone in the lab shared interest in that topic. Bill Conner was eager to study the reproductive biology of the moth and he proposed looking into courtship. I remember asking whether he really wanted to do this, since moth courtship was being studied in so many other labs. But he persisted and I relented. Bill had a good track record. As an undergraduate, he had done research on mosquitoes in the laboratory of the legendary George Craig of Notre Dame University, a laboratory that had already spawned a whole galaxy of entomological stars, including James Truman and H. Frederick Nijhout, so I never doubted that whatever Bill chose to do would pan out.

Moths, as a rule, court at dusk, and the ritual involves the chemical attraction of the flying male to the stationary female. There is no way that the sexes could find each other by scent alone if both were on the wing. It is difficult enough to locate an odor source that is stationary. In moths, it is the female that emits the chemical attractant that signals that she is receptive. The chemical she uses for that purpose is by definition a pheromone, a substance that conveys messages between members of the same species.

Bill's initial tests involved caging individual virgin female *Utetheisa* from our colonies in small screened containers and placing these containers outdoors besides stands of *Crotalaria mucronata* at the Archbold

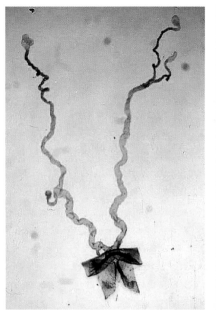
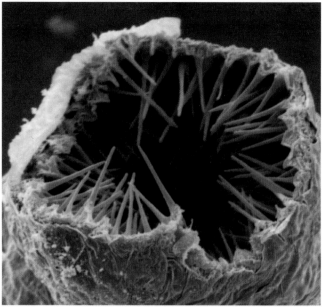
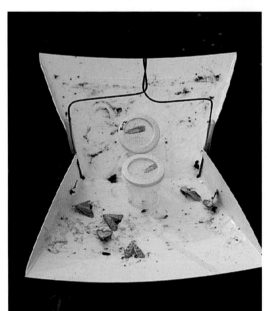
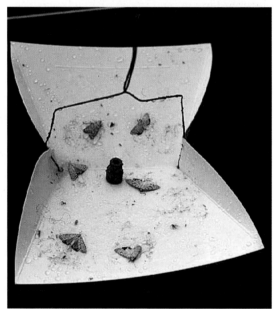

Top left: The two pheromonal glands of the *Utetheisa ornatrix* female, dissected together with a portion of the abdominal tip on which they open. Top right: A view into the inside of the female's pheromone gland. Bottom left: A sticky trap baited with two caged virgin *Utetheisa* females. A number of males have been attracted and have become stuck in the glue of the trap. Bottom right: The same trap, but now baited with a sample of the triene (the chemical was trickled into the rubber cup at the center).

THE SWEET SMELL OF SUCCESS

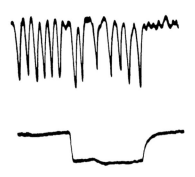

A *Utetheisa ornatrix* electroantenno-gram (EAG) response to a "scenting" female (above) and to a sample of the triene. Downward deflection of the baseline is an indication of stimulus detection by the antennal nerve.

Station. By positioning himself close to such cages he was able to keep visual track of approaching males, and what he found is that these would make their appearance at a fixed time, in the half hour following the hour after sunset. I have myself observed males in the field flying to caged females, and it is quite a remarkable sight. The males arrive with great punctuality, always from downwind, and along a straight aerial trajectory. Although somewhat at the mercy of air turbulence, they are remarkably adept at steering the course, even if momentarily blown from the intended path.

To characterize the female pheromone we resorted to some standard procedures. We confined virgin females to glass chambers and flushed these with air that we then passed into a trap containing a chemical absorbent. We then extracted the absorbent with solvent, fractionated the resulting extract by gas chromatography, and tested the ensuing fractions for electroantennogram activity. The electroantennogram technique, known as the EAG technique, offers a simple way for testing whether a given chemical sample stimulates the antennal nerve of an insect. Antennae are an insect's nose and therefore an appropriate organ for assessment of chemical sensitivity. The EAG technique involves isolating an insect antenna and hooking it up to electrodes so that you can eavesdrop on the response of the antennal nerve as stimuli are presented to the antenna in the form of chemical puffs. We systematically stimulated antennae of *Utetheisa* males with fractions of our extract and found the antennal response to be strongest to a fraction that Jerry's associates found to contain an unsaturated hydrocarbon, Z, Z, Z-3, 6, 9-heneicosatriene. For obvious reasons we adopted a shortened name for the compound. In acknowledgement of its three double bonds, we called it the triene. Jerry's group synthesized the compound.

Synthetic triene proved highly active as an antennal stimulant in EAG tests, and it lured males in tests we did in the field. We placed some of the compound in rubber cups that we affixed to "sticky traps"—glue-covered trays that we hung near *Crotalaria* stands outdoors—and found that males were attracted and became stuck in these traps. Caged virgin females that we used in control tests with the same kind of traps also brought in males.

We had obviously isolated the right compound, but learned later that the attractant pheromone of *Utetheisa* sometimes contains two additional

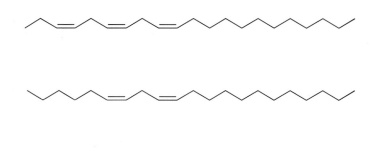

The three components of the sex attractant of the female *Utetheisa ornatrix:* from top to bottom, triene, diene, and tetraene.

hydrocarbons, differing from the triene in having, respectively, 2 and 4 double bonds. We called these compounds diene and tetraene, respectively. Their presence is not consistent in the pheromone and we do not know whether they enhance the attractancy of the mixture, although we did establish that both are active in EAG tests.

It became a rather easy task to locate the glands that produce the attractant pheromone. They are two tubular structures that open separately near the abdominal tip. Ordinarily highly coiled, they have been straightened out for photographic purposes in the photo on page 368.

Bill had noticed that when the females were scenting—when they were "calling" males—their abdomens underwent a throbbing action, in which they extruded, with rhythmic precision, what is in fact two of their terminal abdominal segments. The openings of the glands, which are on the membranes between these segments, were automatically exposed with each extrusion. The extrusion rate was in the range of one to two per second, roughly at a par with a human heartbeat. We debated what this might mean and speculated that it could indicate that the female emits her pheromone in pulses. A temporal patterning had never been demonstrated for an aerial chemical signal and we were intrigued by the idea.

I'll never forget the expression of joy on Bill's face when he greeted me one morning with a succinct statement. "She pulses," he said. "I've got proof." Bill had taken females that were visibly scenting and placed them in an airstream a few centimeters upwind from an EAG preparation. The electrical response of the antennal nerve showed a beautiful oscillation, essentially an on-and-off rhythm, matching the throbbing rate of the female's abdomen. A more elegant proof of the discontinuous pheromonal emission could not have been obtained. Control tests with pieces of filter paper bearing triene, positioned upwind at the same site as the females, yielded sustained antennal responses.

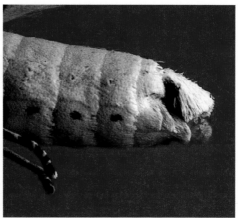

Top: The abdomen of a *Utetheisa ornatrix* female during pulsed pheromone emission. The process is characterized by periodic protrusion of the abdominal tip, as shown on the right. The pheromone gland openings are exposed by such protrusion. Bottom: Simulation of a temporally patterned aerial chemical signal, imitative of *Utetheisa*'s pheromone plume. Air laden with a visible marker (titanium tetrachloride) is pulsed at the rate of two "puffs" per second from the glass capillary tube shown at the left (windspeed = approximately 11 centimeters per second). The pulses remain discrete over the distance, about 60 centimeters, spanned by the photograph.

   The female glands have some special features. They lack compressor muscles and are overlain by secretory cells only. Internally they are beset with spines and therefore only partly compressible. They are air-filled, although their entire inner surface is coated with a thin film of secretion.

   We can only speculate on how the glands operate. But since they are tubular, partly compressible, and (because of their inner spines) elastically reexpansible, we could imagine them ventilating mechanically in the manner of lungs. We concluded that pulsation is achieved simply by rhythmic changes in abdominal blood pressure. Such changes might be the natural concomitants of the visible abdominal motions that character-

THE SWEET SMELL OF SUCCESS

ize the scenting process and they could effect the regular compression and decompression of the glands. Air drawn into the glands during gland decompression, by contacting the pheromonal film within, would become saturated with pheromone vapor, and on subsequent compression would be expelled as a pheromonal puff. Ongoing "inhalation" and "exhalation" would generate the discontinuous pheromonal output. Essential to this mechanism are the internal spines of the glands. These ensure not only that the evaporative film of pheromone within the glands is appropriately large, but that the glands will reexpand under their own elasticity during decompression.

We wondered what *Utetheisa* might accomplish by pulsing its attractant pheromone and briefly entertained the idea that by so doing the female was sending a species-specific code. In other words, we thought that the female was identifiable to the male both by the chemistry of the pheromone and by the discontinuity of the pheromonal output. But given air turbulence, it seemed unlikely that the puffs could remain physically discrete, and the code discernible, at any meaningful distance from the source. And indeed, when we simulated the pheromone delivery system of *Utetheisa* by generating a pulsed plume of air labeled with a visible marker (titanium tetrachloride), we found that beyond a distance of about 1 meter from the source the pulses were no longer detectable as such, even at moderate wind speeds.

Another possibility that occurred to us is that the female, by pulsing the pheromone, was providing orientation cues by which she might be more easily located. Male moths, when they detect the scent of a female, initially respond simply by flying upwind. The pulsation, we thought, might tell the male that he is close to target and might help prevent him from overshooting his goal. But we had no evidence to suggest that the pulsation provided such cues.

It is known now that *Utetheisa* is not alone among moths in pulsing its attractant pheromone. Pulsation rates are generally similar in moths that pulse, which in itself negates the possibility that by pulsing the females are sending out diagnostic codes. The most likely possibility is that pulsation helps the female economize on the amount of pheromone released. Pheromone production, after all, does not come free. But neither does the pumping action that effects the pulsation. It would be interesting to know to what extent the metabolic savings accrued through pulsed emission are offset by the muscular costs of the pumping.

We ourselves did not pursue these questions. Instead, we became interested in the behavior that comes into play once the male encounters the female. Mating, it turned out, was by no means the immediate or inevitable sequel to the encounter. *Utetheisa,* before mating, engages in "pillow talk." We decided that it might be worth listening in.

■ **AT THE ARCHBOLD STATION,** Bill Conner had arranged to videotape the courtship of *Utetheisa* under seminatural conditions. The technique involved releasing males at a site outdoors and luring these to an individual stationary female that was kept under observation with a video camera. The tests were carried out during the period of semidarkness to darkness when *Utetheisa* ordinarily court, which necessitated illuminating the scene with infrared lamps and monitoring events with a special video camera sensitive to infrared light (the same type of camera used to detect nocturnal intruders in commercial establishments).

The females in the tests were coaxed to take up a resting position on an upright wire perch placed directly in front of the video camera. The coaxing procedure was facilitated by rubbing the perch beforehand with freshly macerated *Crotalaria* leaves. *Utetheisa* females ordinarily "call" males while positioned on *Crotalaria,* and the pretreatment of the perch probably helped make the females feel at home. Males were released simultaneously at distances of 2 meters in a circle around the female. The observers, together with the video monitor and the video tape recorder, were positioned several meters outside the circle.

Tests with "normal" *Utetheisa*—with individuals captured in the wild—had a high success rate. Ten courtship sequences that were videotaped all culminated in copulations. The events proceeded as follows: The male approaches the female from downwind; the male hovers beside the female and contacts her with his antennae and legs; the male flexes his abdomen abruptly and, in a brief action lasting about a third of a second, thrusts the abdominal tip toward the female; the female raises her wings, thereby exposing her abdomen; the male lands alongside the female, makes genital contact, and proceeds to copulate.

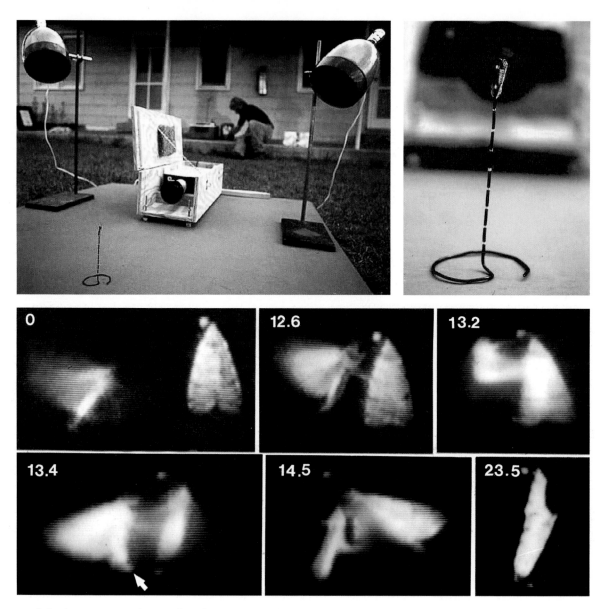

Top left: The experimental set-up for videotaping *Utetheisa ornatrix* courtship. The female moth is positioned on the upright wire stand, placed in front of the boxed video camera. Two infrared lights are providing illumination. The observer, with the video monitor, is in the background. For photographic purposes the picture was taken in daylight. Top right: A close-up of the female on the wire perch. Bottom: A courtship sequence, photographed directly from the screen of the video monitor (the numbers give the time in seconds from the first frame shown). The male approaches the female from downwind (0); the male contacts the female (12.6); the male flexes his abdomen and thrusts his abdominal tip toward the female while everting some sort of structure (13.2); the male straightens his abdomen while the structure (arrow) is still everted (13.4); the male makes genital contact with the female (14.5); and finally the male mates with her (23.5).

THE SWEET SMELL OF SUCCESS

Top left: A male *Utetheisa ornatrix* brushing his everted coremata against a female in courtship. Top right: Everted coremata; the "hairs" on the brushes are modified scales. Bottom left: Three corematal scales, transected to show their hollow insides. Bottom right: An enlarged view of the corematal scale surface, showing the minute pores through which hydroxydanaidal is presumed to seep out from the scale's hollow.

The event that captured our attention was that momentary abdominal thrust effected by the male, which seemed always to precede copulation. Although image resolution was far from satisfactory on the videotapes, it seemed that at the moment of thrusting some sort of structure became everted from the male's abdominal tip. In parallel observations we made on *Utetheisa* that courted in normal light in the laboratory, we thought we could see that these structures consist of a pair of brushes. We tried to capture the moment of thrusting in photographs and eventually succeeded. Thrusting was indeed the maneuver by which the male wiped a pair of eversible brushes against the female. The brushes, which consist of tufts of modified scales, were known. They had previously been described and named coremata (singular, corema). Moreover, in another species of *Utetheisa* they had been shown to be glandular and to produce a compound, hydroxydanaidal, whose structure was such as to suggest that it was derived from pyrrolizidine alkaloid. But the function of the brushes, including the possible communicative role of hydroxydanaidal, remained unknown.

The coremata are easily removed surgically. All you have to do is squeeze the male's abdomen so as to cause the brushes to be partly everted and to snip them off with microscissors when they protrude. Males thus treated are fully viable but they proved less acceptable to females. Bill videotaped 11 courtship sequences with such males and found 5 to be rejected. Being corematectomized did not keep the males from being attracted to the females, and it did not prevent them from executing the abdominal thrusts that ordinarily accompany the corematal eversions. But the females seemed uninterested. They showed reluctance to present the abdomen by raising their wings, and on occasion made attempts to evade the males (by moving around the perch to the opposite side). The unacceptability of the males could not have been due to side effects from the surgery itself. Males that were sham-operated—males that were physically handled like corematectomized males but had scales removed from beside the coremata rather than from the coremata themselves—fared normally (8 of 9 were accepted).

Hydroxydanaidal.

From Jerry's lab came the news that the coremata of our species of *Utetheisa* produced hydroxydanaidal as well (or HD, as we shall call the compound). Moreover, the supposition that HD might be produced from pyrrolizidine alkaloid received instant support when we found that our (−) *Utetheisa,* which had no dietary access to such alkaloid, had no trace of HD in their coremata.

Bill wasted no time in testing (−) *Utetheisa* in his courtship tests. Such *Utetheisa* offered the opportunity of checking on the success of males that lacked HD without lacking the organs by which they ordinarily administer the compound in courtship. The (−) males showed the same limited success as corematectomized males. They showed decreased ability to induce female wing raising, increased tendency to trigger female evasion, and reduced incidence of mating (7 of 28 were rejected).

A simple additional test demonstrated that HD had signal value—that it had true pheromonal capacity. Scenting females that were stimulated in the laboratory by being stroked with everted coremata responded more often by raising their wings if the coremata bore HD. Stimulation involved brushing the coremata a fixed number of times over a specific body region of the female. Isolated male abdomens, inflated with air to induce corematal eversion, were used to administer the stimulus. The coremata of field-collected males were more effective by far than coremata from (−) males, and HD itself, when added to the coremata of (−) males, rendered such coremata potently stimulative. It was clear that HD could help the male gain access to the female.

Anatomically, the coremata are essentially thin-walled sacs, internally furnished with scales, and ordinarily kept withdrawn by invagination. The sacs can be abruptly everted, as they are in courtship, which causes the scales to be splayed. The scales are covered with a thin film of oily fluid (HD is an oil), presumably produced by secretory cells associated with the base of the scales. The scales are hollow and bear minute pores. We think the oil is secreted into the hollow of the scales and seeps through the pores onto the scale surface. Permanently wetted, the coremata are thus kept ready for action.

Although the story was coming together I was bothered by the fact that we did not really have an explanation for what the male was "saying" with its brushes. Was HD simply the male's way of announcing its arrival—of saying to the female, "I am here, I've heeded your call, raise your wings, and let's mate"? Or might he be telling her something more subtle about himself? I had an idea that I dismissed at first but that wouldn't go away (I remember it first occurred to me when I was watching sea otters off the California coast, near Carmel.) We knew from some of the early chemical work that *Utetheisa* in the wild differ greatly in body alkaloid content. I didn't know the reasons for this, but speculated that *Utetheisa* might differ in their larval ability to obtain the chemicals from their food plants. It could all relate to competition for seeds, I thought. Larvae might differ in their ability to locate seeds or to gain access to seeds (some larvae could be more pushy than others in the pods), or they could differ in their ability to absorb alkaloids from the food. These abilities could be under genetic control. If so, might not HD be an indicator of the male's competitive ability? Might this molecule, which we knew to be derived from an alkaloid, be used by the male to tell the female how much alkaloid he possessed and therefore, indirectly, how good he was in competing for the chemical? And did it not make sense for the female to show preference for males able to offer proof of such a capacity, since the trait could be heritable? Should the female not favor a male that "smelled sweetly of success?"

In the paper in which we published the results of our courtship studies we advanced these speculations, not realizing that we were short on facts, and not nearly imaginative enough.

■ IF WE WERE RIGHT in our suggestion that the male used HD to advertise his alkaloid content and that his alkaloid content was a reflection of his larval alkaloid-acquiring ability, one would assume two relationships to hold true for the male: (1) a quantitative relationship between the amount of alkaloid ingested as a larva and the amount absorbed from the gut and retained in the body into the adult stage, and (2) a quantitative relationship between the amount of alkaloid acquired and the amount of HD produced in the coremata.

Both proportionalities were confirmed. We fed *Utetheisa* larvae on diets containing increasing concentrations of *Crotalaria* seeds (and therefore of alkaloid) and found that the more seeds they ingested, the more alkaloid they stored. And the more alkaloid they stored, the more HD they produced in their coremata.

Dave Dussourd was in the midst of his graduate studies and, wanting to expand his research beyond the study of vein-cutting insects, decided to look into the defensive chemistry of *Utetheisa* eggs. It was not long before he made a major discovery. The (−) *Utetheisa* females, he found, could lay (+) eggs. Paradoxical as that seemed, the explanation was straightforward. All such females had to do was mate with (+) males. Such males transmitted alkaloid to the females with the sperm package (the spermatophore) and the females bestowed some of this gift on the eggs. The females did not use only their mate's alkaloid for the purpose. If they themselves contained alkaloid (which they were bound to, in varying measure, under natural conditions) they contributed some of their own alkaloid as well, so that the eggs received biparental protection.

We did an experiment by which we determined how much alkaloid is contributed to the eggs by each parent. We raised males and females on diets containing different pyrrolizidine alkaloids, monocrotaline in case of the females and usaramine in case of the males, then paired the sexes and—with the help of Jerry's associates—analyzed the eggs for alkaloid content. The eggs contained primarily monocrotaline, which indicates that the mother had been the chief donor, but the father's contribution amounted to fully one third of the total. Predation tests done with ladybird beetles showed that this lesser amount, of paternal origin, made a difference. Eggs that received alkaloid from the father only, from crossings where the mother had been alkaloid-free, were significantly less acceptable to the beetles than eggs that were entirely alkaloid-free.

We were forced to modify our view of the coremata's role. HD, it seemed, could serve not only as a proclamation of alkaloid load and of a genetic capacity, but also as a direct announcement of a nuptial gift. By way of HD the male could be providing a measure of how much alkaloid he holds in store for his mate, and the female could be exercising mate choice on the basis of that promise. We postulated that the magnitude of the male's alkaloidal offering—the amount of alkaloid he bestows upon the female—should be proportional to his body alkaloid content, and found that to be the case. Since we knew the corematal HD titer to be a reflection of the male's body alkaloid content, it followed that HD does indeed provide a measure of the male's alkaloidal gift.

We wondered whether the female herself derives benefit from the male's gift and were able to show that she does. And the benefit is immediate. The (−) females, which we knew to be vulnerable to attack by lycosid spiders, lost that vulnerability when they mated with a (+) male. We proved this experimentally by pairing the *Utetheisa* in Ithaca and testing the females with spiders at the Archbold Station. We mailed the mated females to Florida by overnight express so that we would be able to test them as soon after mating as possible, but came to realize that overnight delivery was not early enough. The females were already fully unacceptable by the time we received them in Florida. In fact, we completed the experiment by taking lycosid spiders back to Ithaca, where we were able to offer them females that had literally just completed mating. And sure enough, as early as 5 minutes after uncoupling from the male, a female *Utetheisa* could already prove distasteful. We injected some (−) females with monocrotaline, in amounts comparable to what they would receive at mating, and found that they were already judged unacceptable by lycosids within 5 minutes after injection. The alkaloidal gift is evidently put to virtually immediate use by the female. And it can convey lasting protection. Even senescent females that we offered to lycosids on day 18 after receipt of their alkaloidal gift proved unacceptable to the spiders (the lifespan of adult *Utetheisa* is about 3 weeks). Over time after mating, as the female bestows alkaloid on the eggs, she evidently does not exhaust her supply of the chemicals. She retains the alkaloid in sufficient quantity for her own protection, thereby exercising a strategy that appears intended to protect the egg carrier as well as the eggs. Additional tests with *Nephila* showed that the male's alkaloidal gift can protect female *Utetheisa* against this spider as well.

In a sense it could be argued that our experiments with (−) *Utetheisa* were not natural, inasmuch as such females have little chance of occurring in nature. Female *Utetheisa* are likely always to contain at least some quantity of self-acquired alkaloid. But that quantity is variable, and it can be low, as when the larvae had access to leaves primarily, rather than seeds. The male's gift may therefore constitute an important supplement, which during times of alkaloid shortage could be the bonus that "makes the difference."

These experiments on nuptial acquisition of defensive capacity by the female had been fun for me because they provided me with the opportunity to converse in Spanish again. Even better still, they provided me with a chance to speak Uruguayan Spanish, which has a character all its own. My team had been joined by two Uruguayan graduate students, a married couple, Carmen Rossini and Andrés González, both trained in chemistry and biology and both eager to cast their lot with *Utetheisa*. They were the best of research partners. Blessed with the benefits of academic and parental effort, they are now back in Uruguay, with their daughter, Paulita, and freshly received Ph.D. degrees, promoting the virtues of chemical ecology in Montevideo's educational circles. It had meant a great deal to me to have young Uruguayans in training in my lab, particularly ones willing eventually to return to Uruguay and make a go of it despite the limited opportunities for research available in their country. Uruguay had sheltered my family during World War II and introduced me to the world of bugs. Carmen and Andrés brought back the

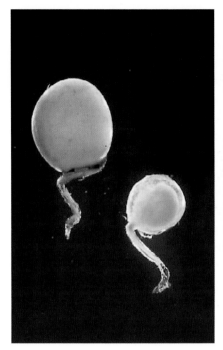

memories and unwittingly rekindled my feeling of indebtedness to their country.

When male *Utetheisa* mate, they lose, on average, 10 percent of their body mass. The loss is to the benefit of the female, in which it registers as an equivalent gain in mass. Copulation takes upward of 9 hours in *Utetheisa*. It is during that time that the male transmits the spermat-ophore, and it is this package that accounts for the mass transferred at mating. Does alkaloid plus sperm account for the entirety of the sper-matophore? Most certainly not. A substantial fraction of the spermat-ophore is made up of nutrient, as seems to be the case generally in Lepidoptera. Although we have never analyzed this nutrient chemically in *Utetheisa,* we know from indirect evidence that the female puts the nu-trient to use. Craig LaMunyon, who also got his Ph.D. by befriending *Utetheisa,* showed that mating a second time enables female *Utetheisa* to increase their egg production by about 15 percent. In fact, *Utetheisa* fe-males are promiscuous and they appear able to boost egg production after later matings as well. Mating is thus to be viewed as a means by which the female *Utetheisa* stocks up, not just on sperm and alkaloid, but on nutri-ent, which she is able to invest in egg production.

Frequency of mating can be determined with precision in female

Utetheisa. The reason for this is that for every spermatophore she receives, she retains the colla, its narrow twisted "stalk," which survives the breakdown undergone by the remainder of the structure. All you need to do to check into the mating history of a female is to dissect her bursa (the pouch that receives the spermatophores) and count the number of colla within. Data published by others told us that females mated on average with four to five males over their lifetime. We found that the number could be higher. In one population at the Archbold Station females had on average 11 colla; the record holder had a total of 23.

By mating frequently the female is provided with an ongoing supply of nutrient for egg production and alkaloid for egg defense. She is also provided with a diversity of sperm, but there is evidence that she does not utilize sperm from all partners. Craig LaMunyon showed elegantly that twice-mated females use sperm from one male only; by checking on the biochemical traits of the offspring, he was able to rule out that sperm from both male partners had been utilized. Moreover, he demonstrated that it was not a matter of being the first or second mate that determined whether a given male's sperm was used. The deciding factor was the size of the spermatophore—it was the sperm from the larger spermatophore that had the competitive advantage. We don't know that the rules of the game are the same for later matings, but see no reason why sperm of larger spermatophores might not consistently have the edge. We are also uncertain as to how the competition plays itself out between sets of *Utetheisa* sperm. Do the sperm from different males somehow fight it out inside the female? Or does the female exercise some degree of control over the selection process? LaMunyon found that if the female is anesthetized, and therefore presumably unable to put into action whatever muscles in her inner reproductive chambers ordinarily control the process by which one set of sperm is favored over another, the selection process breaks down. But either way, whatever the mechanism of sperm sorting, it is certain that the "winning" fathers that produce the larger spermatophores are themselves physically larger. Physical size in male *Utetheisa* correlates positively with alkaloid content as well, and therefore also with both alkaloid-donating capacity and corematal HD content. By favoring males with a high HD content, female *Utetheisa* are therefore choosing males with a large body size. To expand the perspective a bit, since spermatophore size can be expected to be an indicator of spermatophore nutrient content, by favoring large males, the females are also providing for receipt of larger nutrient gifts. Large males are therefore preferable because they are more generous donors of both alkaloid and nutrient. The question is whether they were also the source of better genes. They obviously would be if body size is a heritable trait.

And indeed it is. In careful experiments, involving matings of male and female *Utetheisa* of known body mass, and determinations of correlations of the offspring's and parents' body mass, another student devotee of *Utetheisa*, Vikram Iyengar, showed that body mass is heritable in both sexes, and therefore under genetic control.

What this meant in *Utetheisa* is that by choosing a larger male, the female is assured that her offspring will themselves be larger and, as a consequence, potentially have increased fitness. Larger sons, by virtue of being larger, could be expected to be more successful in courtship, and larger daughters (as we knew from separate experiments) could be expected to lay more eggs. Vik looked into the fitness of the progeny of his controlled pairings, and found that both predicted advantages held true. Sons of larger fathers (or, for that matter, of larger mothers) were at an advantage in being chosen in courtship, and daughters were more fecund.

In very elegant experiments, Vik also showed that it is by HD alone, rather than its correlates, that the female *Utetheisa* appraises the male. In other words, the female judges the male by the intensity of his pheromonal scent only, rather than by his body size or alkaloid content. He was able to demonstrate that females fail to differentiate between males that differ in body mass or alkaloid content if such males lack HD, but does discriminate between males that are size-matched and alkaloid-free if one of the males has been experimentally endowed with HD. He also showed that females can discriminate between males that differ in incremental quantities of HD. The male's pheromonal scent, therefore, is indeed a chemical yardstick.

There is much that we still hope to learn about *Utetheisa*, although we do feel we have come to understand some subtleties of its sexual strategy. It is clear that the female chooses her partner on the basis of his scent, and that by favoring the more strongly scented, she provides for her own increased fecundity and defensibility, as well as for the increased defensibility of the eggs and the improved genetic quality of the offspring. We would like to know more about the males. How often do they mate, and do they show any discrimination against mated females? The evidence that we do have suggests that they are indeed promiscuous, but that they mate with any female that accepts them. "Any port in a storm" seems to be their guiding principle. And what about when males have very little alkaloid? Do they then "lie" by producing exaggerated quantities of HD? We have little evidence, but what we do know suggests that the males are honest "salesmen."

We are in love with *Utetheisa*. The moth has introduced us to levels of complexity of insect life we never imagined could exist. It taught us to ask questions, and it was generous with its answers—generous enough to crown five graduate students with Ph.D.s. It has led us also to look into others species, and to the discovery of other sexual strategies, involving in some cases mate appraisal and gift giving, as in *Utetheisa*. The courtship of *Neopyrochroa* discussed in the last chapter was one such side venture. But there were others.

■ **ARCTIID MOTHS** include many colorful species, but *Cosmosoma myrodora* must be one of the most beautiful. It occurs at the Archbold Station and it is there, and at his current academic home, Wake Forest University, in Winston-Salem, North Carolina, that Bill Conner, together with Ruth Boada, one of his students, studied the moth. He made his field observations in Florida, but succeeded in raising the moth in his laboratory, and it was there that he studied its courtship.

Cosmosoma, like *Utetheisa*, relies on pyrrolizidine alkaloid for defense, but it does not obtain the chemical from its larval food plant. That plant, *Mikania scandens*, a member of the Asteraceae, is totally free of pyrrolizidine alkaloid. In *Cosmosoma*, only the male procures the alkaloid, and it does so as an adult. It is attracted to plants such as *Eupatorium capillifolium* (Asteraceae), which contains pyrrolizidine alkaloids, and it imbibes alkaloid by lapping up excrescent fluid from the plant's surface with its proboscis. *Cosmosoma* males can be collected by luring them to senescent *Eupatorium* plants, and it may be the alkaloid itself that draws them to the plant tissue. In the laboratory, male *Cosmosoma* eagerly feed on crystalline pyrrolizidine alkaloid.

By presenting *Cosmosoma* males to *Nephila* spiders, Bill showed that the acquired alkaloid pays off for the moths. Whereas males are palatable be-

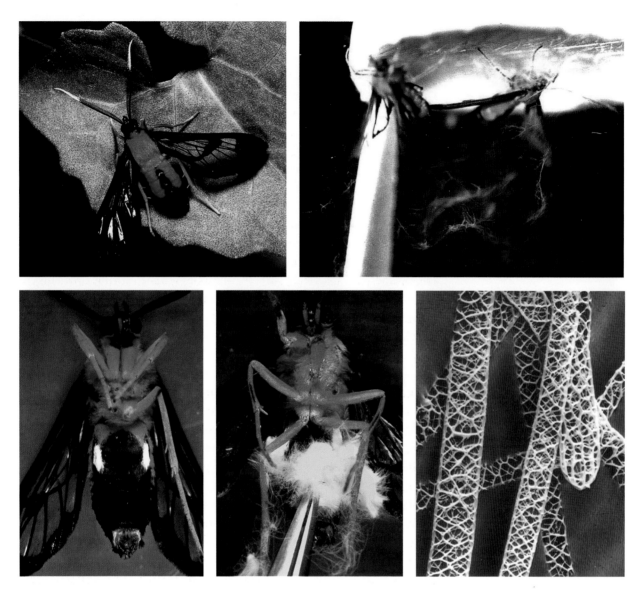

Top left: A *Cosmosoma myrodora* (male) on its larval food plant, *Mikania scandens.* Top right: Courting *Cosmosoma,* photographed in a wind tunnel; the male (left) has just ejected his flocculent, which is visible as a diffuse cloud. Bottom left: Ventral view of a male; the flap that covers the abdominal pouch is the area between the white markings. Bottom middle: The same, with flocculent being teased out of the pouches with forceps. Bottom right: A close-up view of the filaments of flocculent.

fore alkaloid acquisition, they are cut from the web after the enrichment. The parallel with *Utetheisa* is striking.

The *Cosmosoma* male shares the alkaloid with the female, and he does so in a remarkable way. At the base of the abdomen, on the ventral surface, the male has a pair of large pouches, ordinarily invisible from the outside, within which he keeps tucked away a cottony mass of fine cuticular filaments. There are thousands of these fibers in the pouches, which literally bulge as a result. Working in collaboration with Bill, Frank Schroeder in Jerry's lab showed that following ingestion of the alkaloid, the entire cottony mass, or flocculent as Bill termed it, becomes impregnated with the alkaloid. The male, it turns out, can expel the flocculent, and he does so explosively, directly upon the female, during courtship. It is as if he were festooning her. He waits until he is within her immediate vicinity, having presumably been drawn to her by her attractant pheromone, and then "lets go." The effect is decorative, but the consequences are defensive. Coated with filaments, the female is left garmented in alkaloid, and the garment protects her against *Nephila*. Females are palatable to the spider before courtship, but they are substantially less acceptable after being with an alkaloid-fed male or after being experimentally decorated with alkaloid-laden flocculent.

The male also transfers alkaloid to the female with the sperm package, alkaloid that the female subsequently allocates in part to the eggs.

We know of no other instance of nuptial bestowment of defensive filaments by a male insect. Nor do we know whether the flocculent has signal value in *Cosmosoma*. We cannot rule out, for instance, that the female gauges male worth by assessing the alkaloid content of the flocculent. She could, by such assessment, conceivably obtain a measure of the alkaloidal gift that might be forthcoming from the male with the sperm package, and choose a mate on the basis of that criterion.

At high magnification the *Cosmosoma* filaments are stunningly beautiful. They are modified scales, and appear to be built for lightness, flexibility, and strength. They are flattened rather than cylindrical, and highly sculpted. The sculpting could have the special function of providing for a large evaporative surface for dissemination of the alkaloid.

■ **GIFT GIVING** of pyrrolizidine alkaloid, from male to female by seminal infusion, is not restricted to arctiid moths among the Lepidoptera, but

THE SWEET SMELL OF SUCCESS

occurs also in a group of butterflies known as the danaines (subfamily Danainae). One of these is the queen butterfly, *Danaus gilippus,* a milkweed feeder and close relative of the familiar monarch butterfly.

Years ago, before I became seriously involved with *Utetheisa,* a group of us worked on the queen, and it was the queen that spurred our interest in the defensive use of pyrrolizidine alkaloids by insects. The year was 1966 and we were paid a visit by Lincoln Brower, the world's expert on the monarch, who had just finished making a movie of the courtship antics of the queen. The queen courts in daytime. Males and females spot one another on the wing, by sight, as butterflies generally do, without use of a chemical attractant. But chemistry does come into play at close range in the queen once the male comes upon the female. He then flutters about her, and as he does he everts a pair of coremata from the abdominal tip, brushlike structures very much like those of *Utetheisa,* but quite clearly of separate evolutionary origin. Lincoln suggested that we look into the role of the coremata and he proposed that we enlist an undergraduate honor student of his, Thomas Pliske, in the effort. We followed his advice, and in due course, thanks to Tom's collaboration with Jerry's lab, were able to show that the queen's coremata play an important role in male acceptance. Jerry's group showed that the coremata are glandular and that they produce a compound, danaidone, that bears a fundamental similarity to pyrrolizidine alkaloid. An Australian investigator, J. A. Edgar, and his collaborators showed that close relatives of the queen butterfly do indeed derive corematal compounds from ingested pyrrolizidine alkaloids, which left little doubt that danaidone itself was also of alkaloidal origin. We know now that the queen male obtains its pyrrolizidine alkaloid in *Cosmosoma* fashion, from excrescences that it imbibes from the surface of pyrrolizidine alkaloid–containing plants. Pliske had shown that possession of danaidone increases the male's chance of acceptance in courtship. Males lacking the compound were decidedly less successful, but they could be made more acceptable if danaidone was experimentally added to the coremata.

The similarities to *Utetheisa* were striking, but they did not stop there. Dave Dussourd, while still a graduate student, showed that there is a transfer of alkaloid from the male to the female in the queen, and that the female transmits the gift to the eggs. The transfer occurs with amazing efficiency. The male transmits 63 percent of his acquired alkaloid to the female and she bestows 90 percent of the amount received to the eggs.

I would like to believe that there is a basic theme emerging from these

The queen butterfly, *Danaus gillipus*. Top left: A mating pair. Top right: A male feeding on crystalline monocrotaline. Bottom left: The everted coremata of the male. Bottom right: A close-up view of coremetal hairs. Note the covering of fine particles. These are coated with danaidone, and are transferred to the surface of the female (including the female's antennae) during courtship.

THE SWEET SMELL OF SUCCESS

Danaidone.

studies pertaining to male salesmanship in courtship and the egg provisioning that is the subject of the self-advertisement. *Neopyrochroa, Utetheisa,* and the queen all represent proven examples, but cases in which males offer proof during precopulatory interaction of possession of defensive chemicals for transmission with the sperm may be more widespread than suspected. Entirely different insects may also engage in the strategy, and the defensive chemicals at stake could well include many more than pyrrolizidine alkaloid and cantharidin. The insect spermatophore could indeed be a vehicle for the most diverse types of substances. And, as the following example will show, the chemical within its wrappings need not be organic, but can be the simplest of inorganic commodities.

ANYONE who has ever seen butterflies in the act of puddling will not forget the experience. The phenomenon is particularly common in the tropics, where butterfly abundance is at its peak, but it is part of butterfly behavior in temperate areas as well. Puddling involves swarms of butterflies aggregating on wet soil for the purpose of drinking. Thousands of butterflies may congregate for the purpose. Side by side they drink, each a colorful addition to what is often a veritable carpet of jewels. Puddling can take place at the edge of lakes and streams, but it occurs most often beside puddles, hence its name. I have seen puddling aggregations in Uruguay and Panama, and quite spectacular ones, mostly of pierid butterflies, in the Arizona desert. Modest versions of it, sometimes with involvement of only a single species, occur in the countryside around Ithaca. Although it is not generally known, moths puddle as well. People tend to be oblivious of the nocturnal doings of moths. Naturalists collect moths at lights and therefore miss out on much of their behavior, but those who track about with headlamps do at times come upon puddlers in the night. Both butterflies and moths drink by use of the proboscis, the long tubular "soda straw" that ordinarily serves them for uptake of nectar and other nutrients. When not in use, the proboscis is kept coiled and hidden from view. During puddling it is fully extended, with the tip dipped into water or wet soil.

What I found most interesting about puddling is that puddlers are, overwhelmingly, males. There had to be an explanation for why the behav-

A butterfly *(Agraulis vanillae)* puddling. Note the filamentous proboscis.

ior was sex-biased, but none had been provided. Considering what we were discovering about seminal gift giving in insects, I was intrigued by the idea that by puddling males might be procuring a commodity that subsequently found its way to the eggs. Puddling seemed like an ideal next project. I was therefore delighted when our group was enriched by the addition of a graduate student with a passion for butterflies. Scott Smedley loved butterflies aux naturelles. He didn't catch them to pin them in boxes. He wanted to study them alive. And he shared an interest in puddling. The best suggestion then put forth about puddling was that it served to procure salt, or more specifically sodium ions, and there was evidence that some butterflies, in choice experiments, favored sodium-containing over sodium-free fluids. Moreover, in one species of skipper butterfly it was shown that the male loses sodium at mating, while the female is left sodium-enriched. But direct uptake of sodium through puddling had not been demonstrated, and there was no hint as to what females might do with sodium gifts.

There are some 120,000 described species of Lepidoptera in the world, and Scott thought he'd choose one for intensive study. He scanned the literature and, on the basis of reports that it was a good puddler, picked *Gluphisia septentrionis,* a moth of the family Notodontidae. He would not regret the choice. *Gluphisia* turned out to be a prodigious drinker and the best of experimental partners. It did everything "right," and although

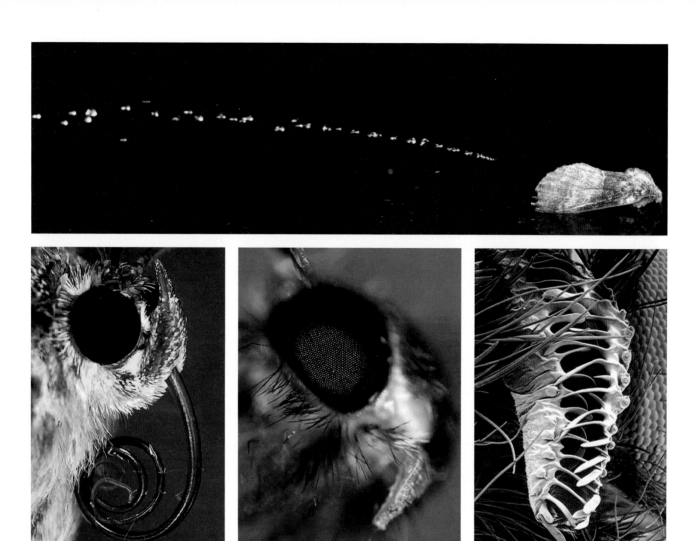

Top: A male *Gluphisia septentrionis* ejecting anal squirt while puddling. Bottom left: The head of an unidentified butterfly, showing the coiled proboscis that is typical of Lepidoptera. Bottom middle: The head of a male *Gluphisia septentrionis*, showing the short beak present in the species in lieu of a proboscis. Bottom right: An enlarged view of the *Gluphisia* beak, showing the cleft through which fluid is imbibed. Note the interdigitating projections that guard against intake of unwanted particulate matter.

sometimes difficult to obtain in appropriate numbers, provided us with all the answers we wanted.

Gluphisia males drink readily in the lab. All you need to do is confine them individually in plastic containers and add an artificial puddle. You may then need to coax the moths a bit, so that they make contact with the water, but when they do, they usually begin to puddle at once. Methodically, and sometimes uninterruptedly for as long as over 2 hours, they take in fluid and pump it through the body. The result is an ongoing voidance of liquid, in the form of anal jets, discharged at a rate of almost 20 per minute, to distances of 0.4 meters or more. The moths' performance is no less spectacular when they are puddling outdoors. On average, the puddling male passes nearly 14 milliliters of fluid per puddling session, by way of jets that amount individually to 12 percent of body mass. The all-time record was that of a male that ejected 38.4 milliliters of fluid, by way of 4,325 individual jets, over a period of just under 3.5 hours. For an 80-milligram *Gluphisia* male this amounts to over 600 times the body mass. The human equivalent, in imaginary urine output, is 45,500 liters, passed at a rate of 3.8 liters per second. Fraternity guzzling pales by comparison.

By determining the difference in sodium concentration in imbibed and ejected fluids Scott showed that puddling does indeed result in uptake of sodium. In a particularly neat experiment Scott made these determinations for fluid samples that he collected in nature. He located *Gluphisia* males that were puddling along dirt roads in rural Pennsylvania, and brought back samples of both puddle water (taken from directly in front of the puddler) and of the anal jets (caught by placing receptacles in the trajectory of the squirts). There could be no question. The ejected fluid was sodium-impoverished.

Further proof of uptakes was obtained by analysis of the body parts of *Gluphisia*. Males that had not puddled contained on average only 2.3 micrograms of sodium per individual. After puddling, they contained 17.0 micrograms. The duration that a moth spends at the puddle turned out, very logically, to depend on the concentration of sodium in the puddle water. Moths drinking a more concentrated solution (0.1 millimolar sodium) spent less than an hour puddling, while those given a more dilute source (0.01 millimolar sodium) puddled for over twice as long.

Gluphisia is specialized for rapid drinking. In lieu of the typical elongated lepidopteran proboscis, it has a short recurved beak, opening by way of an elongate frontal cleft. This gaping orifice is ideally suited for quick intake, and it is even shielded by an interdigitating arrangement of projections that guard against uptake of unwanted particulate matter. The male *Gluphisia* also has a specialized intestine. The anterior portion of the hindgut, where the sodium absorption is likely to take place, is longer and wider in the male than in the female, and in the male also bears a much folded inner lining. The male is thereby provided with a vastly expanded intestinal surface for ionic uptake.

Why is *Gluphisia* so intent on lapping up sodium? Its special appetite may be the consequence of a dietary deficiency. As a larva, *Gluphisia* feeds on quaking aspen *(Populus tremuloides)*, a species that contains sodium at concentrations (2.9 parts per million, dry weight) far lower than is average for trees.

For the female *Gluphisia,* which shares the ionic shortage of the male, the quest for sodium is linked to mating. As Scott showed in meticulous detail, the female obtains sodium by seminal transfer from the male. Moreover, this gift enables the female to remedy the ionic shortfall of the offspring. The female *Gluphisia* does indeed salt her eggs. Of the 17 micrograms of sodium that the male sequesters by puddling, 10 micrograms are on average transmitted to the female, of which about 5 micrograms are bestowed by her upon the eggs. Thus about one third of the sodium initially procured by the male makes its way to the offspring. The spermatophore in *Gluphisia* is therefore a well-salted package. Although it amounts to only 9 percent of male body mass, it conveys over half of a puddler male's body sodium.

And what about "sweet whisperings"? Does the female *Gluphisia* have the means to gauge the male's sodium content during courtship? To our knowledge she does not. We found no evidence that after puddling the males are somehow able and driven to proclaim that they are now "worth their salt."

Epilogue

How is it, I am often asked, that I make discoveries? I always feel a bit awkward about answering the question, because I do not have a particular method. The truth is that I spend a fair amount of time looking around. I already knew as a boy that if I wanted to see things happen—if I wanted to win the revelatory lottery of nature—I had to buy a lot of tickets. So it was in my youth that I formed the habit of taking exploratory walks, whenever possible and as often as possible, for the sole purpose of "eavesdropping" on nature. Naturalists thrive on such walks, driven by curiosity and the hope of witnessing chance events. Taken at face value, such events may not amount to much. But they may "connect" to what you already know, to previous observations stored away in your memory, and thus take on added meaning. There has to be a constant readiness to make such connections. Every tidbit of new information, no matter how trivial, has the potential of amounting to more than a speck of color. Properly assigned to the pointillist canvas that constitutes your inner view of the natural world, the new speck adds dimension to the vision.

Not everyone shares the predisposition constantly to place observation in context. I used to teach a course at the Archbold Station, entitled "Exploration, Discovery, and Follow-up," in which students designed a research project based on leads they obtained on exploratory walks at the Station. Although all were gifted and genuinely interested in nature, few had that capacity—so magically evident in Bob Silberglied, Ian Baldwin, Fotis Kafatos, and Mark Deyrup—of being able to conceptualize, to derive a greater reality from simple observation. I have been extremely lucky in having nature reveal itself on occasion through chance events in my presence. I can remember as if it were yesterday witnessing for the first time *Utetheisa* being cut from a spider web, or *Chrysopa* dressing itself as an aphid, or *Ammophila* carrying a "flower," and I yearn for future occasions when I may again be granted unexpected glimpses into the workings of nature. One of the great joys of returning to your natural haunts time and again, is that you have the opportunity of grasping the broader image. Observations tend then to become cumulative, to be evocative and revelatory in ways that are not possible until you begin to feel at home in the area. For the naturalist, in fact, feeling at home means having achieved a biological appreciation of a region.

An *Arion* slug (3×) and its eggs (6×).

I have a mental construct that leads me to ask four simple questions when I come upon something in nature that piques my curiosity, whether it is a structural feature of an animal or plant or a behavioral peculiarity. First I ask myself what the structure or behavior is all about. What is its function? In other words, what is its adaptive significance? That to me is always the fundamental question. I then pose the comparative query. What do other organisms have in lieu of such a structure or behavior? In other words, how are they specialized to achieve the same ends? And finally I ask two questions pertaining to origins. How is the feature likely to develop, to be formed embryologically? And, ultimately, how is it likely to have evolved, to have been molded over the course of generations? Mere reflection about such questions can help put a discovery in context.

Lest it be thought that I am entirely dependent upon the fortuitous, let me say that I also approach some problems on the basis of logic, or biorationality, as I am fond of saying. Some examples, based on recent experiences, will serve to illustrate the point.

I have always been fascinated by slugs, and by their ability to survive under hostile conditions. Living in soil, as so many of them do, entails special risks. Predators alone are a major hazard, in the form of ants, carabid beetles, centipedes, and spiders, and there is really no way that you can imagine slugs being spared exposure to such enemies. In particular you wonder how slug eggs manage to survive, given that they are not only immobile, but gelatinous and soft. It seemed to me a foregone conclusion that slug eggs are chemically protected. As a chemical ecologist, I found the temptation irresistible to put the idea to the test, and Frank Schroeder,

Miriamin.

a superb chemist from Jerry's group, was easily persuaded to join in the venture. The project, logically conceived, paid off. The eggs of a slug (a species of *Arion*) turned out to contain an interesting new isoprenoid compound, characterized as a polyoxygenated geranylgeraniol derivative, which we found to be potently deterrent to insects. Since the compound was new, we thought we would give it a name, and chose miriamin in honor of Miriam Rothschild, dear friend and naturalist extraordinaire. Miriam was delighted to be so honored, and found it in no way demeaning to be recognized via a slug product.

Biorationality told me also that the slugs themselves had to be protected, and I found that they did indeed have a remarkable way of coping with the likes of ants. There is a simple experiment anyone can do to activate this defense. Look for a slug, and when you find it, poke it gently with a toothpick. A pine needle or leaf stalk will do as well. As long as you keep the stick motionless, nothing will happen. But if you wiggle the stick, the slug will set in motion a coagulation mechanism, whereby the slime in the immediate vicinity of the contact point is converted into a rubbery blob that clings to the tip of the stick. The mechanism is wonderfully effective because it keeps an enemy from piercing the body wall of the slug. Ants are literally muzzled when they bite into a slug. They are thwarted the moment they bear down with their mandibles, and as they back away, are left with their mouthparts encased in coagulated slime.

I still don't know how the mechanism works but I can imagine that some sort of coagulation or macromolecular cross-linking is involved. Dan Aneshansley and I have data based on the response of slugs to localized application of mild electrical stimuli that show the coagulation to be triggered in a fraction of a second—literally in less time than it takes an ant to clamp down with the mandibles. And we have learned that in some slugs the coagulation is accompanied by the visible injection of crystalline material into the slime from specialized integumental cells. But basically

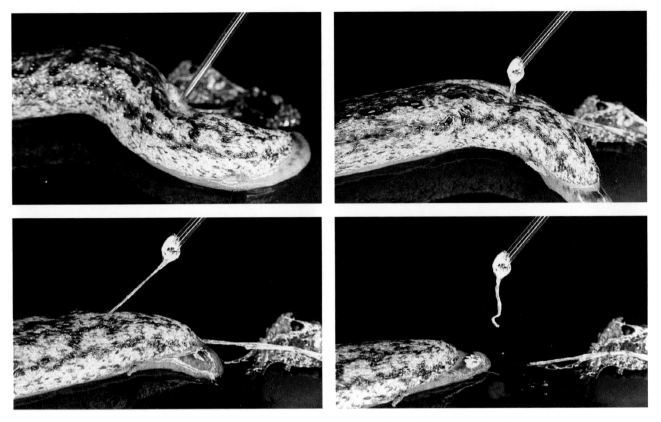

Slime coagulation in a slug. The experimenter presses a glass rod against the back of the slug and wiggles it, causing a localized plug of slime to form. The plug clings to the glass rod when the rod is pulled away.

An ant backing away from a slug it has just bitten. Coagulated slime is sticking to its mouthparts, and it is pulling a column of slime from the slug.

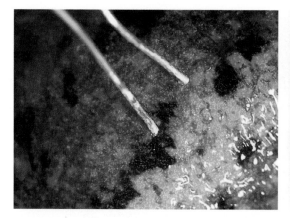 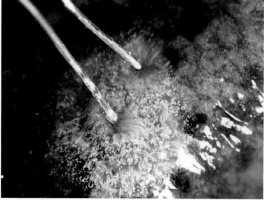

Electrical induction of slime coagulation in the slug *Philomycus carolinianus*. In this species, the coagulation is triggered when crystal-laden integumental cells void their white contents into the slime. In the right photo the emptying of these crystal-laden cells has been induced by mild electric shock delivered by way of the two electrodes shown.

we are still far from being able to explain the coagulation mechanism in molecular terms. Nature provides numerous examples of soil-dwelling animals with slimy investitures. Biorationality predicts that these slimes may also be worth investigating, not only to see how they confer protection, but to determine whether there are applied benefits to be derived from an understanding of their composition and physical structure.

Another project that was engendered by logic rather than chance observation led to the discovery of highly unusual secretory products in certain beetle pupae. The pupa is a relatively helpless stage in the life cycle of an insect. Unable to walk or fly, it is potentially vulnerable to any number of predators. Not surprisingly, many pupae are protected. Some are concealed underground, others are camouflaged, and still others are enclosed in cocoons. Quite unusual are pupae that defend themselves actively by biting. They don't use mandibles for the purpose, but alternative structures present on the thorax or abdomen and fashioned as jaws. Such devices, activated by body motions when the pupa is touched, can effectively deter predators. Many beetle pupae have such devices, which, to judge from their multiplicity of form, must have evolved independently in a number of beetle lineages.

Some years ago I took a fancy to ladybird beetles (family Coccinellidae) and got interested in their defenses. Their pupae in particular drew my attention. Conspicuously colored and positioned visibly on vegetation, they seemed ready made for the taking. I took a brush and proceeded to stroke

The pupa of the moth *Urodus parvula*. The cocoon, although loosely spun, doubtless protects the pupa against predation.

them, in hopes they might mistake me for an ant. They did, and whenever I touched them they retaliated by activating what I was to discover were extremely effective biting devices on the back of their abdomens. These devices take the form of four deep clefts, ordinarily held agape when the pupa is at rest with its body recumbent against the substrate. Disturbance, however, causes the pupa to straighten up, with the result that all clefts are snapped shut. Stimulating the pupa with a single hair suffices to trigger the response, as does exposure of the pupa to individual ants. No sooner did an ant brush its antennae against a pupa's back than the pupa flipped upright and "bit." Ants that were pinched fled instantly.

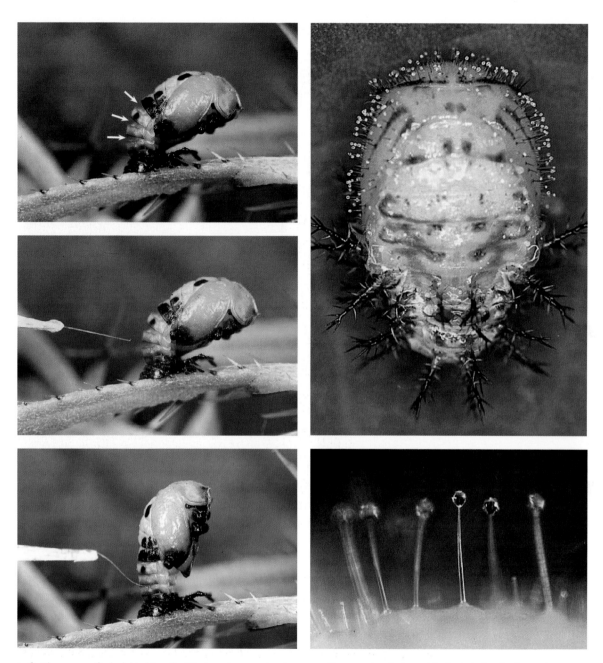

Left: The pupa of a ladybird beetle *(Cycloneda sanguinea)* responding to stimulation with the bristle of a fine paintbrush. The jawlike devices on the back of the pupa are ordinarily held agape (arrows). Poking the jaws with the bristle causes the pupa to flip upward, with the result that the bristle is "bitten." Top right: Dorsal view of a pupa of the Mexican bean beetle *(Epilachna varivestis)*. Notice the glandular hairs that fringe the pupa. Bottom right: An enlarged view of the glandular hairs of the *E. varivestis* pupa.

While experimenting with these "jaws," I remembered working earlier with a coccinellid beetle whose pupa seemed to lack biting devices. The coccinellid was a well-known pest species, the Mexican bean beetle, *Epilachna varivestis,* which Karen Hicks and I, together with Jerry's group, had found to be well protected as an adult, by virtue of chemicals in its blood. We had characterized one of these chemicals, an interesting tropane alkaloid, which we found to be deterrent to spiders, and had published these findings. I checked and found I had remembered correctly. There were no pupal snapping devices in *Epilachna*. Was it because *Epilachna* didn't need them? Surely not. *Epilachna* pupated out in the open like any coccinellid beetle and was bound to be plagued by the same set of enemies as coccinellids generally. Biorationality told me that *Epilachna* had to have an alternative defense.

I took a close look and found that the *Epilachna* pupa was indeed different in that it was covered with microscopic glandular hairs, tiny secretory structures consisting of slender stalks with droplets of secretion at the tip. The sides and back of the pupa were densely covered with these hairs, so that it would be virtually impossible for a predator even to inspect an *Epilachna* pupa without coming in contact with the secretion. This was an exciting finding. I teamed up with Scott Smedley and we were able to show that the hairs are indeed deterrent to ants. Mere contact with a few hairs caused ants to engage in frenzied cleansing activities. Other species of *Epilachna*, such as the squash beetle, *E. borealis,* also had pupal secretory hairs, as did species of the related coccinellid genus *Subcoccinella*. There was evidently a nice chemical problem here. The secretion could be collected in pure form simply by taking up one droplet after another in glass capillary tubing, so it was only a matter of time before we gathered enough material from three of the species for analysis. At the hands of three of Jerry's best associates, Kevin McCormick, Frank Schroeder, and Athula Attygalle, the chemistry panned out beautifully. The secretion turned out to contain an entirely new group of compounds that we called azamacrolides, which the beetles are able to synthesize from simple precursors, and which they produce as variants, the polyazamacrolides, with up to enormously large ring structures. Time will tell whether these novel substances have potentially applicable properties other than insect deterrency. I highlight them here not only to exemplify discovery through

An azamacrolide (left) and a polyazamacrolide (right).

biorational reasoning, but to illustrate that chemical prospecting in the world of insects can still bring real rewards.

I am often asked whether there is much left to be discovered in the natural world, and whether we can count on future generations remaining curious about nature. These are separate questions and I have no qualms about answering the first. I think the potential for discovery of novelty in nature is still enormously high. How could it be otherwise? With the technical tools now available, exploration can be carried out at all levels of biological organization, with the result that individuals of the most diverse backgrounds can join the ranks of the naturalist explorer. Think of it. Most species have not even been discovered yet, let alone examined for their biological characteristics. There is general agreement that the 1.5 million species so far described amount to less (probably *far* less) than half the total in existence. Think of the implications. There may be millions of unique biological entities awaiting discovery, each with its own particular habits, its own way of interacting with mates, enemies, pathogens, and symbionts. The opportunities for discovery in the decades ahead—of biological structures, of mechanisms and functions, and of course of new molecules—are literally limitless. Potentially, at least, natural history could be at the threshold of a golden age. Whether the potential is realized will depend, of course, on whether we have the good sense to preserve what is left of the natural world.

Will we remain curious enough about nature to maintain the exploratory momentum? Will the collective urge to discover keep natural history alive? Although I don't have the answer to this question, I despair when I think of the possibility that curiosity may be on the wane. Without curiosity, without a passion for discovery, nature cannot endure. And without nature, curiosity will fade. Think of the consequences if it came to the worst. Of what purpose would be our intellect in a world without nature? Of what purpose our senses, our eyes and ears? Imagine the riches foregone, the myriad stories left untold.

I cannot help feeling that, ultimately, curiosity will be sustained. It is so fundamentally human to thirst for knowledge and to turn to nature for visions of the unknown. Will we be wise enough to put limits on our encroachment upon nature? Perhaps we can keep the hope alive that we will eventually succeed in doing so. It might help if naturalists of all persuasions closed ranks to redefine the rules by which we coexist with the living world. Can love of insects make a difference? I am not sure. But I would like to believe that it does.

Bibliography

GENERAL READINGS

Ackerman, D. 1990. *A Natural History of the Senses.* New York: Random House.

Agosta, W. C. 1992. *Chemical Communication: The Language of Pheromones.* New York: Scientific American Library.

Angier, N. 1995. *The Beauty and the Beastly.* Boston: Houghton Mifflin.

Berenbaum, M. R. 1989. *Ninety-nine Gnats, Nits, and Nibblers.* Urbana: University of Illinois Press.

———1993. *Ninety-Nine More Maggots, Mites, and Munchers.* Urbana: University of Illinois Press.

———1995. *Bugs in the System.* Reading, Mass.: Perseus Books.

Bettini, S. 1978. *Arthropod Venoms.* Berlin: Springer-Verlag.

Borror, D. J., and R. E. White. 1970. *A Field Guide of the Insects of America North of Mexico.* Boston: Houghton Mifflin.

Blum, M. S. 1981. *Chemical Defenses of Arthropods.* New York: Academic Press.

Conniff, R. 1996. *Spineless Wonders.* New York: Henry Holt.

Covell, C. V., Jr. 1984. *A Field Guide to the Moths.* Boston: Houghton Mifflin.

Cott, H. B. 1957. *Adaptive Coloration in Animals.* 2nd ed. London: Methuen.

Dawkins, R. 1986. *The Blind Watchmaker.* New York: Norton.

Dethier, V. G. 1962. *To Know a Fly.* San Francisco: Holden-Day.

Deyrup, M. D. 2000. *Florida's Fabulous Insects.* Tampa, Fla.: World Publications.

Edmunds, M. 1974. *Defence in Animals.* New York: Longman.

Eisner, T., and J. Meinwald. 1995. *Chemical Ecology: The Chemistry of Biotic Interaction.* Washington, D.C.: National Academy Press.

Evans, D. L., and J. O. Schmidt, ed. 1990. *Insect Defenses.* Albany: State University of New York Press.

Evans, H. E. 1968. *Life on a Little-Known Planet.* New York: E. P. Dutton.

——1985. *The Pleasures of Entomology*. Washington, D.C.: Smithsonian Institution Press.

Futuyma, D. J. 1986. *Evolutionary Biology*. 2nd ed. Sunderland, Mass.: Sinauer Associates.

Gullan, P. J., and P. S. Cranston. 2000. *The Insects*. 2nd ed. London: Blackwell Science.

Heinrich, B. 1979. *Bumblebee Economics*. Cambridge, Mass.: Harvard University Press.

Hölldobler, B., and E. O. Wilson. 1994. *Journey to the Ants*. Cambridge, Mass.: Harvard University Press.

Hoyt, E., and T. Schultz. 1999. *Insect Lives*. New York: John Wiley and Sons.

Leopold, A. 2001. *A Sand County Almanac*. New York: Oxford University Press.

Nijhout, H. F. 1998. *Insect Hormones*. Princeton: Princeton University Press.

Opler, P. A., and V. Malikul. 1992. *Field Guide to the Eastern Butterflies*. New York: Houghton Mifflin.

Rosenthal, G. A., and M. R. Berenbaum. 1991. *Herbivores: Their Interactions with Secondary Plant Metabolites,* vol. 1. 2nd ed. San Diego: Academic Press.

——1992. *Herbivores: Their Interactions with Secondary Plant Metabolites, vol. 2.* 2nd ed. San Diego: Academic Press.

Sondheimer, E., and J. B. Simeone. 1970. *Chemical Ecology*. New York: Academic Press.

Tilden, J. W., and A. C. Smith. 1986. *A Field Guide to Western Butterflies*. Boston: Houghton Mifflin.

Von Frisch, K. 1967. *The Dance Language and Orientation of Bees*. Cambridge, Mass.: The Belknap Press of Harvard University Press.

Waldbauer, G. 2000. *Millions of Monarchs, Bunches of Beetles: How Bugs Find Strength in Numbers*. Cambridge, Mass.: Harvard University Press.

——2003. *What Good Are Bugs? Insects in the Web of Life*. Cambridge, Mass.: Harvard University Press.

White, R. E. 1983. *A Field Guide to the Beetles*. Boston: Houghton Mifflin.

Wickler, W. 1968. *Mimicry in Plants and Animals*. London: Weidenfeld and Nicolson.

Wigglesworth, V. B. 1964. *The Life of Insects*. Cleveland: World Publishing Co.

Wilson. E. O. 1971. *The Insect Societies*. Cambridge, Mass.: The Belknap Press of Harvard University Press.

——1984. *Biophilia*. Cambridge, Mass.: Harvard University Press.

——1992. *The Diversity of Life*. Cambridge, Mass.: The Belknap Press of Harvard University Press.

BIBLIOGRAPHY

————1994. *Naturalist*. Washington, D.C.: Island Press.

————2002. *The Future of Life*. New York: Alfred A. Knopf.

1 BOMBARDIER

Aneshansley, D. J., T. Eisner, J. M. Widom, and B. Widom. 1969. Biochemistry at 100°C: explosive secretory discharge of bombardier beetles *(Brachinus)*. *Science* 165: 61–63.

Barry, D. 1991. *Dave Barry Talks Back*. New York: Crown Trade Paperbacks.

Butenandt, A., and P. Karlson. 1954. Über die Isolierung eines Metamorphosehormons der Insekten in kristallisierter Form. *Zeitschrift für Naturforschung* 9b: 389–391.

Butenandt, A., R. Beckmann, D. Stamm, and E. Hecker. 1959. Über den Sexuallockstoff des Seidenspinners *Bombyx mori* Reindarstellung und Konstitution. *Zeitschrift für Naturwissenschaften* B14: 283–384.

Darwin, C. 1958. *The Autobiography of Charles Darwin and Selected Letters,* ed. F. Darwin. New York: Dover Publications.

Dean, J. 1980. Effect of thermal and chemical components of bombardier beetle chemical defense: glossopharyngeal response in two species of toads *(Bufo americanus, Bufo marinus)*. *Journal of Comparative Physiology* 135: 51–59.

————1980. Encounters between bombardier beetles and two species of toads *(Bufo americanus, Bufo marinus)*: speed of prey-capture does not determine success. *Journal of Comparative Physiology* 135: 41–50.

Dean, J., D. J. Aneshansley, H. E. Edgerton, and T. Eisner. 1990. Defensive spray of the bombardier beetle: a biological pulse jet. *Science* 248: 1219–1221.

Dialogues of Entomology. 1819. London: R. Hunter.

Edwards, J. S. 1961. The action and composition of the saliva of an assassin bug *Platymeris rhadamanthus* Gaerst. (Hemiptera: Reduviidae). *Journal of Experimental Biology* 38: 61–77.

Eisner, T. 1958. The protective role of the spray mechanism of the bombardier beetle, *Brachynus ballistarius* Lec. *Journal of Insect Physiology* 2: 215–220.

————1958. Spray mechanism of the cockroach *Diploptera punctata*. *Science* 128: 148–149.

————1970. Chemical defense against predation in arthropods. In E. Sondheimer and J. B. Simeone, ed., *Chemical Ecology*. New York: Academic Press.

————1972. Chemical ecology: on arthropods and how they live as chemists. *Verhandlungsbericht der Deutschen Zoologischen Gesellschaft* 65: 123–137.

Eisner, T. and D. J. Aneshansley. 1999. Spray aiming in the bombardier beetle:

photographic evidence. *Proceedings of the National Academy of Sciences USA* 96: 9705–9709.

Eisner, T., and D. Blumberg. 1959. Quinone secretion: a widespread defensive mechanism of arthropods. *The Anatomical Record* 134: 558–559.

Eisner, T., and L. M. Dalton. 1993. Emetic voidance of stomach lining induced by massive beetle ingestion in a beluga whale. *Journal of Chemical Ecology* 19: 1833–1836.

Eisner, T., and J. Meinwald. 1966. Defensive secretions of arthropods. *Science* 153: 1341–1350.

Eisner, T., D. J. Aneshansley, J. Yack, A. B. Attygalle, and M. Eisner. 2001. Spray mechanism of crepidogastrine bombardier beetles (Carabidae; Crepidogastrini). *Chemoecology* 11: 209–219.

Fieser, L. F., and M. I. Ardao. 1956. Investigation of the chemical nature of gonyleptidine. *Journal of the American Chemical Society* 78: 774–781.

Karlson, P., and M. Lüscher. 1956. "Pheromones": a new term for a class of biologically active substances. *Nature* 183: 55.

Pavan, M. 1959. Biochemical aspects of insect poisons. *Proceedings of the Fourth International Congress of Biochemistry, Vienna* 12: 15–36.

Rolander, D. 1754. Die Schussfliege. *Der Königlichen Schwedischen Akademie der Wissenschaften Abhandlungen* 12: 298–302.

Roth, L. M., and B. Stay. 1958. The occurrence of *para*-quinones in some arthropods, with emphasis on the quinone-secreting tracheal glands of *Diploptera punctata* (Blattaria). *Journal of Insect Physiology* 1: 305–318.

Rothschild, M. 1961. Defensive odours and Müllerian mimicry among insects. *Transactions of the Royal Entomological Society of London* 113: 101–121.

Schildknecht, H. 1957. Zur Chemie des Bombardierkäfers. *Angewandte Chemie* 69: 62.

Schildknecht, H., and K. Holoubek. 1961. Die Bombardierkäfer und ihre Explosionschemie. *Angewandte Chemie* 73: 1–7.

Schildknecht, H., E. Maschwitz, and U. Maschwitz. 1968. Die Explosionschemie der Bombardierkäfer (Coleoptera, Carabidae) III. Mitteilung: Isolierung and Charakterisierung der Explosionskatalisatoren. *Zeitschrift für Naturforschung* 23b: 1213–1218.

Stratton, J. A. 1991. Harold Eugene Edgerton. *Proceedings of the American Philosophical Society* 135: 443–450.

Thomson, R. H. 1971. *Naturally Occurring Quinones.* New York: Academic Press.

Williams, C. M. 1952. Morphogenesis and the metamorphosis of insects. *Harvey Lectures* 47: 126–155.

Wilson, E. O. 1958. A chemical releaser of alarm and digging behavior in the ant *Pogonomyrmex badius* (Latreille). *Psyche* 65: 41–51.

———1959. Source and possible nature of the odor trail of fire ants. *Science* 129: 643–644.

Wilson, E. O., N. I. Durlach, and L. M. Roth. 1958. Chemical releasers of necrophoric behavior in ants. *Psyche* 65: 108–114.

Zwingle, E. 1987. The man who made time stand still. *National Geographic* 172: 464–483.

2   VINEGAROONS AND OTHER WIZARDS

Chadha, M. S., T. Eisner, and J. Meinwald. 1961. Defense mechanisms of arthropods. IV. *Para*-benzoquinones in the secretion of *Eleodes longicollis*. Lec. (Coleoptera: Tenebrionidae). *Journal of Insect Physiology* 7: 46–50.

Eisner, H. E., D. W. Alsop, and T. Eisner. 1967. Defense mechanisms of arthropods. XX. Quantitative assessment of hydrogen cyanide production in two species of millipedes. *Psyche* 74: 107–117.

Eisner, H. E., T. Eisner, and J. J. Hurst. 1963. Hydrogen cyanide and benzaldehyde produced by millipedes. *Chemistry and Industry* 1963: 124–125.

Eisner, T. 1961. Demonstration of simple reflex behavior in decapitated cockroaches. *Turtox News* 39: 196–197.

———1962. Survival by acid defense. *Natural History* 71: 10–19.

———1966. Beetle's spray discourages predators. *Natural History* 75: 42–47.

Eisner, T., F. McHenry, and M. M. Salpeter. 1964. Defense mechanisms of arthropods. XV. Morphology of the quinone-producing glands of a tenebrionid beetle (*Eleodes longicollis* Lec.). *Journal of Morphology* 115: 355–400.

Eisner, T., D. Alsop, K. Hicks, and J. Meinwald. 1978. Defensive secretions of millipeds. In S. Bettini, ed., *Handbook of Experimental Pharmacology*, vol. 48, *Arthropod Venoms*. Berlin: Springer-Verlag.

Eisner, T., J. Meinwald, A. Monro, and R. Ghent. 1961. Defense mechanisms of arthropods. I. The composition and function of the spray of the whipscorpion, *Mastigoproctus giganteus* (Lucas) (Arachnida, Pedipalpida). *Journal of Insect Physiology* 6: 272–298.

Eisner, T., H. E. Eisner, J. J. Hurst, F. C. Kafatos, and J. Meinwald. 1963. Cyanogenic glandular apparatus of a millipede. *Science* 139: 1218–1220.

Guldensteeden-Egeling, C. 1882. Über Bildung von Cyanwasserstoffsäure bei einem Myriapoden. *Pflüger's Archiv für die gesamte Physiologie* 28: 576–579.

Happ, G. M. 1968. Quinone and hydrocarbon production in the defensive glands of *Eleodes longicollis* and *Tribolium castaneum* (Coleoptera: Tenebrionidae). *Journal of Insect Physiology* 14: 1821–1837.

Hurst, J. J., J. Meinwald, and T. Eisner. 1964. Defense mechanisms of arthropods. XII. Glucose and hydrocarbons in the quinone-containing secretion of *Eleodes longicollis. Annals of the Entomological Society of America* 57: 44–46.

Jones, T. H., W. E. Conner, J. Meinwald, H. E. Eisner, and T. Eisner. 1976. Benzoyl cyanide and mandelonitrile in the cyanogenetic secretion of a centipede. *Journal of Chemical Ecology* 2: 421–429.

Meinwald, J., K. F. Koch, J. E. Rogers, Jr., and T. Eisner. 1966. Biosynthesis of arthropod secretions. III. Synthesis of simple *p*-benzoquinones in a beetle *(Eleodes longicollis). Journal of the American Chemical Society* 88: 1590–1592.

Meinwald, Y. C., and T. Eisner. 1964. Defense mechanisms of arthropods. XIV. Caprylic acid: an accessory component of the secretion of *Eleodes longicollis. Annals of the Entomological Society of America* 57: 513–514.

Noirot, C., and A. Quennedey. 1974. Fine structure of insect epidermal glands. *Annals of the Entomological Society of America* 19: 61–80.

Peschke, K., and T. Eisner. 1987. Defensive secretion of a beetle *(Blaps mucronata):* physical and chemical determinants of effectiveness. *Journal of Comparative Physiology* 161: 377–388.

Rossini, C., A. B. Attygalle, A. González, S. R. Smedley, M. Eisner, J. Meinwald, and T. Eisner. 1997. Defensive production of formic acid (80%) by a carabid beetle *(Galerita lecontei). Proceedings of the National Academy of Sciences USA* 94: 6792–6797.

Thiele, H. U. 1977. *Carabid Beetles in Their Environments.* Berlin: Springer-Verlag.

3 WONDERS FROM WONDERLAND

Attygalle, A. B., D. J. Aneshansley, J. Meinwald, and T. Eisner. 2001. Defense by foot adhesion in a chrysomelid beetle *(Hemisphaerota cyanea):* characterization of the adhesive oil. *Zoology* 103: 1–6.

Blest, A. D. 1957. The function of eyespot patterns in the Lepidoptera. *Behaviour* 11: 209–256.

Common, I. 1990. *Moths of Australia.* Carlton, Victoria: Melbourne University Press.

Dani, F. R., S. Cannoni, S. Turillazi, and E. D. Morgan. 1996. Ant repellent effect of the sternal gland secretion of *Polistes dominulus* (Christ) and *P. sulcifer* (Zimmerman) (Hymenoptera: Vespidae). *Journal of Chemical Ecololgy* 22: 37–48.

Darwin, C. 1898. *Insectivorous Plants*. New York: Appleton.

Davidson, B. S., T. Eisner, and J. Meinwald. 1991. 3,4-Didehydro-ß-ß··-caroten-2-one, a new carotenoid from the eggs of the stick insect *Anisomorpha buprestoides*. *Tetrahedron Letters* 32: 5651–5654.

De Brunhoff, L. 1981. *Babar's Anniversary Album*. New York: Random House.

Duelli, P. 1984. Oviposition. In M. Canard, Y. Séméria, and T. R. New, ed., *Biology of Chrysopidae*. The Hague: Dr. W. Junk.

Eisner, T. 1964. Catnip: its raison d'être. *Science* 146: 1318–1320.

———1965. Defensive spray of a phasmid insect. *Science* 148: 966–968.

———1985. Still more on bird attacks. *New England Journal of Medicine* 313: 1232–1233.

Eisner, T., and D. J. Aneshansley. 2000. Defense by foot adhesion in a beetle (*Hemisphaerota cyanea*). *Proceedings of the National Academy of Sciences USA* 97: 6568–6573.

Eisner, T., and M. Eisner. 2000. Defensive use of a fecal thatch by a beetle larva (*Hemisphaerota cyanea*). *Proceedings of the National Academy of Sciences USA* 97: 2632–2636.

Eisner, T., and Y. C. Meinwald. 1965. Defensive secretion of a caterpillar (*Papilio*). *Science* 150: 1733–1735.

Eisner, T., and J. Shepherd. 1965. Caterpillar feeding on a sundew plant. *Science* 150: 1608–1609.

———1966. Defense mechanisms of arthropods. XIX. Inability of sundew plants to capture insects with detachable integumental outgrowths. *Annals of the Entomological Society of America* 59: 868–870.

Eisner, T., M. Eisner, and M. Deyrup. 1996. Millipede defense: use of detachable bristles to entangle ants. *Proceedings of the National Academy of Sciences USA* 93: 10848–10851.

Eisner, T., E. van Tassell, and J. E. Carrel. 1967. Defensive use of a "fecal shield" by a beetle larva. *Science* 158: 1471–1473.

Eisner, T., D. Alsop, K. Hicks, and J. Meinwald. 1978. Defensive secretions of millipeds. In S. Bettini, ed., *Handbook of Experimental Pharmacology*, vol. 48, *Arthropod Venoms*. Berlin: Springer-Verlag.

Eisner, T., M. Eisner, D. J. Aneshansley, C.-L. Wu, and J. Meinwald. 2000. Chemical defense of the mint plant, *Teucrium marum* (Labiatae). *Chemoecology* 10: 211–216.

Eisner, T., A. F. Kluge, M. I. Ikeda, Y. C. Meinwald, and J. Meinwald. 1971. Sesquiterpenes in the osmeterial secretion of a papilionid butterfly, *Battus polydamas*. *Journal of Insect Physiology* 17: 245–250.

Eisner, T., A. B. Attygalle, W. E. Conner, M. Eisner, E. MacLeod, and J. Meinwald. 1996. Chemical egg defense in a green lacewing *(Ceraeochrysa smithi)*. *Proceedings of the National Academy of Sciences USA* 93: 3280–3283.

Eisner, T., T. E. Pliske, M. Ikeda, D. F. Owen, L. Vázquez, H. Pérez, J. G. Franclemont, and J. Meinwald. 1970. Defense mechanisms of arthropods. XXVII. Osmeterial secretions of papilionid caterpillars *(Baronia, Papilio, Eurytides)*. *Annals of the Entomological Society of America* 63: 914–915.

Jolivet, P. H. and M. L. Cox, ed. 1996. *Chrysomelidae Biology*, vol. 2, *Ecological Studies*. Amsterdam: SPB Academic Publishing.

Jolivet, P. H., M. L. Cox, and E. Petitpierre, eds. 1994. *Novel Aspects of the Biology of Chrysomelidae*. Dordrecht, Netherlands: Kluwer.

Meinwald, J., M. S. Chadha, J. J. Hurst, and T. Eisner. 1962. Defense mechanisms of arthropods. IX. Anisomorphal, the secretion of a phasmid insect. *Tetrahedron Letters* 1962: 29–33.

Masters, W. M. 1979. Insect disturbance stridulation: its defensive role. *Behavioral Ecology and Sociobiology* 5: 187–200.

————1979. Irradiance modulation used to examine sound-radiating motion in insects. *Science* 203: 57–60.

Meinwald, J., G. M. Happ, J. Labows, and T. Eisner. 1966. Cyclopentanoid terpene biosynthesis in a phasmid insect and in catmint. *Science* 151: 79–80.

Meinwald, J., T. H. Jones, T. Eisner, and K. Hicks. 1977. New methylcyclopentanoid terpenes from the larval defensive secretion of a chrysomelid beetle *(Plagiodera versicolora)*. *Proceedings of the National Academy of Sciences USA* 74: 2189–2193.

Pennisi, E. 2002. Biology reveals new ways to hold on tight. *Science* 296: 250–251.

Sax, K. 1960. *Standing Room Only*. Boston: Beacon Press.

Seifert, G. 1966. Häutungverursachende Reize bei *Polyxenus*. *Zoologischer Anzeiger* 177: 258–263.

Yarbus, A. L. 1967. *Eye Movements and Vision*. New York: Plenum.

4 MASTERS OF DECEPTION

Cott, H. B. 1957. *Adaptive Coloration in Animals*. 2nd ed. London: Methuen.

Darlington, P. J. 1938. Experiments on mimicry in Cuba, with suggestions for future study. *Transactions of the Royal Entomological Society of London.* 87: 681–695.

Eisner, T,. and F. C. Kafatos. 1962. Defense mechanisms of arthropods. X. A pheromone promoting aggregation in an aposematic distasteful insect. *Psyche* 69: 53–61

Eisner, T., F. C. Kafatos, and E. G. Linsley. 1962. Lycid predation by mimetic adult Cerambycidae (Coleoptera). *Evolution* 16: 316–324.

Eisner, T., K. Hicks, M. Eisner, and D. S. Robson. 1978. "Wolf-in-sheep's-clothing" strategy of a predaceous insect larva. *Science* 199: 790–794

Eisner, T., D. F. Wiemer, L. W. Haynes, and J. Meinwald. 1978. Lucibufagins: defensive steroids from the fireflies *Photinus ignitus* and *P. marginellus* (Coleoptera: Lampyridae). *Proceedings of the National Academy of Sciences USA* 75: 905–908.

Eisner, T., M. A. Goetz, D. E. Hill, S. R. Smedley, and J. Meinwald. 1997. Firefly "femmes fatales" acquire defensive steroids (lucibufagins) from their firefly prey. *Proceedings of the National Academy of Sciences USA* 94: 9723–9728.

Goetz, M., D. F. Wiemer, L. W. Haynes, J. Meinwald, and T. Eisner. 1979. Lucibufagines, Partie III. Oxo-11-et oxo-12-bufalines, steroïdes défensifs des lampyres *Photinus ignitus* et *P. marginellus* (Coleoptera: Lampyridae). *Helvetica Chimica Acta* 62: 1396–1400.

Goetz, M. A., J. Meinwald, and T. Eisner. 1981. Lucibufagins, IV. New defensive steroids and a pterin from the firefly *Photinus pyralis* (Coleoptera: Lampyridae). *Experientia* 37: 679–680.

González, A., J. F. Hare, and T. Eisner. 2000. Chemical egg defense in *Photuris* firefly "femmes fatales." *Chemoecology* 9: 177–185.

González, A., F. Schroeder, J. Meinwald, and T. Eisner. 1999. *N*-methylquinolinium 2-carboxylate, a defensive betaine from *Photuris versicolor* fireflies. *Journal of Natural Products* 62: 378–380.

González, A., F. C. Schroeder, A. B. Attygalle, A. Svatos, J. Meinwald, and T. Eisner. 1999. Metabolic transformation of acquired lucibufagins by firefly "femmes fatales." *Chemoecology* 9: 105–112.

Knight, M., R. Glor, S. R. Smedley, A. González, K. Adler, and T. Eisner. 1999. Firefly toxicosis in lizards. *Journal of Chemical Ecology* 25: 1981–1986.

Linsley, E. G., T. Eisner, and A .B. Klots. 1961. Mimetic assemblages of sibling species of lycid beetles. *Evolution* 15: 15–29.

Lloyd, J. E. 1965 Aggressive mimicry in *Photuris:* firefly femmes fatales. *Science* 149: 653–654.

———1975. Aggressive mimicry in *Photuris* fireflies: signal repertoires by femmes fatales. *Science* 187: 452–453.

Meinwald, J., D. F. Wiemer, and T. Eisner. 1979. Lucibufagins. 2. esters of 12-

Oxo-2ß,5ß,11a-trihydroxybufalin, the major defensive steroids of the firefly *Photinus pyralis* (Coleoptera: Lampyridae). *Journal of the American Chemical Society* 101: 3055–3060.

Meinwald, J., J. Smolanoff, A. C. Chibnall, and T. Eisner. 1975. Characterization and synthesis of waxes from homopterous insects. *Journal of Chemical Ecology* 1: 269–274

Moore, B. P., and W. V. Brown. 1981. Identification of warning odour components, bitter principles, and antifeedants in an aposematic beetle: *Metriorrhynchus rhipidius* (Coleoptera: Lycidae). *Insect Biochemistry* 11: 493–499.

5 AMBULATORY SPRAY GUNS

Benfield, E. F. 1972. A defensive secretion of *Dineutes discolor* (Coleoptera: Gyrinidae). *Annals of the Entomological Society of America* 65: 1324–1327.

Carrel, J. E. 1984. Defensive secretion of the pill millipede *Glomeris marginata*. *Journal of Chemical Ecology.* 10: 41–51.

Carrel, J. E., and T. Eisner. 1984. Spider sedation induced by defensive chemicals of milliped prey. *Proceedings of the National Academy of Sciences USA* 81: 806–810.

Carrel, J. E., J. P. Doom, and J. P. McCormick. 1985. Arborine and methaqualone are not sedative in the wolf spider *Lycosa ceratiola* Gertsch and Wallace. *The Journal of Arachnology* 13: 269–271

Dethier, V. G. 1980. Food aversion learning in two polyphagous caterpillars, *Diacrisia virginica* and *Estigmene congrua*. *Physiology and Entomology* 5: 321–325.

Eisner, T. 1968. Mongoose and millipedes. *Science* 160: 1367.

Eisner, T., and D. J. Aneshansley. 1982. Spray aiming in bombardier beetles: jet deflection by the Coanda effect. *Science* 215: 83–85.

———2000. Chemical defense: aquatic beetle *(Dineutes hornii)* vs. fish *(Micropterus salmoides)*. *Proceedings of the National Academy of Sciences USA* 97: 11313–11318.

Eisner, T., and J. A. Davis. 1967. Mongoose throwing and smashing millipedes. *Science* 155: 577–579.

Eisner, T., I. Kriston, and D. J. Aneshansley. 1976. Defensive behavior of a termite *(Nasutitermes exitiosus)*. *Behavioral Ecology and Sociobiology* 1: 83–125.

Eisner, T., A. F. Kluge, J. E. Carrel, and J. Meinwald. 1971. Defense of phalangid: liquid repellent administered by leg dabbing. *Science* 173: 650–652.

Ernst, E. 1959. Beobachtungen beim Spritzakt der Nasutitermes-Soldaten. *Revue Suisse Zoologie* 66: 289–295.

Gelperin, A. 1976. Complex associative learnings in small neural networks. *Trends in Neurosciences* 9: 323–328.

Kerfoot, W. C., and A. Sih. 1987. *Predation.* Hanover, N.H.: University Press of New England.

Kriston, I., J. A. L. Watson, and T. Eisner. 1977. Non-combative behaviour of large soldiers of *Nasutitermes exitiosus* (Hill): an analytical study. *Insectes Sociaux* 24: 103–111.

Meinwald, J., K. Opheim, and T. Eisner. 1972. Gyrinidal: a sesquiterpenoid aldehyde from the defensive glands of gyrinid beetles. *Proceedings of the National Academy of Sciences USA* 69: 1208–1210.

Meinwald, Y. C., J. Meinwald, and T. Eisner. 1966. 1,2-Dialkyl-4 (3H)-quinazolinones in the defensive secretion of a millipede *(Glomeris marginata). Science* 154: 390–391

Miller, J., L. Hendry, and R. Mumma. 1975. Norsesquiterpenes as defensive toxins of whirligig beetles (Coleoptera: Gyrinidae). *Journal of Chemical Ecology.* 1: 59–82.

Prestwich, G. D. 1979. Chemical defense by termite soldiers. *Journal of Chemical Ecology* 5: 459–480.

———1983. The chemical defenses of termites. *Scientific American* 249: 78–87.

Reba, I. 1966. Applications of the Coanda effect. *Scientific American* 214: 84–92.

Schildknecht, H., and W. F. Wenneis. 1966. Über Arthropoden-(Insekten) Abwehrstoffe. XX. Strukturaufklärung des Glomerins . *Zeitschrift für Naturforschung* 21b: 552.

Wagner, D. 1946. *Umhlanga: A Story of the Coastal Bush of South Africa.* Durban, South Africa: Knox.

6 TALES FROM THE WEBSITE

Blackledge, T. A. 1998. Signal conflict in spider webs driven by predators and prey. *Proceedings of the Royal Society of London* 265: 1991–1996.

Blackledge, T. A., and J. W. Wenzel. 2001. Silk mediated defense by an orb web spider against predatory mud-dauber wasps. *Behaviour* 138: 155–171.

Eisner, T., and S. Camazine. 1983. Spider leg autotomy induced by prey venom injection: an adaptive response to "pain"? *Proceedings of the National Academy of Sciences USA* 80: 3382–3385.

Eisner, T,. and J. Dean. 1976. Ploy and counterploy in predator-prey interactions:

orb-weaving spiders versus bombardier beetles. *Proceedings of the National Academy of Sciences USA* 73: 1365–1367.

Eisner, T., and S. Nowicki. 1983. Spider web protection through visual advertisement: role of the "stabilimentum." *Science* 219: 185–187.

Eisner, T., and J. Shepherd. 1966. Defense mechanisms of arthropods. XIX. Inability of sundew plants to capture insects with detachable integumental outgrowths. *Annals of the Entomological Society of America* 59: 868–870.

Eisner, T., R. Alsop, and G. Ettershank. 1964. Adhesiveness of spider silk. *Science* 146: 1058–1061.

Eisner, T., M. Eisner, and M. Deyrup. 1991. Chemical attraction of kleptoparasitic flies to heteropteran insects caught by orb-weaving spiders. *Proceedings of the National Academy of Sciences USA* 88: 8194–8197.

Herberstein, M. E., C. L. Craig, J. A. Coddington, and M. A. Elgar. 2000. The functional significance of silk decorations of orb-web spiders: a critical review of the empirical evidence. *Biological Review* 75: 649–669.

Horton, C. C. 1980. A defensive function for the stabilimenta of two orb weaving spiders (Araneae, Araneidae). *Psyche* 87: 13–20.

Kerr, A. M. 1993. Low frequency of stabilimenta in orb webs of *Argiope appensa* (Araneae: Araneidae) from Guam: an indirect effect of an introduced avian predator? *Pacific Science* 47: 328–337.

Masters, W. M., and T. Eisner. 1990. The escape strategy of green lacewings from orb webs. *Journal of Insect Behavior* 3: 143–157.

Roeder, K. D. 1967. *Nerve Cells and Insect Behavior*. Cambridge, Mass.: Harvard University Press.

Seah, W. K., and D. Li. 2001. Stabilimenta attract unwelcome predators to orb-webs. *Proceedings of the Royal Society of London* 268: 1553–1558.

Schoener, T. W., and D. A. Spiller. 1992. Stabilimenta characteristics of the spider *Argiope argentata* on small islands: support of the predator-defense hypothesis. *Behavioral Ecology and Sociobiology* 31: 309–318.

7 THE CIRCUMVENTERS

Berenbaum, M. 1983. Coumarins and caterpillars: a case for coevolution. *Evolution* 37: 163–179.

Brodie, E. D., and E. D. Brodie, Jr. 1990. Tetrodotoxin resistance in garter snakes: an evolutionary response of predators to dangerous prey. *Evolution* 44: 651–659.

Chadha, M. S., T. Eisner, and J. Meinwald. 1961. Defense mechanisms of arthropods. IV. *Para*-benzoquinones in the secretion of *Eleodes longicollis* Lec. (Coleoptera: Tenebrionidae). *Journal of Insect Physiology* 7: 46–50.

Dussourd, D. E. 1993. Foraging with finesse: caterpillar adaptations for circumventing plant defenses. In N. E. Stamp and T. M. Casey, ed., *Ecological and Evolutionary Constraints on Foraging*. New York: Chapman and Hall.

———1999. Behavioral sabotage of plant defense: do vein cuts and trenches reduce insect exposure to exudate? *Journal of Insect Behavior* 12: 501–515.

Dussourd, D. E., and T. Eisner. 1987. Vein-cutting behavior: insect counterploy to the latex defense of plants. *Science* 237: 898–901.

Dussourd, D. E., and A. M. Hoyle. 2000. Poisoned plusiines: toxicity of milkweed latex and cardenolides to some generalist caterpillars. *Chemoecology* 10: 11–16.

Eisner, T. 1966. Beetle's spray discourages predators. *Natural History* 75: 43–47.

———1970. Chemical defense against predation in arthropods. In E. Sondheimer and J. B. Simeone, ed., *Chemical Ecology*. New York: Academic Press.

Eisner, T., I. T. Baldwin, and J. Conner. 1993. Circumvention of prey defense by a predator: ant lion vs. ant. *Proceedings of the National Academy of Sciences USA* 90: 6716–6720.

Eisner, T., M. Eisner, and E. R. Hoebeke. 1998. When defense backfires: detrimental effect of a plant's protective trichomes on an insect beneficial to the plant. *Proceedings of the National Academy of Sciences USA* 95: 4410–4414.

Eisner, T., D. Alsop, K. Hicks, and J. Meinwald. 1978. Defensive secretions of millipeds. In S. Bettini, ed., *Handbook of Experimental Pharmacology*, vol. 48, *Arthropod Venoms*. Berlin: Springer-Verlag.

Eisner, T., M. Eisner, A. B. Attygalle, M. Deyrup, and J. Meinwald. 1998. Rendering the inedible edible: circumvention of a millipede's chemical defense by a predaceous beetle larva (Phengodidae). *Proceedings of the National Academy of Sciences USA* 95: 1108–1113.

Hölldobler, B., and E. O. Wilson. 1990. *The Ants*. Cambridge, Mass.: The Belknap Press of Harvard University Press.

Hulley, P. E. 1988. Caterpillar attacks plant mechanical defence by mowing trichomes before feeding. *Ecological Entomology* 13: 239–241.

Hurst, J. J., J. Meinwald, and T. Eisner. 1964. Defense mechanism of arthropods. XII. Glucose and hydrocarbons in the quinone-containing secretion of *Eleodes longicollis*. *Annals of the Entomological Society of America* 57: 44–46.

Kim, S. 1994. Food poisoning: fish and shellfish. In K. R. Olson, ed., *Poisoning and Drug Overdose*, vol. 1. Norwalk, Conn.: Appleton and Lange.

Mosher, H. S. 1986. The chemistry of tetrodotoxin. *Annals of the New York Academy of Sciences* 479: 32–43.

Mosher, H. S., and F. A. Fuhrman. 1984. Occurrence and origin of tetrodotoxin. In E. P. Ragelis, ed., *Seafood Toxins*. Washington, D.C.: American Chemical Society.

Nishida, R. 2002. Sequestration of defensive substances from plants by Lepidoptera. *Annual Review of Entomology* 47: 57–92.

Roesel von Rosenhof, A. J. 1755. *Insekten Belustigung*, vol. 3. Nuremberg: J. J. Fleischmann.

Roy, J., and JM. Bergeron. 1990. Branch-cutting behavior by the vole *(Microtus pennsylvanicus)*. *Journal of Chemical Ecology* 16: 735–741.

Shure, D. J., L. A. Wilson, and C. Hochwender. 1989. Predation on aposematic efts of *Notophthalmus viridescens*. *Journal of Herpetology* 23: 437–439.

Tiemann, D. L. 1967. Observations on the natural history of the western banded glowworm *Zarhipis integripennis* (Le Conte) (Coleoptera: Phengodidae). *Proceedings of the California Academy of Sciences* 35: 235–264.

Tune, R., and D. E. Dussourd. 2000. Specialized generalists: constraints on host range in some plusiine caterpillars. *Oecologia* 123: 543–549.

Vietmeyer, N. D. 1984. The preposterous puffer. *National Geographic* 166: 260–270.

Von Frisch, K. 1974. *Animal Architecture*. New York: Harcourt Brace Jovanovich.

Zalucki, M. P., L. P. Brower, and A. Alonso. 2001. Detrimental effects of latex and cardiac glycosides on survival and growth of first-instar monarch butterfly larvae *Danaus plexippus* feeding on the sandhill milkweed *Asclepius humistrata*. *Ecological Entomology* 26: 212–224.

8 THE OPPORTUNISTS

Daly, J. W., H. M. Garraffo, T. F. Spande, C. Jaramillo, and A. S. Rand. 1994. Dietary source for skin alkaloids of poison frogs (Dendrobatidae)? *Journal of Chemical Ecology* 20: 943–955.

Daly, J. W., S. I. Secunda, H. M. Garraffo, T. F. Spande, A. Wisnieski, and J. F. Cover, Jr. 1994. An uptake system for dietary alkaloids in poison frogs (Dendrobatidae). *Toxicon* 32: 657–663.

De Geer, K. 1776. *Abhandlungen zur Geschichte der Insekten*. Leipzig: Müllers Buch-und Kunsthandlung.

DeVol, J. E., and R. D. Goeden. 1973. Biology of *Chelinidea vittiger* with notes on its host-plant relationship and value in biological weed control. *Entomologist* 2: 231–240.

Donkin, R. A. 1977. Spanish red: an ethnological study of cochineal and the Opuntia cactus. *Transactions of the American Philosophical Society* 67: 4–84.

Eisner, T. 1990. De Geer's pioneering phytochemical observation. *Journal of Chemical Ecology* 16: 2489–2492.

————1994. Integumental slime and wax secretion: defensive adaptations of sawfly larvae. *Journal of Chemical Ecology* 20: 2743–2749.

Eisner, T., and D. J. Aneshansley. 1983. Adhesive strength of the insect-trapping glue of a plant *(Befaria racemosa)*. *Annals of the Entomological Society of America* 76: 295–298.

Eisner, T., M. Eisner, and E. R. Hoebeke. 1998. When defense backfires: detrimental effect of a plant's protective trichomes on an insect beneficial to the plant. *Proceedings of the National Academy of Sciences USA* 95: 4410–4414.

Eisner, T., S. Nowicki, M. Goetz, and J. Meinwald. 1980. Red cochineal dye (carminic acid): its role in nature. *Science* 208: 1039–1042.

Eisner, T., J. E. Carrel, E. van Tassell, E. R. Hoebeke, and M. Eisner. 2001. Construction of a defensive trash packet from sycamore leaf trichomes by a chrysopid larva (Neuroptera: Chrysopidae). *Proceedings of the Entomological Society of Washington* 104: 437–446.

Eisner, T., J. S. Johnessee, J. Carrel, L. B. Hendry, and J. Meinwald. 1974. Defensive use by an insect of a plant resin. *Science* 184: 996–999.

Eisner, T., R. Ziegler, J. L. McCormick, M. Eisner, E. R. Hoebeke, and J. Meinwald. 1994. Defensive use of an acquired substance (carminic acid) by predaceous insect larvae. *Experientia* 50: 610–615.

Fleming, S. 1983. The tale of the cochineal: insect farming in the New World. *Archaeology* September/October: 68–69, 79.

Greene, E. 1989. A diet-induced developmental polymorphism in a caterpillar. *Science* 243: 643–646.

Leonard, M. D. 1928. *A List of the Insects of New York, with a List of the Spiders and Certain Allied Groups*. Ithaca: Cornell University Press.

Morrow, P. A., T. E. Bellas, and T. Eisner. 1976. *Eucalyptus* oils in the defensive oral discharge of Australian sawfly larvae (Hymenoptera: Pergidae). *Oecologia* 24: 193–206.

Sickerman, S. L., and J. K. Wangberg. 1983. Behavioral responses of the cactus bug, *Chelinidea vittiger* Uhler, to fire damaged host plants. *The Southwestern Entomologist* 8: 263–267.

Treiber, M. 1979. Composites as host plants and crypts for *Synchlora aerata* (Geometridae). *Journal of the Lepidopterists' Society* 33: 239–244.

BIBLIOGRAPHY

9 THE LOVE POTION

Blodgett, S. L., J. E. Carrel, and R. A. Higgins. 1991. Cantharidin content of blister beetles (Coleoptera: Meloidae) collected from Kansas alfalfa and implications for inducing cantharidiasis. *Environmental Entomology* 20: 776–780.

Carrel, J. E., and T. Eisner. 1974. Cantharidin: potent feeding deterrent to insects. *Science* 183: 755–757.

Eisner, T., M. Goetz, D. Aneshansley, G. Ferstandig-Arnold, and J. Meinwald. 1986. Defensive alkaloid in blood of Mexican bean beetle *(Epilachna varivestis)*. *Experientia* 42: 204–207.

Eisner, T., S. R. Smedley, D. K. Young, M. Eisner, B. Roach, and J. Meinwald. 1996. Chemical basis of courtship in a beetle *(Neopyrochroa flabellata)*: cantharidin as "nuptial gift." *Proceedings of the National Academy of Sciences USA* 93: 6499–6503.

Eisner, T., S. R. Smedley, D. K. Young, M. Eisner, B. Roach, and J. Meinwald. 1996. Chemical basis of courtship in a beetle *(Neopyrochroa flabellata)*: cantharidin as precopulatory "enticing" agent. *Proceedings of the National Academy of Sciences USA* 93: 6494–6498.

Eisner, T., J. Conner, J. E. Carrel, J. P. McCormick, A. J. Slagle, C. Gans, and J. C. O'Reilly. 1990. Systemic retention of ingested cantharidin by frogs. *Chemoecology* 1: 57–62.

Fey, F. 1954. Beiträge zur Biologie der canthariphilen Insekten. *Beiträge zur Entomologie* 4: 180–187.

Görnitz, K. 1937. Cantharidin als Gift und Anlockungsmittel für Insekten. *Arbeiten für physikalische angewandte Entomologie* 4: 116–159.

Happ, G., and T. Eisner. 1961. Hemorrhage in a coccinellid beetle and its repellent effect on ants. *Science* 134: 329–331.

Holz, C., G. Streil, K. Dettner, J. Dütemeyer, and W. Boland. 1994. Intersexual transfer of a toxic terpenoid during copulation and its paternal allocation to developmental stages: quantification of cantharidin in cantharidin-producing oedemerids (Coleoptera: Oedemeridae) and canthariphilous pyrochroids (Coleoptera: Pyrochroidae). *Zeitschrift für Naturforschung* 49c: 856–864.

Leonard, M. D. 1928. *A List of the Insects of New York, with a List of the Spiders and Certain Allied Groups*. Ithaca: Cornell University Press.

McCormick, J. P., and J. E. Carrel. 1987. Cantharidin biosynthesis and function in meloid beetles. In G. D. Prestwich and G. J. Blomquist, ed., *Pheromone Biochemistry* . New York: Academic Press.

Meynier, J. 1893. Empoisonnement par la chair de grenouilles infestées par des insectes du genre *Mylabris* de la famille des Meloides. *Archives de Medicine et de Pharmacie Militaires* 22: 53–56.

Prischam, D. A., and C. A. Sheppard. 2002. A world view of insects as aphrodisi-acs, with special reference to Spanish fly. *American Entomologist* 48: 208–220.

Say, T. 1826. Descriptions of new species of coleopterous insects, inhabiting the United States. *Journal of the Academy of Natural Sciences of Philadelphia* 5: 237–284.

Schütz, C., and K. Dettner. 1992. Cantharidin-secretion by elytral notches of male anthicid species (Coleoptera: Anthicidae). *Zeitschrift für Naturforschung* 47c: 290–299.

Sierra, J. R., W.-D. Woggon, and H. Schmid. 1976. Transfer of cantharidin (1) during copulation from the adult male to the female *Lytta vesicatoria* (Span-ish flies). *Experientia* 32: 142–144.

Smedley, S. R., C. L. Blankespoor, Y. Yang, J. E. Carrel, and T. Eisner. 1996. Pred-atory response of spiders to blister beetles (family Meloidae). *Zoology* 99: 211–217.

Yosef, R., J. E. Carrel, and T. Eisner. 1996. Contrasting reactions of loggerhead shrikes to two types of chemically defended insect prey. *Journal of Chemical Ecology* 22: 173–181.

Young, D. K. 1975. A revision of the family Pyrochroidae (Coleoptera: Heteromera) for North America based on the larvae, pupae, and adults. *Con-tributions of the American Entomological Institute* 11: 1–39.

———1984. Field records and observations of insects associated with cantharidin. *The Great Lakes Entomologist* 17: 195–199.

———1984. Field studies of cantharidin orientation by *Neopyrochroa flabellata* (Coleoptera: Pyrochroidae). *The Great Lakes Entomologist* 17: 133–135.

10 THE SWEET SMELL OF SUCCESS

Bogner, F., and T. Eisner. 1991. Chemical basis of egg cannibalism in a caterpil-lar *(Utetheisa ornatrix)*. *Journal of Chemical Ecology* 17: 2063–2075.

———1992. Chemical basis of pupal cannibalism in a caterpillar *(Utetheisa ornatrix)*. *Experientia* 48: 97–102.

Boppré, M. 1990. Lepidoptera and pyrrolizidine alkaloids: exemplification of complexity in chemical ecology. *Journal of Chemical Ecology* 16: 165–185.

Brower, L. P. 1969. Ecological chemistry. *Scientific American* 1969: 28–29.

Brower, L. P., J. V. Z. Brower, and F. P. Cranston, 1965. Courtship behavior of the queen butterfly, *Danaus gilippus berenice*. *Zoologica* 50: 1–39.

Bull, L. B., C. C. J. Culvenor, and A. T. Dick. 1968. *The Pyrrolizidine Alkaloids: Their Chemistry, Pathogenicity and Other Biological Properties.* Amsterdam: North-Holland Publishing Co.

Conner, W. E., B. Roach, E. Benedict, J. Meinwald, and T. Eisner. 1990. Courtship pheromone production and body size as correlates of larval diet in males of the arctiid moth, *Utetheisa ornatrix. Journal of Chemical Ecology* 16: 543–552.

Conner, W. E., T. Eisner, R. K. Vander Meer, A. Guerrero, and J. Meinwald. 1981. Precopulatory sexual interaction in an arctiid moth *(Utetheisa ornatrix):* role of a pheromone derived from dietary alkaloids. *Behavioral Ecology and. Sociobiology* 9: 227–235.

Conner, W. E., R. Boada, F. C. Schroeder, A. González, J. Meinwald, and T. Eisner. 2000. Chemical defense: bestowal of a nuptial alkaloidal garment by a male moth upon its mate. *Proceedings of the National Academy of Sciences USA* 97: 14406–14411.

Conner, W. E., T. Eisner, R. K. Vander Meer, A. Guerrero, D. Ghiringelli, and J. Meinwald. 1980. Sex attractant of an arctiid moth *(Utetheisa ornatrix):* a pulsed chemical signal. *Behavioral Ecology and Sociobiology* 7: 55–63.

Dusenbery, D. B. 1989. Calculated effect of pulsed pheromone release on range of attraction. *Journal of Chemical Ecology* 15: 971–977.

Dussourd, D. E., C. A. Harvis, J. Meinwald, and T. Eisner. 1989. Paternal allocation of sequestered plant pyrrolizidine alkaloid to eggs in the danaine butterfly, *Danaus gilippus. Experientia* 45: 896–898.

———1991. Pheromonal advertisement of a nuptial gift by a male moth *Utetheisa ornatrix. Proceedings of the National Academy of Sciences USA* 88: 9224–9227.

Dussourd, D. E., K. Ubik, C. Harvis, J. Resch, J. Meinwald, and T. Eisner. 1988. Biparental defensive endowment of eggs with acquired plant alkaloid in the moth *Utetheisa ornatrix. Proceedings of the National Academy of Sciences USA* 85: 5992–5996.

Edgar, J. A., C. C. Culvenor, and G. S. Robinson. 1973. Hairpencil dihydrophrrolizidines of Danainae from the New Hebridwes. *Journal of the Australian Entomological Society* 12: 144–150.

Edgar, J. A., C. C. Culvenor, and L. W. Smith. 1971. Dihydropyrrolizidine derivatives in hairpencil secretion of danaid butterflies. *Experientia* 27: 761–762.

Eisner, T. 1982. For love of nature: exploration and discovery at biological field stations. *BioScience* 32: 321–326.

Eisner, T., and M. Eisner. 1991. Unpalatability of the pyrrolizidine alkaloid containing moth, *Utetheisa ornatrix,* and its larva, to wolf spiders. *Psyche* 98: 111–118.

Eisner, T., and J. Meinwald. 1987. Alkaloid-derived pheromones and sexual selection in Lepidoptera. In G. D. Prestwich and G. J. Blomquist, ed., *Pheromone Biochemistry.* Orlando: Academic Press.

———1995. The chemistry of sexual selection. *Proceedings of the National Academy of Sciences USA* 92: 50–55.

Eisner, T. C. Rossini, A. González, V. K. Iyengar, M. V. S. Siegler, and S. Smedley. 2002. Paternal investment in egg defense. In M. Hilker and T. Meiners, ed., *Chemoecology of Insect Eggs and Egg Deposition.* Oxford: Blackwell Publishing.

Eisner, T., M. Eisner, C. Rossini, V. K. Iyengar, B. L. Roach, E. Benedikt, and J. Meinwald. 2000. Chemical defense against predation in an insect egg. *Proceedings of the National Academy of Sciences USA* 97: 1634–1639.

González, A., C. Rossini, M. Eisner, and T. Eisner. 1999. Sexually transmitted chemical defense in a moth *(Utetheisa ornatrix)*. *Proceedings of the National Academy of Sciences USA* 96: 5570–5574.

Grant, A. J., R. J. O'Connell, and T. Eisner. 1989. Pheromone-mediated sexual selection in the moth *Utetheisa ornatrix:* olfactory receptor neurons responsive to a male-produced pheromone. *Journal of Insect Behavior* 2: 371–385.

Hare, J. F., and T. Eisner. 1993. Pyrrolizidine alkaloid deters ant predators of *Utetheisa ornatrix* eggs: effects of alkaloid concentration, oxidation state, and prior exposure of ants to alkaloid-laden prey. *Oecologia* 96: 9–18.

———1995. Cannibalistic caterpillars *(Utetheisa ornatrix)* fail to differentiate between eggs on the basis of kinship. *Psyche* 102: 27–33.

Hartmann, T., and L. Witte. 1995. Chemistry, biology and chemoecology of the pyrrolizidine alkaloids. In S. W. Pelletier, ed., *Alkaloids: Chemical and Biological Properties* Oxford: Pergamon Press.

Iyengar, V. K., and T. Eisner. 1999. Female choice increases offspring fitness in an arctiid moth *(Utetheisa ornatrix)*. *Proceedings of the National Academy of Sciences USA* 96: 15013–15016.

———1999. Heritability of body mass, a sexually selected trait, in an arctiid moth *(Utetheisa ornatrix)*. *Proceedings of the National Academy of Sciences USA* 96: 9169–9171.

———2002. Parental body mass as a determinant of egg size and egg output in an arctiid moth *(Utetheisa ornatrix)*. *Journal of Insect Behavior* 15: 309–318.

Iyengar, V. K., H. K. Reeve, and T. Eisner. 2002. Paternal inheritance of female moths' mating preference. *Nature* 419: 830–832.

Iyengar, V. K., C. Rossini, and T. Eisner. 2001. Precopulatory assessment of male quality in an arctiid moth *(Utetheisa ornatrix)*: hydroxydanaidal is the only criterion of choice. *Behavioral Ecology and Sociobiology* 49: 283–288.

Iyengar, V. K., C. Rossini, E. R. Hoebeke, W. E. Conner, and T. Eisner. 1999. First record of the parasitoid *Archytas aterrimus* (Diptera: Tachinidae) from *Utetheisa ornatrix* (Lepidoptera: Arctiidae). *Entomological News* 110: 144–146.

Jain, S. C., D. E. Dussourd, W. E. Conner, T. Eisner, A. Guerrero, and J. Meinwald. 1983. Polyene pheromone components from an arctiid moth *(Utetheisa ornatrix)*: characterization and synthesis. *Journal of Organic Chemistry* 48: 2266–2270.

LaMunyon, C. W. 1997. Increased fecundity, as a function of multiple mating, in an arctiid moth, *Utetheisa ornatrix*. *Ecological Entomology* 22: 69–73.

LaMunyon, C. W., and T. Eisner. 1993. Post copulatory sexual selection in an arctiid moth *(Utetheisa ornatrix)*. *Proceedings of the National Academy of Sciences USA* 90: 4689–4692.

———1994. Spermatophore size as determinant of paternity in an arctiid moth *(Utetheisa ornatrix)*. *Proceedings of the National Academy of Sciences USA* 91: 7081–7084.

Mattocks, A. R. 1986. *Chemistry and Toxicology of Pyrrolizidine Alkaloids*. London: Academic Press.

Meinwald, J., Y. C. Meinwald, and P. H. Mazzocchi. 1969. Sex pheromone of the queen butterfly: chemistry. *Science* 164: 1174–1175.

Nishida, R. 2002. Sequestration of defensive substances from plants by Lepidoptera. *Annual Review of Entomology* 47: 57–92.

Pliske, T. E., and T. Eisner. 1969. Sex pheromone of the queen butterfly: biology. *Science* 164: 1170–1172.

Rossini, C., A. González, and T. Eisner. 2001. Fate of an alkaloidal nuptial gift in the moth *Utetheisa ornatrix*: systemic allocation for defense of self by the receiving female. *Journal of Insect Physiology* 47: 639–647

Rossini, C., E. R. Hoebeke, V. K. Iyengar, W. E. Conner, M. Eisner, and T. Eisner. 2000. Alkaloid content of parasitoids reared from pupae of an alkaloid-sequestering arctiid moth *(Utetheisa ornatrix)*. *Entomological News* 111: 287–290.

Schulz, S., W. Francke, M. Boppré, T. Eisner, and J. Meinwald. 1993. Insect pheromone biosynthesis: stereochemical pathway of hydroxydanaidal production from alkaloidal precursors in *Creatonotos transiens* (Lepidoptera, Arctiidae). *Proceedings of the National Academy of Sciences USA* 90: 6834–6838.

Smedley, S. R., and T. Eisner. 1995. Sodium uptake by "puddling" in a moth. *Science* 270: 1816–1818.

————1996. Sodium: a male moth's gift to its offspring. *Proceedings of the National Academy of Sciences USA* 93: 809–813.

Storey, G. K., D. J. Aneshansley, and T. Eisner. 1991. Parentally provided alkaloid does not protect eggs of *Utetheisa ornatrix* (Lepidoptera: Arctiidae) against entomopathogenic fungi. *Journal of Chemical Ecology* 17: 687–693.

Stowe, M. K. 1988. Chemical mimicry. In K. C. Spencer, ed., *Chemical Mediation of Coevolution*. San Diego: Academic Press.

Stowe, M. K., J. H. Tumlinson, and R. R. Heath. 1987. Chemical mimicry: bolas spiders emit components of moth prey species sex pheromones. *Science* 236: 1635–1637.

Trigo, J. R., K. S. Brown, L. Witte, T. Hartmann, L. Ernst, and L. E. S. Barata. 1996. Pyrrolizidine alkaloids: different acquisition and use patterns in Apocynaceae and Solanaceae feeding ithomiine butterflies (Lepidoptera: Nymphalidae). *Biological Journal of the Linnean Society* 58: 99–123.

11 EPILOGUE

Attygalle, A. B., K. D. McCormick, C. L. Blankespoor, T. Eisner, and J. Meinwald. 1993. Azamacrolides: a family of alkaloids from the pupal defensive secretion of a ladybird beetle *(Epilachna varivestis)*. *Proceedings of the National Academy of Sciences USA* 90: 5204–5208.

Eisner, T. 1982. For love of nature: exploration and discovery at biological field stations. *BioScience* 32: 321–326.

————2001. Chemical prospecting: the new natural history. In M. J. Novacek, ed., *The Biodiversity Crisis: Losing What Counts*. New York: The New Press.

Eisner, T., and M. Eisner. 1992. Operation and defensive role of "gin traps" in a coccinellid pupa *(Cycloneda sanguinea)*. *Psyche* 99: 265–273.

Rossini, C., A. González, J. Farmer, J. Meinwald, and T. Eisner. 2000. Antiinsectan activity of epilachnene, a defensive alkaloid from the pupae of Mexican bean beetles *(Epilachna varivestis)*. *Journal of Chemical Ecology* 26: 391–397.

Schroeder, F. C., A. González, T. Eisner, and J. Meinwald. 1999. Miriamin, a defensive diterpene from the eggs of a land slug *(Arion sp.)*. *Proceedings of the National Academy of Sciences USA* 96: 13620–13625.

Schröder, F. C., J. J. Farmer, S. R. Smedley, T. Eisner, and J. Meinwald. 1998. Absolute configuration of the polyazamacrolides, macrocyclic polyamines produced by a ladybird beetle. *Tetrahedron Letters* 39: 6625–6628.

Schroeder, F. C., J. J. Farmer, S. R. Smedley, A. B. Attygalle, T. Eisner, and J. Meinwald. 2000. A combinatorial library of macrocyclic polyamines produced by a ladybird beetle. *Journal of the American Chemical Society* 122: 3628–3634.

Schröder, F., J. J. Farmer, A. B. Attygalle, J. Meinwald, S. R. Smedley, and T. Eisner. 1998. Combinational chemistry in insects: a library of defensive macrocyclic polyamines. *Science* 281: 428–431.

Schroeder, F. C., S. R. Smedley, L. K. Gibbons, J. J. Farmer, A. B. Attygalle, T. Eisner, and J. Meinwald. 1998. Polyazamacrolides from ladybird beetles: ring-size selective oligomerization. *Proceedings of the National Academy of Sciences USA* 95: 13387–13391.

Wilcove, D. S., and T. Eisner. 2000. The impending extinction of natural history. *Chronicle of Higher Education* September 15: B24.

Acknowledgments

Thanks in infinite measure go to my wife, Maria, for sharing the adventure. Her companionship and indefatigable spirit brought joy to the exploration, and her keen scientific instinct helped shape the questions. Without her encouragement this book would not have come to light.

Thanks are due also to our daughters, Yvonne, Vivian, and Christina, for sharing our love of nature, for helping, and for putting up with Daddy's "zoo."

None of the work discussed in this book could have been done without the close collaboration of the many graduate students, postdoctoral fellows, and undergraduate research students who enlivened my laboratory over the years. I have made it a point to acknowledge their contributions throughout the text, although such mention does not always do justice to the full measure of their individual input.

Essential also has been the close partnership I have had with Jerrold Meinwald and his associates. Without their enthusiastic collaboration, the various molecules that are at the root of the stories discussed here would have remained uncharacterized. Jerry's friendship and scientific wisdom added immensely to the sheer enjoyment that we all derived from working together.

Very special thanks are due to Edward O. Wilson, fellow entomophile and conservationist, for his enduring friendship and for defining, through his science, as no other scientist has, the vision, the goals, and the possibilities.

I have also been extremely fortunate in having talented technical and secretarial help over the years. Two assistants of immense ability who by their mere presence graced the lab were Karen Hicks and the late Rosalind Alsop. Of my former secretaries, Ruth Roberts and Carol Kautz became valued friends. At present, management of the office is in the very competent hands of Janis Strope, whose organizational skills have on many occasions kept me from drifting off course. For maintaining our insects in culture, and persuading them that our insectary is the real world, I am indebted to Janice Schlesinger.

I am grateful to a number of people, including my current graduate students, Alex Bezzerides, Lynn Fletcher, Jackie Grant, Vick Iyengar, Josh Ladau, and Shannon Olsson, and several friends, Daniel Aneshansley, Mark Deyrup, David Liittschwager, Susan Middleton, Charles Pearman, Frank Schroeder, and Carol Skinner, for commenting on parts of the manuscript. Jerry Meinwald was particularly helpful in passing judgment on the chemical sections. Four close friends, May Berenbaum, Helen Ghiradella, John Hildebrand, and Melody Siegler, were kind enough to critique the entire text. Their input led to corrections of fact and interpretation. My friend and colleague Richard Hoebeke, upon whom I have come to depend in many matters entomological, provided some of the identifications of insects portrayed in these pages.

I am grateful also to a number of people whose friendship has been an inspiration over the years: Diane Ackerman, Kraig Adler, Natalie Angier, Ian Baldwin, Michael Bean, May Berenbaum, Ron Booker, the late William L. Brown Jr., the late Frank M. Carpenter, the late Mont Cazier, Kenneth Christiansen, William E. Conner, Dale Corson, the late Vincent Dethier, Mark Deyrup, Jared Diamond, Paul and Anne Ehrlich, the late Howard E. Evans, Turid Forsythe, Rodrigo Gámez, Carl Gans, John Hildebrand, the late Howard Hinton, Roald Hoffmann, Berthold Hölldobler, Dan Janzen, Fotis Kafatos, Rosalind Lasker, John Law, E. Gort Linsley, Simon Levin, Martin Lindauer, Tom Lovejoy, Hubert Markl, Mitch Masters, Norman Myers, Peter Narins, Richard D. O'Brien, William Provine, Peter Raven, the late Kenneth Roeder, Wendell Roelofs, Louis Roth, Miriam Rothschild, the late Carl Sagan, the late Robert E. Silberglied, Andy Sinauer, Noel Snyder, the late Adrian Srb, Asher Treat, Ari van Tienhoven, Charles Walcott, Rüdiger Wehner, David Wilcove, and the late Carrol M. Williams.

My 45 years at Cornell have been happy ones, in large measure because of the many personal ties forged over time. To countless Cornellians, including the staff and my colleagues from the Department of Neurobiology and Behavior, and the thousands of students who over the years attended my courses in introductory biology, animal behavior, chemical ecology, chemical communication, and entomology, I extend my thanks for brightening the days.

To the Cornell Library and its extraordinary staff, to the Cornell Plantations, the Johnson Museum, the Laboratory of Ornithology, and the entire staff of University Photography go thanks for countless favors rendered. I am grateful also to my journalist friend Roger Segelken for converting what I often cloaked in obscurity into press releases accessible to all.

For 44 years I have been supported uninterruptedly by the same research grant from the National Institutes of Health. I was incredulous when I received that first phone call, while in the field in Portal, Arizona, in 1959, telling me that my grant application was approved, and I still marvel that my request for renewal should have been funded in full at 5-year intervals ever since. I am appreciative of the support received through the years from the Lalor Foundation, the National Science Foundation, and the Society of the Sigma XI. Particularly helpful in recent years has been support for my chemical ecological work from the Johnson and Johnson Company.

Over the years I have been involved with various films dealing wholly or in part with my work. I am particularly grateful to the following directors and camera operators: Caroline Weaver, Rodger Jackman (*Secret Weapons*, BBC, London, 1983); Robin Brown (*Nature Watch*, ATV Network, London, 1981); John Rubin (*The Ruling Class*, WQED, Pittsburgh, 1992); David Suzuki, Vishnu Mathur, and Rudolf Kovanic (*The Bugman of Ithaca*, CBC, Toronto, 1990); Nick Upton, Kevin Flay (*Beetlemania*, Green Umbrella for BBC, London, 1996); Nancy Block (*Bug Stories*, Discovery Channel, Toronto, 1997–2000); Christopher and Lotte Sykes (*Seven Wonders of the World*, Christopher Sykes Productions for BBC, London, 1997).

I am immeasurably indebted to the late Richard Archbold, visionary founder of the station that bears his name, who early on recognized the hidden value of the Florida scrub and the need for its exploration and preservation. My indebtedess extends to the past directors of the station, James Layne, Jim Wolfe, and John Fitzpatrick, as well as to the present director, Hilary Swain, and to the senior station affiliates, Mark Deyrup, Fred Lohrer, and Eric Menges, for converting the station into an exploratory paradise and center of conservationist influence.

It is also a very special pleasure to acknowledge my debt to my late parents, and to my sister, who provided the most accommodating milieu a young naturalist could have hoped for. No matter how cluttered my room, or how menacing the caged vermin, there was no questioning of my activities. My bugs were serious business and my interest in them was assumed to take priority over school. My debt to family is intimately linked to my debt to music. Ours was a music-oriented family. My father, a good pianist, led the way, with the result that I too became committed to the keyboard. I met Maria at the piano, and to this day we make sure always to have an electronic keyboard along when we travel into the field, so as not to miss the opportunity to play 4-hands. Through the years I have been utterly devoted to playing chamber music, particularly, but not exclusively, with scientist-musicians. Unforgettable have been the evenings of ensemble playing and occasional performing with Jelle Atema, Helen Ghiradella, Jane Houston, Jerry Meinwald, and Noel Snyder.

I also wish to acknowledge how pleasant it has been to work with the staff of Harvard University Press throughout the publication process. I am much indebted to Michael Fisher, Executive Editor for Science and Medicine, for providing encouragement and help all along, and to Sara Davis for many favors rendered. Annamarie Why and David Foss deserve credit, respectively, for designing the book and for guiding it through the production stages. Nancy Clemente edited the entire manuscript, displaying at all times the friendly disposition and flawless judgment for which she is famous.

To my friend James McConkey I extend thanks for calling to my attention the quote from Martin Rees that serves as the book epigraph.

Illustration Credits

Some of the illustrations, listed next below by source, have appeared previously in papers published originally in the journal *Science*. These are reproduced here with permission from the American Association for the Advancement of Science:

Page 16	Eisner, T. 1958. *Science* 128: 148–149.
Page 163 (bottom)	Eisner, T., H. E. Eisner, J. J. Hurst, F. C. Kafatos, and J. Meinwald. 1963. *Science* 139: 1218–1220.
Pages 219 (middle left), 220, 221	Eisner, T., R. Alsop, and G. Ettershank. 1964. *Science* 146: 1058–1061.
Pages 86, 87 (top), 88	Eisner, T. 1965. *Science* 148: 966–968.
Page 113	Eisner, T., and J. Shepherd. 1965. *Science* 150: 1608–1609.
Pages 21, 69 (bottom left and right), 70 (A), 87 (bottom), 263 (left)	Eisner, T., and J. Meinwald. 1966. *Science* 153: 1341–1350.
Page 214	Eisner, T., and J. A. Davis. 1967. *Science* 155: 577–579.
Pages 29, 34	Aneshansley, D. J., T. Eisner, J. M. Widom, and B. Widom. 1969. *Science* 165: 61–63.
Pages 187, 188 (right)	Eisner, T., A. F. Kluge, J. E. Carrel, and J. Meinwald. 1971. *Science* 173: 650–652.
Page 311 (top left, bottom left and right)	Eisner, T., J. S. Johnessee, J. Carrel, L. B. Hendry, and J. Meinwald. 1974. *Science* 184: 996–999.

Pages 159, 160 (bottom left), 162, 163	Eisner, T., K. Hicks, M. Eisner, and D. S. Robson. 1978. *Science* 199: 790–794.
Page 319 (top left)	Eisner, T., S. Nowicki, M. Goetz, and J. Meinwald. 1980. *Science* 208: 1039–1042.
Pages 180 (bottom left and right), 181, 182	Eisner, T., and D. J. Aneshansley. 1982. *Science* 215: 83–85.
Pages 223 (bottom left, middle, and right), 225 (left), 226	Eisner, T., and S. Nowicki. 1983. *Science* 21:, 185–187.
Pages 283 (bottom left and right), 285 (bottom right), 286 (top left and right, bottom right), 287 (right)	Dussourd, D. E., and T. Eisner. 1987. *Science* 237: 898–901.
Pages 35, 36, 38	Dean, J., D. J. Aneshansley, H. E. Edgerton, and T. Eisner. 1990. *Science* 248: 1219–1221.
Page 391 (top and bottom right)	Smedley, S. R., and T. Eisner. 1995. *Science* 270: 1816–1818

I am also grateful for permission to reproduce the following figures, which appeared initially in the *Proceedings of the National Academy of Sciences:*

Pages 230 (bottom left and right), 231 (bottom right)	Eisner, T., and J. Dean. 1976. *Proc. Nat. Acad. Sci. USA* 73: 1365–1367.
Pages 250 ,251, 252 (top right, bottom left and right)	Eisner, T., and S. Camazine. 1983. *Proc. Nat. Acad. Sci. USA* 80: 3382–3385.
Pages 266 (top right), 270 (right), 271, 272	Eisner, T. , I. T. Baldwin, and J. Conner. 1993. *Proc. Nat. Acad. Sci. USA* 90: 6716–6720.
Page 209 (bottom left)	Carrel, J. E, and T. Eisner. 1984. *Proc. Nat. Acad. Sci. USA* 81: 806–810.
Pages 353 (bottom left), 356 (left), 358 (top left), 360 (bottom left), 368 (bottom left and right), 375 (top left and right)	Eisner, T., and J. Meinwald. 1995. *Proc. Nat. Acad. Sci. USA* 92: 50–55.
Page 314 (top left and right)	Meinwald, J., and T. Eisner. 1995. *Proc. Nat. Acad. Sci. USA* 92: 14–18.
Pages 339 (top right), 340, 342	Eisner, T., S. R. Smedley, D. K. Young, M. Eisner, B. Roach, and J. Meinwald. 1996. *Proc. Nat. Acad. Sci. USA* 93: 6494–6498.

Pages 343, 345	Eisner, T., S. R. Smedley, D. K. Young, M. Eisner, B. Roach, and J. Meinwald. 1996. *Proc. Nat. Acad. Sci. USA* 93: 6499–6503.
Pages 104 (top), 106 (bottom), 107	Eisner, T., M. Eisner, and M. Deyrup. 1996. *Proc. Nat. Acad. Sci. USA* 93: 10848–10851.
Pages 116, 118	Eisner, T., A. B. Attygalle, W. E. Conner, M. Eisner, E. MacLeod, and J. Meinwald. 1996. *Proc. Nat. Acad. Sci. USA* 93: 3280–3283.
Pages 142, 148, 150 (top and bottom left)	Eisner, T., M.A. Goetz, D. E. Hill, S. R. Smedley, and J. Meinwald. 1997. *Proc. Nat. Acad. Sci. USA* 94: 9723–9728.
Pages 301 (top left, bottom left, middle, and right), 302 (top right, bottom left), 303 (top left, bottom left and right)	Eisner, T., M. Eisner, and R. E. Hoebeke. 1998. *Proc. Nat. Acad. Sci. USA* 95: 4410–4414.
Pages 274, 277, 278 (right), 279	Eisner, T., M. Eisner, A. B. Attygalle, M. Deyrup, and J. Meinwald. 1998. *Proc. Nat. Acad. Sci. USA* 95: 1108–1113.
Page 401 (left column, bottom right)	Schroeder, F. C., S. R. Smedley, L. K. Gibbons, J. J. Farmer, A. B. Attygalle, T. Eisner, and J. Meinwald. 1998. *Proc. Nat. Acad. Sci. USA* 95: 13387–13391.
Pages 31 (B-E), 32	Eisner, T. and D. J. Aneshansley. 1999. *Proc. Nat. Acad. Sci. USA* 96: 9705–9709.
Pages 128, 130, 131, 132 (top left), 135 (top), 136, 137, 138	Eisner, T., and D. J. Aneshansley. 2000. *Proc. Nat. Acad. Sci. USA* 97: 6568–6573.
Page 396	Schroeder, F. C., A. González, T. Eisner, and J. Meinwald. 1999. *Proc. Nat. Acad. Sci. USA* 96: 13620–13625.
Pages 190, 192, 193, 195 (bottom left and right)	Eisner, T., and D. J. Aneshansley. 2000. *Proc. Nat. Acad. Sci. USA* 97: 11313–11318.
Pages 120(bottom), 121 (top), 122, 123, 124, 125, 126, 127	Eisner, T, and M. Eisner. 2000. *Proc. Nat. Acad. Sci. USA* 97: 2632–2636.
Page 385	Conner, W. E., R. Boada, F. C. Schroeder., A. González, J. Meinwald, and T. Eisner. 2000. *Proc. Nat. Acad. Sci. USA* 97: 14406–14411.

Thanks go also to the following publishers for permission to reproduce figures from some of their journals:

The American Entomology Society

Page 207 (top)	Eisner, T., and M. Eisner. 2002. *Ent. News* 113: 6–10

Birkhäuser Verlag AG

Page 388 (top right)	Dussourd, D. E., C. A. Harvis, J. Meinwald, and T. Eisner. 1989. *Experientia* 45: 896–898.
Page 366 (middle right)	Bogner, F., and T. Eisner. 1992. *Experientia* 48: 97–102.
Page 325	Eisner, T, R. Ziegler, J. L. McCormick, M. Eisner, E. R. Hoebeke, and J. Meinwald. 1994. *Experientia* 50: 610–615.

Elsevier Science

Page 19	Eisner, T. 1958. *J. Ins. Physiol.* 2: 215–220
Page 51	Eisner, T., J. Meinwald, A. Monro, and R. Ghent. 1961. *J. Ins. Physiol.* 6: 272–298

Entomological Society of America

Page 184 (bottom)	Eisner, T., A. F. Kluge, J. C. Carrel, and J. Meinwald. 1972. *Ann. Ent. Soc. Amer.* 65: 765–766.

John Wiley & Sons

Pages 69 (bottom left), 70	Eisner, T., F. McHenry, and M. M. Salpeter. 1964. *J. Morph.* 115: 355–400.

Kluwer Academic / Plenum Publishing

Pages 246, 247 (bottom left), 248	Masters, W. M., and T. Eisner. 1990. *J. Ins. Behav.* 3: 143–157
Page 68 (bottom left)	Jones, T. H., W. E. Conner, J. Meinwald, H. E. Eisner, and T. Eisner. 1976. *J. Chem. Ecol.* 2: 421–429
Page 316	Eisner, T. 1990. *J. Chem. Ecol.* 16: 2489–2492.
Page 366 (middle left)	Bogner, F., and T. Eisner. 1991. *J. Chem. Ecol.* 17: 2063–2075.
Page 309 (top left and right)	Eisner, T. 1994. *J. Chem. Ecol.* 20: 2743–2749.

Springer Verlag

Pages 199, 200 (top), 201, 202, 204	Eisner, T., I. Kriston, and D. J. Aneshansley. 1976. *Behav. Ecol. and Sociobiol.* 1: 83–125
Page 314 (top right and bottom)	Morrow, P. A., T. E. Bellas, and T. Eisner. 1976. *Oecologia* 24: 193–206.
Pages 368 (top right, bottom left and right), 369, 371	Conner, W. E., T. Eisner, R. K. Vander Meer, A. Guerrero, D. Ghiringelli, and J. Meinwald. 1980. *Behav. Ecol. Sociobiol.* 7: 55–63.
Pages 374 (bottom 2 rows), 375 (top left, bottom left and right)	Conner, W. E., T. Eisner, R. K. Vander Meer, A. Guerrero, and J. Meinwald. 1981. *Behav. Ecol. Sociobiol.* 9: 227–235

Urban & Fischer Verlag

Page 135 (bottom)	Attygalle, A. B., D. J. Aneshansley, J. Meinwald, and T. Eisner. 2001. *Zoology* 103: 1–6.

Sources of other illustrations are:

Pages 23, 179: Beetle outlines courtesy of Frances Fawcett.

Page 42: Stamp © 1999 by the U.S. Postal Service. Reproduced with permission. All rights reserved.

Page 78: Courtesy of the Archbold Biological Station; photo by John A. Wagner.

Page 79 (bottom): Courtesy of the Archbold Biological Station; photo by Reed Bowman.

Page 88: Drawing courtesy of Frances A. McKittrick.

Page 99: Courtesy of the Oeffentliche Kunstsammlung Basel, Kunstmuseum. Photo by Martin Bühler.

Page 100: From *The Travels of Babar* by Jean de Brunhoff, copyright © 1934, renewed 1962 by Random House, Inc. Used by permission of Random House Children's Books, a division of Random House, Inc.

Page 102: Photos courtesy of Ian Common, from *Moths of Australia*, Melbourne University Press, 1990; reprinted by permission of the publisher.

Page 136: Drawing courtesy of Susan Pulakis.

Page 152: Photo courtesy of Frank DiMeo.

Page 214: Drawings by M. A. Menadue.

Page 299 (left): Photo courtesy of Sandy Podulka, Cornell University.

Page 385: Top left photo courtesy of William E. Conner; top right photo courtesy of Nickolay Hristov.

Page 390: Photo courtesy of Scott Smedley.

Margaret Nelson was kind enough to provide the figures on pages 24, 39, 54, 65, 66, 71, 130, 178, 220, 221, 226, and 243 (top). The electron micrographs on pages 84, 106, 107, 180 (top), 248 (top), 279, and 301 (bottom) were graciously provided by Maria Eisner. Frank Schroeder kindly drafted the chemical formulas.

The picture on page ii is of a female aphid giving birth, and the scanning electron micrograph on page xii is of an ant that has been entangled by the defensive bristles of a polyxenid millipede. The pictures on the chapter opening pages are: *Schinia gloriosa,* a noctuid moth (Prologue); a *Brachinus* bombardier beetle (Chapter 1); a *Zygaena* moth (Chapter 2); *Plagiodera versicolora,* a chrysomelid beetle larva (Chapter 3); *Lycus fernandezi,* a lycid beetle (Chapter 4); an *Iridomyrmex humilis* ant under attack by *Nasutitermes exitiosus* termites (Chapter 5); a female of the spider *Argiope florida,* with the much smaller male, in her web (Chapter 6); *Tetraopes tetrophthalmus,* a cerambycid beetle (Chapter 7); an *Ammophila* wasp, carrying a *Synchlora* caterpillar (Chapter 8); a *Neopyrochroa flabellata* beetle feeding on cantharidin crystals (Chapter 9); the moth *Utetheisa ornatrix* mating (Chapter 10); an unidentified coreid bug on a eucalyptus leaf (Epilogue).

Index